# Applied
# Strength of Materials

### Third Edition

by
## ALFRED JENSEN
Emeritus Professor, Engineering
University of Washington, Seattle

and

## HARRY H. CHENOWETH
Associate Professor, Civil Engineering
University of Washington, Seattle

**McGRAW-HILL BOOK COMPANY**

New York    St. Louis    Dallas    San Francisco    Düsseldorf
Johannesburg    Kuala Lumpur    London    Mexico    Montreal    New Delhi
Panama    Paris    São Paulo    Singapore    Sydney    Tokyo    Toronto

Library of Congress Cataloging in Publication Data

Jensen, Alfred.
    Applied strength of materials.

    1. Strength of materials.  I. Chenoweth,
Harry H., joint author.  II. Title.
TA405.J4  620.1'12  74-18384
ISBN 0-07-032471-9

APPLIED STRENGTH OF MATERIALS

    3 4 5 6 7 8 9 0    DODO    7 8 3 2 1 0 9 8

*The editors for this book were Donald E. Hepler and Alice V. Manning, the designer was Tracy Glasner, and its production was supervised by Valerie Klima and Laurence Charnow. It was set in Laurel by York Graphic Services, Inc.*

By  Alfred Jensen and
     Harry H. Chenoweth

Applied Engineering Mechanics, Third Edition
Applied Strength of Materials, Third Edition
Statics and Strength of Materials, Second Edition

# Contents

**Preface**

## 1 Stress and Deformation

## 2 Engineering Materials and Their Properties

### 3   Riveted and Welded Joints. Thin-walled Pressure Vessels

### 4   Torsion

### 5   Centroids and Moments of Inertia of Areas

### 6   Shear and Moment in Beams

## 10    Statically Indeterminate Beams

## 11    Combined Stresses

## 12    Columns

## 13   Special Topics

# Tables

# Preface

The ultimate purpose of the study of engineering mechanics and strength of materials is to gain the basic knowledge required for the successful design of machine parts and of structural members, parts, and connections.

In this book virtually all the subject matter normally taught in the usual college course in strength of materials has been included. A distinguishing feature of the book, however, is that the developments of all design formulas have been accomplished without the use of calculus. In these developments the authors have relied on thorough explanations of existing physical relationships and of the easily understood relationships of the action and reaction. This simplified approach is similar to that used in the authors' "Applied Engineering Mechanics," to which this book is a companion volume. The book is, therefore, especially suitable for college courses in which the use of calculus is not required and for courses offered in technical institutes and in industrial training programs, since an understanding of ordinary high school mathematics is sufficient.

The general presentation of the subject matter and the more than six hundred problems are the result of the authors' experience in teaching during more than a quarter-century. The many illustrative problems have been worked out in unusual detail, both to relieve instructors of time-consuming explanations and to serve as guides to students in setting up their problem solutions.

As in "Applied Engineering Mechanics," each chapter is composed of several articles, each of which presents additional theory, a new concept, or a different aspect, and contains one or more completely solved illustrative problems. These are followed by a large number of problems arranged in order of difficulty, more than one-third of which are supplied with answers. Classroom instructors may, upon request to the authors, obtain a list of answers to the remaining problems. At the end of each chapter is found a summary of its important points and formulas, giving students a quick review of its essential parts. The summary in turn is followed by a considerable number of review problems and questions chosen to test the student's grasp of subject matter.

A departure in presentation of the subject matter is the strong emphasis placed on the actual design of structural members, parts, and connections. This emphasis is of particular advantage to the many students who do not later take additional courses in design. It is also very helpful to those already in industry who have need for such a book for reference and self-study.

Because relatively few students have the opportunity to study separately the methods of manufacture and the properties of the various so-called engineering materials, a brief chapter discussing these items has been included. The subject matter of this chapter is informative rather than technical and is intended to give students only an elementary understanding of some of the most commonly used engineering materials.

Included in Chap. 12, on columns, are the latest formula and design procedures for the design of timber columns, recommended by the National Lumber Manufacturers Association. The older and more conservative formulas recommended by the Forest Products Laboratory are also included.

Special subjects such as (1) axial stresses in members of two materials, (2) eccentrically loaded riveted joints, (3) design of laterally unsupported steel beams, (4) statically indeterminate beams, and (5) eccentrically loaded columns have been included in the firm belief that students should not be unfamiliar with them. Any or all of these subjects may, of course, be omitted without disadvantage to the rest of the subject matter.

In this third edition the following new materials have been added: (1) variation of stress with inclination of the exposing cutting plane and principal stresses, and (2) prestressed beams. In addition, many problems have been changed and new problems have been added. The latest edition of the "Manual of Steel Construction" by the American Institute of Steel Construction (AISC) has been followed where practical.

Despite all efforts by authors, publisher, and printer, some errors invariably creep into a new edition. The authors will be grateful to all who may discover errors for information concerning them.

Appreciation is extended to the American Institute of Steel Construction for permission to reproduce from their handbook many of the tables in the Appendix.

**Alfred Jensen and Harry H. Chenoweth**

# CHAPTER 1

# Stress and
# Deformation

## 1-1 Scope of Text

The title of this text, "Applied Strength of Materials," unfortunately does not disclose very well the actual contents between its covers, especially so to the student and reader not already fairly well acquainted with the subject matter. Such a reader would, quite naturally, perhaps, assume that at least the major part of the text was devoted to discussions of the actual testing of various materials to determine their strengths, together, possibly, with extensive experimental test data and the final conclusions drawn from these data.

As a matter of fact, less than 5 percent of this text deals with the actual testing of materials to determine strength and other properties. The balance is devoted (1) to discussions and explanations of the relationship existing between the externally applied forces and the internally induced stresses in various types of structural members and parts, such as bolts, rivets, shafts, pressure tanks, beams, columns, etc., and (2) to explanations of the relationship existing between these same externally applied forces and the resulting deformations, such as the elongation of a steel rod in tension, the shortening of a wood post in compression, the amount of twist in a shaft due to a torque, the deflection of a beam due to transverse loads, etc. The above text material, together with such closely allied subjects as riveted and welded connections, centroids of areas, moments of inertia of areas, analysis of statically indeterminate beams, combined axial and bending stress, and others, normally comprise the content of a textbook on strength of materials.

Other titles, such as "Mechanics of Materials" and "Resistance of Materials,"

1

that have been used by a few authors for many years do not seem to explain the scope of the text material any better. Most authors and professional engineers have, therefore, accepted the title "Strength of Materials," or "Applied Strength of Materials."

### 1-2 Definition of Stress

**Stress** is defined as *resistance to external forces.* It is measured in terms of the *force exerted per unit of area.* In engineering practice, the force is generally given in pounds and the resisting area in square inches. Consequently stress is usually expressed in pounds per square inch, often abbreviated to psi. Stress, then, is produced in all bodies upon which forces act.

In practice the word "stress" is often given two meanings: (1) force per unit area, or *intensity* of stress, generally referred to as **unit stress,** and (2) total internal force within a single member, generally called **total stress.** When the word "stress" is used alone, the context should make it clear which of the two meanings is intended.

### 1-3 The Basic Stresses

Only two basic stresses exist: (1) **normal stresses,** which always act normal (perpendicular) to the stressed surface under consideration, and (2) **shearing stresses,** which act parallel to the stressed surface. (Normal stresses may be either tensile or compressive.) Other stresses either are similar to these basic stresses or are a combination of them. For example, the stresses in a bent beam, in a general way referred to as "bending stresses," actually are a combination of tensile, compressive, and shearing stresses; torsional stress, as encountered in the twisting of a shaft, is a shearing stress.

When the external forces acting on a member are parallel to its major axis and the member is of constant cross section, or substantially so, the resulting internal stresses are likewise parallel to that axis. Such forces are called **axial forces,** and the stresses are referred to as **axial stresses.**

**Tensile Stress.** When a pair of axial forces *pull* on a member, and thus tend to stretch or elongate it, they are said to be **tensile forces,** and they produce axial **tensile stresses** internally in the member on a plane lying perpendicular, or normal, to its axis. An example of such a condition is illustrated in Fig. 1-1*a*. An eyebar is subjected to a pair of tensile forces *P*. If a section is cut out of this bar, the internal tensile stresses are disclosed and, being *tensile* stresses, they are, of course, *pulling* on each of the two remaining ends of the bar as well as on the section removed.

The stressed area considered is a plane surface lying normal to the direction of stress. In general, it is the cross-sectional area of the stressed member, and the stress is generally assumed to be uniformly distributed over that area.

**Compressive Stress.** When a pair of axial forces *push* on a member and shorten or compress it, they are called **compressive forces** and they produce

FIG. 1-1 The basic stresses. (*a*) Normal. (*b*) Shearing.

axial **compressive stresses** internally in the member on a plane perpendicular, or normal, to its axis. This condition is also illustrated in Fig.1-1*a*. A short post is supporting at its top an axial load *P*, which, of course, is counteracted by the reacting force *P* at its base. When a section is cut out of this post, the internal compressive stresses are disclosed, pushing on the two remaining ends of the post as well as on the section removed.

As in the case of tension, the stressed area considered is a plane surface lying normal to the direction of stress. In general, it is the cross-sectional area of the stressed member, and the stress is assumed to be uniformly distributed over that area.

**Shearing Stress.** This type of stress differs from tensile and compressive stresses in that the stressed plane (the shear plane) lies parallel with the direction of stress rather than perpendicular to it, as in the cases of tensile and compressive stresses.

An illustration of shear is seen in Fig. 1-1*b*. The vertical-grained block of wood *A* is clamped tightly into a testing machine. A vertical force *P*, applied to the upper surface of the projecting portion *B* of the block, will tend to shear it off the main body of the block along the shear plane *abcd*. Such action is resisted by the shearing strength of the wood fibers. The unit shearing stresses produced by force *P* on surface *abcd* are indicated by the small arrows.

Two examples of shearing failure of simple riveted connections are shown in Fig. 1-2. In the lap joint on the left, failure will occur on the single plane through the rivet coinciding with the plane of the contracting surfaces. This case is referred to as **single shear**, and the area in shear is the cross-sectional

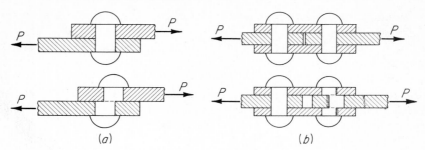

FIG. 1-2 Shear failures of rivets. (*a*) Single shear. (*b*) Double shear.

area of the rivet. In the butt joint on the right, the right rivet is the smallest, and it, therefore, will fail first in shear, while the other rivet will hold. But shear failure must of necessity occur simultaneously on the two planes through the rivet which coincide with the two surfaces of contact between the plates. This case is called **double shear**, and the total shear area is naturally twice the cross-sectional area of the rivet.

In the case of bolts and rivets and in beams, as will be shown later, the forces producing shearing stresses generally act *transversely* to the longitudinal axis of the member. This is in contrast to tensile and compressive forces, which generally act *parallel* to that axis.

In shear, the stressed area considered is a plane surface lying parallel to the direction of stress. In the case of bolts and rivets and in beams, it is generally the cross-sectional area of the member. Depending on the conditions involved, shear stresses may or may not be considered to be uniformly distributed over the stressed area.

**Bearing Stress.** Compressive stress exerted on an *external* surface of a body, as, for example, when one object presses against another, is referred to as bearing stress. Examples of bearing stresses can be found in the middle sketch of Fig. 1-1. Let us assume that we have here a timber post resting on a steel baseplate supported on a concrete footing, which in turn is ultimately supported by the soil. The post *bears* on the plate, the plate *bears* on the footing, and the footing *bears* on the soil. The resulting compressive stresses at the external surfaces of contact between these bodies are *bearing stresses*. In the examples just mentioned, the stressed areas are plane surfaces, all lying normal to the direction of stress. Bearing on curved surfaces, such as occur on bolts and rivets, is discussed in Chap. 3.

### 1-4 Ultimate Stress, Allowable Stress, Factor of Safety, Units
Before a material can safely be used in the making of structural parts or members, a number of its properties must be known. In the case of metals such as steel, cast iron, bronze, aluminum, and others which are widely used as structural members and machine parts we must have reliable information

concerning such properties as strength, elasticity, stiffness, ductility, malleability, toughness, etc.

The production of these metals from the raw materials, and the subsequent processes required for their formation into structural parts, have been so well standardized that the properties found through tests to be possessed by a random sample may safely be assumed to be possessed in like degrees by similar materials produced and processed under similar conditions. Methods used to determine some of these properties are discussed in Chap. 2.

**Ultimate stress,** or strength, is defined as *the greatest unit stress a material can withstand without rupture.* In the practical design of structures and machines, we would not, of course, for obvious reasons of safety, attempt to use the full strength of any material.

**Allowable stress** is that portion of the ultimate strength which may safely be used in design. The terms *working stress* and *safe working stress* have been given the same meaning and both are widely used. However, the term *allowable stress* is most generally accepted and is used in this text.

**Factor of safety** is *the ratio of ultimate stress to allowable stress* and is denoted by the symbol $N$. The appropriate factor of safety for any given design depends upon a number of considerations. Among these are the following:

1. The degree of safety required. That is, what dangers to human life and property are involved?
2. The degree of economy desired. A tendency would exist to use a high percentage of the strength of costly materials.
3. Dependability of the material. Are flaws easily detected? Do apparently similar pieces behave similarly under stress? Does the material give ample warning of impending failure? Hard, brittle materials fail without warning; "soft," tough materials deform and thus give warning.
4. Permanency of design. Lower factors of safety are often permitted in temporary designs.
5. Load conditions: steady, varying, or shock loads?
6. The exactness with which probable loads and stresses can be predicted or determined.
7. Accessibility of parts for inspection and maintenance.
8. Relative difficulty in preventing deterioration of the material, such as decay of timber, rusting of steel, disintegration of concrete, etc.

Designers of structures and machines generally work with a predetermined allowable stress. But in aircraft design, the ultimate strength and a predetermined factor of safety are often used.

Allowable stresses which may safely be used in various designs are, as a rule, based upon extensive laboratory tests made to disclose the various properties of the materials used and upon a vast amount of accumulated experience. Most large cities have established building codes which govern all structural design within their respective municipalities. The Federal government, the Army, the Navy, most states, and many counties all have their separate codes.

The factors of safety suitable in aircraft design are generally determined by the severity of the conditions under which the aircraft is expected to be flown. Machines are usually designed so as to withstand all reasonable wear and tear. A factor of safety of 2 might be sufficient for a temporary structure subjected only to steady loads and endangering neither life nor property, while a factor of safety as high as 10 might be required in a machine which is subjected to unpredictable shock loads and whose failure would endanger life and cause loss of other property. For steady loads, a factor of safety between 3 and 4 is commonly used.

Tables giving properties of various materials and suitable allowable stresses are found throughout the text. A common test to determine the properties of strength and elasticity of steel is described in Art. 1-12.

**Units.** When dealing with quantities such as forces, areas, and volumes, we must establish a system of *units* of these quantities and their accepted abbreviations. The abbreviations shown in Table 1-1 agree with those recommended by the American National Standards Institute (ANSI).

## 1-5 The Direct-Stress Formula

The types of stresses described in the preceding articles are often referred to as **direct stresses,** in order to distinguish them from torsional, bending, and other stresses.

In cases similar to the three illustrated in Fig. 1-1, the reasonable assumption is usually made that the intensity of stress is about the same on all units of stressed area. That is, the maximum and minimum unit stresses do not differ greatly from the average stress. While this assumption in some instances may

**Table 1-1 ENGINEERING UNITS AND ABBREVIATIONS**

| Unit | General usage | Illustrations and solutions | Solutions and answers |
|------|---------------|-----------------------------|----------------------|
| Force | lb, kip, ton, etc. | | |
| Length | in., ft, etc. | | |
| Area | sq in., sq ft, etc. | | |
| Force (or load) per unit of length | lb per in., lb per ft, etc. | lb/in., lb/ft | |
| Force (or load) per unit of area | lb per sq in., lb per sq ft, kips per sq in., etc. | lb/sq in., kips/sq in. | psi, psf, ksi |
| Volume | cu in., cu ft, etc. | | |
| Weight (force) per unit of volume | lb per cu in., lb per cu ft, etc. | lb/cu in., lb/cu ft | pci, pcf |

be incorrect in varying degrees, such instances are usually recognized and appropriate modifications can readily be made.

Using the above-stated assumption, we may now establish a relationship between the average intensity of stress $s$ produced by an externally applied force $P$ on each unit of stressed area $A$. Clearly, the total internal force is the product of the average unit stress and the stressed area, which is $sA$. And this total internal force $sA$ must, according to the law of equilibrium, be equal to the external force $P$ producing it, since the two forces are collinear. Consequently

$$P = sA \quad \text{or} \quad s = \frac{P}{A} \tag{1-1}$$

where  $s$ = average unit stress, psi
       $P$ = total external force or load, lb
       $A$ = stressed area, sq in.

Study of Eq. (1-1) discloses that (1) given $P$ and $A$, we may obtain $s$; (2) given $s$ and $A$, we may obtain $P$; and (3) given $s$ and $P$, we may obtain the required area $A$. Of these cases, (1) and (2) are problems in *analysis;* that is, a given member is analyzed to determine either its unit stress or its load capacity. Case (3) is one of *design;* that is, a member of cross-sectional area $A$ must be selected or designed to withstand the given load $P$.

## 1-6 Simple Structural Members

Many of the structural members used in common engineering practice, such as timber beams and columns, steel bolts, rods, pipes, beams, and columns, and many others, are commercially produced in a wide variety of sizes whose cross-sectional areas and other properties can readily be determined from tabulated information furnished by the manufacturers. A number of such tables are given in the Appendix.

**Threaded Steel Rods.** In many problems involving analysis or design, several details must be considered. For example, if a steel rod is used as a tension member in a truss, as illustrated in Fig. 1-3, its ends are usually threaded to facilitate its being fastened to other members. In such a case, the threading of a 1-in.-diameter rod with 8 threads per inch would reduce its effective diameter from 1 in. to 0.8376 in. at the root of the threads and its area from 0.7854 sq in. to a net **root area** of 0.551 sq in., a reduction of about 30 percent. Such a reduction in area cannot, of course, be ignored by designers and the *root area*, therefore, must be used in the design of threaded rods (see Table E-3).

Two series of threads are in general use: (1) American Standard Coarse Series for structural work and (2) American Standard Fine Series, mostly used in machine work. The material between the threaded ends of a rod is conse-

FIG. 1-3 Unit stresses. Analysis.

quently stressed to only part of its usable strength. To overcome this inefficiency in long rods, **upset ends** are often used; that is, the part to be threaded is enlarged so that its root diameter after threading will not be smaller than the diameter of the unthreaded portion of the rod.

**Standard timber** is cut in rectangular shapes to *rough* cross-sectional dimensions which generally are multiples of 2 in., such as 2 by 4 in., 6 by 10 in., or 12 by 18 in. The actual sizes of surfaced timber are less than these nominal dimensions. Most structural timber is dressed, or surfaced, on all four sides, a treatment denoted by S4S (see Table E-2).

**Structural steel shapes,** such as I-beams, wide-flange sections, channels, angles, and pipe sections, all widely used in structural work, have been standardized as to size and weight per linear foot. The cross-sectional area, weight per foot, and several other useful properties of these sections are given in Tables E-5 to E-15 inclusive.

### 1-7 Problems in Analysis and Design

In engineering practice the term **analysis** includes within its meaning (1) all work relating to the investigation required to estimate correctly all external loads and forces to which a structure or machine which is to be designed may possibly be subjected and (2) the determination of the resulting internal forces in the various members and other component parts. **Applied mechanics,** then, is essentially analysis.

When applied mechanics is combined in such investigations with a knowledge of details and aspects of structural elements, as those of buildings, bridges, and aircraft frames, for example, the process is called **structural analysis.** For economic and other reasons, several differing analyses may be made before one is finally selected from which the actual design may proceed.

**Design** is the final determination of the required sizes of all the various elements, members, and joint details which constitute the structure as a whole and from which determination the actual construction may be accomplished. Design is, then, essentially a problem of *materials,* involving their strength and other properties. The unit stresses which may safely be used in design are called **allowable stresses,** and their values depend upon several properties of the materials in question. The bases for their determination are discussed in Art. 1-4.

## ILLUSTRATIVE PROBLEMS

**1-1.** In Fig. 1-3 is shown a part of a common type of roof truss, constructed mainly of timbers and steel rods. Let us presume that the forces to be resisted by the various members, and the sizes of the members, have already been determined. The timbers are surfaced (S4S), and the steel rod is threaded. Determine (*a*) the average unit compressive stress in the 8- by 8-in. diagonal timber *A* if the total stress in it is 20 kips (1 kip = 1,000 lb); (*b*) the maximum average unit stress in the $^3/_4$-in. threaded steel rod if the total stress in it is 4 kips; (*c*) the average unit bearing stress between the timber and the 4- by 4-in. square steel washer if the hole in it is $^7/_8$ in. in diameter; (*d*) the average unit bearing stress between the brick wall column and the 8- by 10-in. timber if the total load on the column is 15 kips; (*e*) the unit shearing stress along the dashed line in the 8- by 10-in. timber if the horizontal thrust against the steel shoe is 16 kips and $L = 22$ in.

*Solution:* We note that both timbers are surfaced (S4S) and that the steel rod is threaded. Hence from Table E-2 we find the net area of the 8- by 8-in. timber to be 56.25 sq in. ($7\frac{1}{2}$ by $7\frac{1}{2}$ in.). Similarly, from Table E-3, we find the root area of the $^3/_4$-in.-diameter rod to be 0.302 sq in. Consequently

(*a*)   $P = 20,000$ lb    $A = 56.25$ sq in.    $s_c = \dfrac{P}{A} = \dfrac{20,000}{56.25} = 356$ psi    *Ans.*

(*b*)   $P = 4,000$ lb    $A = 0.302$ sq in.    $s_t = \dfrac{P}{A} = \dfrac{4,000}{0.302} = 13,260$ psi    *Ans.*

(*c*) The net area of the 4- by 4-in. washer is $16 - 0.601 = 15.399$ sq in. Hence

$P = 4,000$ lb    $A = 15.399$ sq in.    $s_b = \dfrac{P}{A} = \dfrac{4,000}{15.399} = 260$ psi    *Ans.*

(*d*) The 8- by 10-in. timber is $7\frac{1}{2}$ in. wide and bears 12 in. on the brick wall column, giving a bearing area of 90 sq in. Therefore

$P = 15,000$ lb    $A = 90$ sq in.    $s_b = \dfrac{P}{A} = \dfrac{15,000}{90} = 167$ psi    *Ans.*

(*e*) The horizontal component of the total stress in the 8- by 8-in. timber is 16 kips. This force tends to shear off the part of the timber above the dashed line. The width of the shear area is $7\frac{1}{2}$ in., and its length is 22 in., a total area of 165 sq in. Then, since $P = 16,000$ lb and $A = 165$ sq in.,

$$s_s = \frac{P}{A} = \frac{16,000}{165} = 97 \text{ psi} \qquad Ans.$$

**1-2.** A typical warehouse-floor support is shown in Fig. 1-4. The short timber post $C$ is capped by a section of steel channel in order to provide greater bearing area for the ends of beams $A$ and $B$. These beams are 8 by 20 in. S4S, with the greatest dimension in a vertical plane. The steel baseplate $P$ likewise provides greater bearing area on the concrete footing $F$. The load transmitted to post $C$ from each of beams $A$ and $B$ is 16 kips, making a total load on the post of 32 kips. The weights of the cap, post, plate, and footing may here be neglected.

FIG. 1-4 Unit stresses. Design.

Find (a) the minimum length $L$ of channel required to support the beams if the maximum bearing stress allowed in the timber *perpendicular to the grain* is 320 psi; (b) the minimum available nominal size of square timber post, S4S, required if the unit stress parallel to the grain is not to exceed 1,100 psi; (c) the required size of the square steel baseplate $P$ if the permissible unit bearing stress on the concrete is 500 psi; (d) the minimum dimension $d$ of the square footing $F$ if the allowed pressure on the soil is 8,000 psf (pounds per square foot).

*Solution:* Since the 8-in. beams are surfaced (S4S), they are only $7\frac{1}{2}$ in. wide, and the area in bearing is 7.5L. When the weights of all parts are disregarded, the load $P$ will be 32,000 lb. Then

(a)   $P = 32,000$ lb      $s = 320$ psi      $A = \dfrac{P}{s} = \dfrac{32,000}{320} = 100$ sq in.

$A = 7.5L$   and   $L = \dfrac{100}{7.5} = 13.3$ in.   *Ans.*

(b)   $P = 32,000$ lb      $s = 1,100$ psi      $A = \dfrac{P}{s} = \dfrac{32,000}{1,100} = 29.1$ sq in.

Use 6- by 6-in. post S4S ($A = 30.25$ sq in.).   *Ans.*

(c)   $P = 32,000$ lb      $s = 500$ psi      $A = \dfrac{P}{s} = \dfrac{32,000}{500} = 64$ sq in.

Use 8- by 8-in. baseplate.   *Ans.*

(d)   $P = 32,000$ lb      $s = 8,000$ psf      $A = \dfrac{P}{s} = \dfrac{32,000}{8,000} = 4$ sq ft

Use 2- by 2-ft footing.   *Ans.*

## PROBLEMS

**1-3.** In Prob. 1-1 assume (a) that the diagonal member is 6 by 6 in. S4S and that the total stress in it is 18 kips; (b) that the steel rod is of 1-in. diameter and the total stress in it is 7.5 kips; (c) that the washer is 5 by 5 in. in size with a $1\frac{1}{8}$-in. hole; (d) that timber B is 6 by 10 in. S4S ($9\frac{1}{2}$ in. vertical) and the load on the column is 15 kips; (e) that the horizontal thrust against the steel shoe is 12 kips and $L = 18$ in. Then solve the five parts.

*Ans.* (a) $s_c = 594$ psi;  (b) $s_t = 13,620$ psi;  (c) $s_b = 312.3$ psi;  (d) $s_b = 227.5$ psi; (e) $s_s = 121.2$ psi

**1-4.** Assume in Prob. 1-2 that beams A and B are 6 by 12 in. S4S ($11\frac{1}{2}$ in. vertical), that each beam transmits a load of 8 kips to the post, and that the allowable stresses are as follows: in timber, 200 psi perpendicular to the grain and 900 psi parallel to the grain; on concrete, 500 psi; on soil, 4,000 psf. Then solve the four parts.

**1-5.** A round, threaded steel rod of $1\frac{1}{8}$-in. diameter supports an axial load $P$ in tension. Calculate (a) the maximum unit tensile stress $s_t$ in the rod if the axial load $P$ is 12,000 lb, and (b) the maximum allowed load $P$ on this rod if the unit tensile stress must not exceed 20,000 psi. (c) What are the corresponding answers for a rod with upset ends?

*Ans.* (a) $s_t = 17,300$ psi; (b) $P = 13,860$ lb; (c) $s_t = 12,080$ psi, and $P = 19,880$ lb

**1-6.** Determine the required nominal diameter $d$ of a threaded steel rod to carry an axial load of 20,000 lb in tension if a unit stress of 20,000 psi is permitted. What is the corresponding answer for a rod with upset ends?

*Ans.* $d = 1\frac{3}{8}$-in. threaded rod and $d = 1\frac{1}{8}$-in. upset rod (less than 1 percent over-stress)

**1-7.** A short 8- by 8-in. S4S timber post supports an axial compressive load. Calculate (a) the maximum unit compressive stress $s_c$ in the post if the load $P$ is 40,000 lb and (b) the maximum allowed load $P$ on this post if the permissible stress in it is 1,000 psi. (c) What are the corresponding answers for a post that is rough instead of S4S?

**1-8.** Determine the required nominal size of a short square timber post (S4S) to withstand an axial compressive load of 50,000 lb if the maximum allowable unit stress is 800 psi. What is the corresponding answer for a post that is rough instead of S4S?

*Ans.* 10 by 10 in.; 8 by 8 in.

**1-9.** The threaded steel rod shown in Fig. 1-5 supports a vertical post of a timber earth-retaining wall. (a) Determine the smallest suitable diameter $d$ which may safely be used if the allowable unit stress in the rod is 20,000 psi. (b) Find also the required side dimensions of a square steel-plate washer if stress in bearing perpendicular to the grain of the timber is 257 psi and the hole in the washer is $\frac{3}{4}$ in. in diameter.

*Ans.* (a) $\frac{5}{8}$ in.; (b) 4 by 4 in.

**1-10.** The short steel column C shown in Fig. 1-6 supports an axial compressive load of 100,000 lb and is welded onto a steel baseplate B resting on a concrete footing

FIG. 1-5 Prob. 1-9.      FIG. 1-6 Prob. 1-10.      FIG. 1-7 Prob. 1-13.

*F.* (*a*) Select the lightest (most economical) wide-flange section to use if the unit stress is not to exceed 13,500 psi. (*b*) Determine the required side dimension *b* of the square baseplate *B* if the allowable bearing stress on the concrete is 525 psi. (*c*) Calculate the required side dimension *f* of the base of the square concrete footing if the pressure on the soil is not to exceed 5,000 psf. Disregard weight of column, baseplate, and footing.          *Ans.* (*a*) W 6 × 25; (*b*) 14- by 14-in. plate; (*c*) *f* = 4 ft 6 in.

**1-11.** A square, hollow tube of wrought aluminum alloy of $1\frac{1}{2}$ in. outside dimension with a wall thickness of 0.1 in. is used as a structural member in an airplane fuselage. Calculate the unit stress in this member caused by an axial tensile load of 10,000 lb.          *Ans.* $s_t = 17{,}860$ psi

**1-12.** A piece of schedule 80 steel pipe used as a structural column is to support an axial load of 45,000 lb. If the allowable unit stress in the column is 12,000 psi, what size pipe should be used?          *Ans.* 4-in. nominal diameter

**1-13.** In Fig. 1-7 is shown the cross section of a heavy structural steel column built up of one 12- by $\frac{1}{2}$-in. web plate, two 13- by $\frac{3}{8}$-in. cover plates, and four 6- by 4- by $\frac{5}{8}$-in. angles. If the allowable unit stress in the column is 14,000 psi, what total load *P* can it support?          *Ans.* $P = 548{,}000$ lb

**1-14.** Two steel straps, each 2 in. wide by $\frac{3}{8}$ in. thick, are held together with a $\frac{3}{4}$-in.-diameter unfinished bolt, as shown in Fig. 1-8. (*a*) If the allowable shearing stress in the bolt is 10,000 psi, what maximum axial load *P* can the bolt resist in shear? (*b*) If the bolt hole is of $\frac{7}{8}$-in. diameter and the allowable unit tensile stress in the straps is 12,000 psi, what maximum load *P* is permitted in tension? Possible friction between the straps may be disregarded.

*Ans.* (*a*) *P* = 4,420 lb; (*b*) *P* = 5,060 lb

FIG. 1-8 Probs. 1-14 and 1-15.      FIG. 1-9 Probs. 1-16 and 1-17.      FIG. 1-10 Prob. 1-18.

**1-15.** The tensile load $P$ in Fig. 1-8 is 5,670 lb. Calculate (a) the required diameter $d$ of the bolt and (b) the required width $w$ of the straps. (Select bolt diameter of nearest higher multiple of $\frac{1}{8}$ in.) Assume that bolt-hole diameter is $\frac{1}{8}$ in. larger than bolt and that thickness of strap is $\frac{3}{8}$ in. Allowable stresses: shear, 10,000 psi; tension, 15,000 psi. Disregard possible friction between the straps.

**1-16.** An eyebar is connected to a fitting by means of a $\frac{1}{2}$-in.-diameter finished pin, as shown in Fig. 1-9. (a) What maximum pull $P$ in the bar can this pin resist in shear if its allowable shearing stress $s_s$ is 15,000 psi? (b) If a tensile stress $s_t$ of 20,000 psi is permitted in the round bar, what diameter $d$ will give the bar the same strength in tension as the pin has in shear?

*Ans.* (a) $P = 5,370$ lb; (b) $d = 0.585$ in. (use $d = \frac{5}{8}$ in.)

**1-17.** The pull $P$ on the eyebar in Fig. 1-9 is 3,000 lb. Determine the required diameters $d$ of (a) the pin and (b) the eyebar. Allowable stresses: shear, 15,000 psi; tension, 20,000 psi. (Select diameters of nearest higher multiple of $\frac{1}{16}$ in.)

*Ans.* (a) $d = \frac{3}{8}$ in.; (b) $\frac{7}{16}$ in.

**1-18.** In Fig. 1-10 a 1-in.-diameter square-headed bolt passes through a heavy supporting steel plate. Under the tensile pull $P$ the bolt might fail in three ways: (1) in tension across the 1-in.-diameter shank, (2) in shear on the cylindrical surface formed if the shank of the bolt should pull straight out of its head, and (3) in bearing on the contacting surfaces between the bolt head and the supporting plate. Calculate the average unit tensile, shearing, and bearing stresses resulting from a pull $P$ of 15,000 lb.

## 1-8 Axial Deformation or Strain

The meaning of the term **strain,** as used in engineering practice, is often misunderstood. *Strain is deformation,* although the word is often used by laymen to designate the *force* which produces deformation in some object. For example, in an expression such as "the fast-moving train subjected the bridge to a *terrific strain,*" the word *strain* is undoubtedly used to mean *force* rather than deformation, since, in the case envisioned, the force may be "terrific," while actually the strain or deformation may be comparatively small.

All material bodies which are subjected to external forces, with internal stresses produced as a consequence, are necessarily deformed (strained). For example, a long rod subjected to an axial tensile load would be stretched or elongated, while a column supporting an axial load would be compressed or shortened. The *total deformation* produced in a member is designated by the Greek letter $\delta$ (delta). If the length of the member is $L$, the deformation per unit of length, expressed by the Greek letter $\epsilon$ (epsilon), is

$$\text{Unit deformation} = \frac{\text{total deformation}}{\text{length}} \quad \text{or} \quad \epsilon = \frac{\delta}{L} \quad (1\text{-}2)$$

The quantities total deformation $\delta$ and length $L$ are generally given in inches. Consequently, unit deformation $\epsilon$ will be in inches per inch. Obviously, then, the quantity *unit deformation* carries no mathematical unit, since inches/inches $= 1$. In a perfectly straight bar of homogeneous material and of constant cross section, $\epsilon$ would be the *actual* unit deformation. Since such perfection does not exist, we must realize that, in practice, $\epsilon$ is the average unit deformation.

The amount of deformation produced in a given member by a given force will vary with the stiffness of the material of which the member is made. In some instances, a considerable deformation may not be objectionable; in others, even small deformations may produce serious results. For example, a deflection of as much as 6 in. of a 50-ft airplane wing under certain flight conditions might well correspond with expectations, while so high a deflection of a 50-ft beam in a building would probably cause its condemnation. Also, in many precision machines even exceedingly small deformations of certain parts may render a machine unserviceable.

## 1-9 Elasticity. Elastic Limit

Another important property of a structural material is that, having been deformed by a force, it be capable of complete recovery of its original form upon removal of the deforming force. A material having this property is said to be **elastic.**

Considerable misconception exists regarding the true technical meaning of elasticity. Popularly, a material is thought to be "elastic" if it will withstand a high percentage of deformation without injury. Thus, rubber is considered to be highly elastic. Technically speaking, however, a material is elastic only if it has the ability to return to its original form after removal of a deforming force. Within this meaning, steel and glass both are elastic.

Good examples of the value of the elastic property of steel is found in the wide use of steel springs: the balance and main springs of a watch, the coil springs used in furniture and beds and as wheel springs on many automobiles, and the spring in a door check which automatically shuts the door after it has been opened.

In all these instances, the satisfactory performance of the intended duty of each spring depends upon the complete recovery of its original form after removal of a deforming force. Failure so to recover would render such a spring useless.

However, in order to preserve the elastic property of a material having limited physical strength, deformations and the stresses which accompany the deformations must not exceed a certain limit, appropriately referred to as its **elastic limit.** A material stressed beyond its elastic limit will return only partially to its original form upon complete removal of the deforming force. The deformation remaining is called **permanent set.**

## 1-10 Hooke's Law. Modulus of Elasticity

According to Hooke's law, *stress is proportional to strain.* To this should be added, *within the elastic limit.*

The various properties of materials are now quite readily being determined in testing laboratories under well-standardized conditions, and Hooke's law can easily be verified.

For example, to verify the proportionality of stress to strain, and to determine the elastic limit of so-called "soft" steel, a standard bar is clamped tightly into a testing machine capable of exerting a gradually increasing pull on the bar. At various intervals, the pull on the bar and the resulting elongation are measured simultaneously. Equal increments in stress are found to produce equal increments in deformation up to a stress of about 36,000 psi, thus substantiating the theory of proportionality of stress to strain. For stresses beyond this point, the elongations will increase at a much faster rate than the stresses, thus indicating that the elastic limit has been passed.

The proportionality of stress to strain is expressed as *the ratio of unit stress to unit strain,* or deformation. In a stiff but elastic material, such as steel, we find that a given unit stress produces a relatively small unit deformation. In a softer, but still elastic material, such as bronze, the deformation caused by the same intensity of stress is about twice that of steel, and in aluminum it is three times that of steel.

Consequently the ratio of unit stress $s$ to unit strain $\epsilon$ of any given material, which may be determined experimentally, then gives us a measure of its stiffness or elasticity which we call the **modulus of elasticity** of the material, denoted by the symbol $E$. That is,

$$\text{Modulus of elasticity} = \frac{\text{unit stress}}{\text{unit deformation}} \quad \text{or} \quad E = \frac{s}{\epsilon} \quad (1\text{-}3)$$

We now have a fixed relationship between modulus of elasticity, unit stress, and unit deformation. The values of the moduli of elasticity of most engineering materials are given in various handbooks (see also Tables 1-2, 8-1, and E-1). For example, for steel, $E$ is about 29,000,000 psi; for bronze, $E$ is about 15,000,000 psi; for brass, $E$ is about 13,500,000 psi; for timber, which is somewhat less elastic, $E$ varies from 1,100,000 to 1,760,000 psi; and for concrete, still less elastic, $E$ varies roughly from 1,500,000 to 3,000,000 psi.

In Eq. (1-2) the relationship $\epsilon = \delta/L$ (or $\delta = \epsilon L$) was established, and, from Eq. (1-3), $\epsilon = s/E$. Also, $s = P/A$. We may now formulate an equation to determine the total deformation $\delta$ in a prismatic member subjected to an axial load $P$. That is,

$$\delta = \epsilon L = \frac{s}{E} \cdot L = \frac{P}{A} \cdot \frac{1}{E} L$$

or

$$\delta = \frac{PL}{AE} \quad (1\text{-}4)$$

Table 1-2 **SOME AVERAGE PROPERTIES OF STRUCTURAL MATERIALS (STEADY LOADS), PSI**

| Material | Allowable stress | | | Modulus of elasticity | Shearing modulus of elasticity |
|---|---|---|---|---|---|
| | Tension | Compression | Shear | | |
| Steel, carbon[a] | 22,000 | 22,000 | 14,500 | 29,000,000 | 12,000,000 |
| low alloy | 25,000–30,000 | 25,000–30,000 | 17,000–20,000 | 29,000,000 | 12,000,000 |
| Cast iron, gray | 5,000–8,000 | 20,000–30,000 | 7,500–12,500 | 14,000,000–18,300,000 | 5,600,000–7,400,000 |
| Aluminum alloy | 15,000–22,000 | 15,000–22,000 | 10,000–14,000 | 10,000,000 | 3,800,000 |
| Brass, cast | 8,000 | 15,000 | . . . | 13,500,000 | 5,100,000 |
| Bronze, Tobin | 10,000 | 10,000 | . . . | 15,000,000 | |
| Copper rods, bolts | 8,000 | 8,000 | . . . | 16,000,000 | 6,000,000 |

[a] ASTM A36 steel.

where $\delta$ = total axial deformation, in.

$P$ = total axial load, lb

$L$ = length of member, in.

$A$ = cross-sectional area of member, sq in.

$E$ = modulus of elasticity of material, psi

Equation (1-4) holds only for stresses *below* the elastic limit of the material. A discussion of the determination of elastic limits of various materials is given in Chap. 3.

## ILLUSTRATIVE PROBLEM

**1-19.** Determine the total elongation $\delta$ of a steel rod 1 in. in diameter and 20 ft long produced by a tensile load $P$ of 15,000 lb if the modulus of elasticity $E$ of the steel is 29,000,000 psi.

*Solution:* According to the list of values following Eq. (1-4), the length of 20 ft must be converted to inches. A bar of 1-in. diameter has a cross-sectional area of 0.7854 sq. in. Then, since $P$ = 15,000 lb, $L$ = (20)(12) = 240 in., $A$ = 0.7854 sq in., and $E$ = 29,000,000 psi,

$$\left[\delta = \frac{PL}{AE}\right] \qquad \delta = \frac{(15,000)(240)}{(0.7854)(29,000,000)} = 0.158 \text{ in.} \qquad Ans.$$

## PROBLEMS

**1-20.** A surveyor's steel tape with a cross-sectional area of 0.006 sq in. must be stretched with a pull of 16 lb when in use. If the modulus of elasticity of this steel is 30,000,000 psi, (a) what is the total elongation $\delta$ in the 100-ft tape; (b) what unit tensile stress is produced by the pull?

*Ans.* (a) $\delta = 0.1069$ in.; (b) $s_t = 2,670$ psi

**1-21.** A round, hollow cast-iron column of 8-in. outside diameter and 6-in. inside diameter, 16 ft long, supports an axial compressive load of 220,000 lb. Calculate the total shortening produced in this column if the modulus of elasticity of cast iron is 15,000,000 psi.    *what area is used?*

**1-22.** A $\frac{3}{4}$- by 4-in. steel eyebar 20 ft long between centers of the pins at its ends is used as a diagonal member in a light bridge, as shown in Fig. 1-11. Assume the bar to be of constant cross-sectional area throughout its length of 20 ft. If the total tensile stress in the bar is 60,000 lb and if $E$ is 30,000,000 psi, calculate (a) the unit tensile stress $s_t$; (b) the unit elongation $\epsilon$; (c) the total elongation $\delta$.

*Ans.* (a) $s_t = 20,000$ psi; (b) $\epsilon = 0.000\ 67$ in. per in.; (c) $\delta = 0.16$ in.

FIG. 1-11  Prob. 1-22.

FIG. 1-12  Prob. 1-25.

**1-23.** A vertical timber column 6 by 8 in. of Douglas fir, S4S, and 10 ft 8 in. long is subjected to a vertical axial load of 32,000 lb. Calculate (a) the unit stress in the timber; (b) the unit axial deformation (see Table 8-1 for values of $E$); (c) the total axial deformation.

*Ans.* (a) $s_c = 774$ psi; (b) $\epsilon = 0.000\ 44$ in. per in.; (c) $\delta = 0.0563$ in.

**1-24.** A section of a heavy concrete foundation wall 10 ft high and of uniform thickness is subjected to a vertical load which produces a unit stress $s$ of 500 psi. If $E$ for this concrete is 2,000,000 psi, calculate (a) the unit vertical deformation, and (b) the total vertical deformation.

*Ans.* (a) $\epsilon = 0.000\ 25$ in. per in.; (b) $\delta = 0.03$ in.

**1-25.** The ends of the laminated-wood roof arch shown in Fig. 1-12 are tied together with a horizontal steel rod 90 ft 10 in. long which must withstand a total stress of 60,000 lb. Two turnbuckles are used. All threaded rod ends are upset. (a) Determine the required diameter $d$ of rod if the maximum allowable stress is 20,000 psi. (b) If the unstressed length of the rod is 90 ft 10 in. and there are 4 threads per inch on the upset ends, how many turns of one turnbuckle will bring it back to its unstressed length after it has elongated under full allowable tensile stress? Use $E = 29,000,000$ psi.

**1-26.** A copper aerial wire 160 ft long, weighing 0.0396 lb per ft, and having a cross-sectional area of 0.010 28 sq in., is suspended vertically from its upper end; $E = 15,000,000$ psi. (a) Calculate the total amount $\delta$ this wire will stretch because of its own weight. (b) Find the total elongation $\delta$ if a weight of 50 lb is attached to its lower end. (c) Determine what maximum weight $W$ this wire can safely support at its lower end if the unit stress in it must not exceed its elastic limit of 10,000 psi.    *Ans.* (a) $\delta = 0.0394$ in.; (b) $\delta = 0.662$ in.; (c) $W = 96.46$ lb

**1-27.** A piston rod of a steam engine is of 4-in. diameter and is alternately subjected to equal tensile and compressive stresses of 10,000 psi. If the distance between two points $A$ and $B$ on the unstressed rod is 4 ft 10 in., calculate the minimum and maximum distances between $A$ and $B$ when the rod is fully stressed to 10,000 psi. Assume $E = 29,000,000$ psi.          *Ans.* Min $L = 57.98$ in.; max $L = 58.02$ in.

## 1-11 Shearing Deformation and Poisson's Ratio

A shearing force causes a shearing deformation, just as an axial force causes a shortening or a lengthening of a member. A shearing deformation is angular (see Fig. 1-13). The shearing force $F$ causes the material to deform into the dashed position. The total deformation is $\delta_s$, and this deformation occurs over the length $L$. The deformation per unit of length is $\delta_s/L = \tan \phi$. For very small angles the tangent of the angle is practically equal to the angle expressed in radians. Therefore the shearing unit strain is the angle $\phi$ (Greek letter phi) in radians. Now, just as normal unit stress is related to unit strain by Hooke's law ($s = E\epsilon$), so is shearing unit stress related to shearing unit deformation. The relationship is

$$s_s = G\phi \qquad\qquad (1\text{-}5)$$

where $G$ is the shearing modulus of elasticity, usually in pounds per square inch. Values of $G$ for common metals are given in Table 1-2. The relative motion between the ends of a shaft under torsion is due to shearing deformation. Such deformations will be studied in Chap. 4.

The ratio of unrestrained unit lateral contraction (or expansion) to the unit longitudinal elongation (or contraction) is called **Poisson's ratio** $\mu$. We are aware of the Poisson effect although we may not have associated a name with the phenomenon. Pulling on a rubber band not only lengthens it in the direction of the pull but also visibly contracts the lateral dimensions of the

FIG. 1-13 Shearing deformation.    FIG. 1-14 Poisson's ratio.

cross section. In symbols (see Fig. 1-14) Poisson's ratio is

$$\mu = \frac{\delta'/b}{\delta/L} \tag{1-6a}$$

where $\delta' = |b - b'|$ and $b$ is the original lateral dimension. Since $\delta/L$ is the unit deformation $\epsilon$, Eq. (1-6a) yields

$$\delta' = \mu\epsilon b \tag{1-6b}$$

Numerical values of Poisson's ratio $\mu$ generally vary between 0.25 and 0.35. Values can be found in handbooks or can be determined by Eq. (1-7) if the appropriate moduli of elasticity are known. The shearing modulus $G$ and the modulus of elasticity $E$ are related by

$$2G = \frac{E}{1 + \mu} \qquad \text{or} \qquad \mu = \frac{E}{2G} - 1 \tag{1-7}$$

### ILLUSTRATIVE PROBLEM

1-28. Compute the numerical value of Poisson's ratio for copper. If a 20-in.-long copper rod of $\frac{1}{2}$-in. diameter is subjected to an axial pull $P$ of 800 lb, determine the change in diameter.

*Solution:* From Table 1-2, $E = 16,000,000$ psi and $G = 6,000,000$ psi. Substitution of these values into Eq. (1-7) gives

$$\left[\mu = \frac{E}{2G} - 1\right] \qquad \mu = \frac{16,000,000}{(2)(6,000,000)} - 1 = 0.333 \qquad Ans.$$

The unit tensile stress $s_t$ is

$$\left[s_t = \frac{P}{A}\right] \qquad s_t = \frac{800}{0.196} = 4,080 \text{ psi}$$

The unit elongation can now be found:

$$\left[E = \frac{s}{\epsilon}\right] \qquad \epsilon = \frac{s}{E} = \frac{4,080}{16,000,000} = 0.000\ 255$$

Now from Eq. (1-6b)

$$[\delta' = \mu\epsilon b] \qquad \delta' = (0.333)(0.000\ 255)(0.5) = 0.000\ 0425 \text{ in.} \qquad Ans.$$

### PROBLEMS

1-29. Electrical strain gages mounted on the surface of a cylindrical-shaped block of metal subjected to axial compressive load show a unit longitudinal strain of 0.0012 and a unit lateral expansion of 0.0004. If the block is 2 in. in diameter and the load $P$ is 40,000 lb, determine (a) Poisson's ratio; (b) the modulus of elasticity $E$; and (c) the shearing modulus of elasticity $G$.

1-30. What pressure in psi must be exerted along the edges of a 10-in.-wide aluminum plate if the 10-in. dimension is to remain constant when the longitudinal compressive stress is raised by 10,000 psi?                                    *Ans.* $s = 3,150$ psi

## 1-12  Determination of Allowable Stress and Modulus of Elasticity. The Stress-Strain Diagram

The allowable stresses that are considered as safe to use in design are usually specified by the governmental agency having jurisdiction over the particular design, such as the Federal, state, county, or city government, and the Army, Navy, Department of Commerce, etc. They are published in the form of codes and are readily available to designers.

As previously mentioned, these allowable stresses are based upon extensive laboratory tests together with a vast amount of accumulated experience. Such tests furnish information concerning the strength, elasticity, ductility, etc., of the material tested. One of the most common of these, the tension test of a soft-steel bar, is described below.

**The Stress-Strain Diagram.** In the tension test most commonly used for metals, a testing machine applies a controlled and gradually increasing tensile force to a round bar until rupture finally occurs. The total pull on the bar at any time during the test is measured by means of scales which are a part of the machine. Extensometers capable of measuring as little as 0.0001 in. are used to measure deformations. At stated intervals, the total tensile force on the bar and the total deformation in an 8-in. section of it are measured simultaneously. From these measurements, apparent unit stresses and strains are calculated and are then plotted to give the stress-strain diagram shown in Fig. 1-15. For soft steel, the stress is found to be proportional to the strain

FIG. 1-15 Stress-strain diagram. Soft steel.

FIG. 1-16 Necking of soft-steel bar.

up to a stress of about 30,000 psi, shown as point $A$ on the curve, thus proving Hooke's law. For stresses beyond point $A$, the strain increases at a faster rate than the stress. Consequently $A$ is the **proportional limit.** At a stress of about 33,000 psi, the **elastic limit** of the material is reached. That is, if subjected to stresses beyond this point, the bar will no longer return to its original length after removal of the load. In other words, it has acquired a permanent set.

From a practical standpoint most designers make little or no distinction between the proportional limit and the elastic limit, both generally being known by the latter term. Since permanent set due to stress beyond the elastic limit is generally objectionable, the allowable stress of a material never exceeds its elastic limit but is usually fixed at some reasonable lower point.

Shortly after the elastic limit has been reached (at $B$ on the curve), a sudden elongation takes place, while the load on the bar actually drops. That is, the material suddenly *yields.* The stress at which this occurs is called the **yield point.** However, the bar again immediately picks up ability to resist increasing stress, but the elongations now increase at a much faster rate than the stresses.

As the test is continued, a point of maximum stress—the **ultimate strength** of the material—that the bar is capable of resisting is finally reached at $E$. That is, the ordinate to $E$ on the curve represents the *ultimate strength* of the material. Beyond point $E$, elongations will continue but at gradually decreasing stresses until finally rupture of the bar occurs. After passing the point of ultimate strength at $E$, a more rapid local failure starts and the bar in Fig. 1-16$a$ begins to "neck," as illustrated in Fig. 1-16$b$, with a consequent reduction in cross-sectional area.

All the unit stresses plotted in the stress-strain diagram are based on the original cross-sectional area of the unstressed bar. This is common practice, but it results in a curve that does not give the true stresses. Obviously, as the bar begins to elongate, there is a consequent reduction in its diameter, which becomes measurable soon after the yield point has been passed and, of course, becomes obvious when necking starts. If the actual rather than the original area is used, the actual calculated stresses give a curve following approximately the dashed line in Fig. 1-15.

If, where point $C$ on the curve is reached, the load on the bar should gradually be decreased, and if simultaneous readings of load and elongations are taken as before and are plotted, the points will fall approximately on line $CD$. That is, an elongation equal to $OD$ will remain after complete removal of all load, thus indicating that amount of *permanent set.*

Line *CD* is found to be parallel to *OA*, thereby indicating no change in the modulus of elasticity of the steel. If then loads are again applied as previously, the new points, if plotted, will also fall approximately along line *DC*, thus indicating that the drawing out of the material of the bar has had the effect of increasing both its proportional and its elastic limits. In fact, by such a process of cold drawing, the usable elastic limit of soft steel can be increased almost to its ultimate strength.

In a soft-steel bar, the **percentage of elongation** in the 8-in. gage length might average about 30 percent. The reduction in area at the ruptured section would be about 50 percent. These percentages show that soft steel is a highly *ductile* material. This property of ductility is exceedingly valuable in steel used for buildings, bridges, derricks, etc., since noticeable deformations often would give ample warning of impending failure.

A similar test on a specimen of *hard steel* (0.6 percent carbon) would show it to have greater strength, higher elastic limit, and no yield point, as indicated by the hard-steel curve in Fig. 1-15. Obviously, hard steel is brittle and would give little warning of impending failure.

A number of other materials, such as cast iron, concrete, and wood, commonly used in the design of machines and structures, have elastic properties somewhat less well defined than those of steel. Study of their stress-strain diagrams, as shown in Fig. 1-17, reveals chiefly their lack of proportionality of stress to strain. This lack has not been found to be especially objectionable. It does, however, complicate the matter of determining the modulus of elasticity of these materials. In such cases the slope of a tangent to the curve at the average allowable stress is often used to determine the modulus of elasticity.

While most steels have approximately equal strengths in tension and compression, wood is slightly stronger in tension than in compression. But copper, brass, cast iron, and concrete are much stronger in compression than in tension. Concrete has practically no usable strength in tension, and is therefore generally reinforced with steel rods, which then resist the tensile stresses.

FIG. 1-17 Stress-strain diagram. Gray cast iron, concrete, and wood.

FIG. 1-18 Stress concentrations. (*a*) Circular hole. (*b*) Circular fillets. (*c*) Stress-concentration factors for flat bars.   *Refer also to Machinery's Handbook.*

## 1-13 Stress Concentration

The relationship $s = P/A$ gives the average stress at any cross section of a slender *prismatic* tension member subjected to axial load. However, an abrupt change in cross section (such as often occurs near the end of a member because of rivet holes, for example) changes radically the distribution of stress at that section. As indicated in Fig. 1-18*a*, a concentration of stress, which often exceeds twice the average stress, will occur at the edge of a hole in a bar, the stress depending on the ratio of the radius $r$ of the hole and the net width $b$ of the bar, as shown in Fig. 1-18.

The reasons for such stress concentrations are readily understandable. Let us say that Fig. 1-18*a* shows only a small section of a much longer bar under stress. We may reasonably assume that a short distance below the hole, such as at section *HI*, the stress is fairly uniformly distributed over the cross-sectional area of the bar. If now we think of the stress as *flowing* upward through the resisting material of this member, it becomes clear that the stress in the material directly below the hole, as it approaches *E*, is diverted around the hole through the material at and near edges *D* and *F* and that it then again converges somewhere above *E′*, probably reaching fairly uniform distribution again at section *AB*. This diversion naturally causes the material at and near *D* and *F* to be more highly stressed. The ordinates shown in the diagram represent probable stresses on a plane through *CDFG*.

Figure 1-18*b* shows similar stress concentrations at another type of change in cross section. The ordinates represent stress on a plane through *CD* if flow of stress is downward.

These stress concentrations should be carefully considered by designers, especially in connection with hard and brittle materials, since such materials are apt to begin to crack without much warning when stresses exceed the elastic limit.

In softer metals, such as ordinary structural steel, the material at points of high stress will yield after the elastic-limit stress has been exceeded, thereby shifting to adjacent material the burden of carrying some of the excess stress. Undoubtedly, excessive stresses, yielding, and consequent redistribution of stresses are frequent occurrences in designs using structural materials capable of yielding.

The maximum stress adjacent to a hole or fillet is given by the equation

$$s_{max} = K \frac{P}{A_{min}} \tag{1-8}$$

where $K$ is a stress-concentration factor which can be obtained from the graph in Fig. 1-18$c$.

## ILLUSTRATIVE PROBLEM

**1-31.** A $\frac{1}{2}$-in.-diameter hole is drilled on the center line of a steel strap $1\frac{1}{2}$ in. wide by $\frac{1}{4}$ in. thick, subjected to an axial pull $P = 1,500$ lb. Calculate ($a$) the unit stress $s_t$ in the strap several inches away from the hole, and ($b$) the maximum unit stress adjacent to the hole.

*Solution:* The stress some distance away from the hole is fairly uniformly distributed and is given by the equation

$$\left[ s_t = \frac{P}{A} \right] \qquad s_t = \frac{1,500}{(0.25)(1.5)} = 4,000 \text{ psi} \qquad Ans.$$

The net width $b$ at the hole is $1\frac{1}{2} - \frac{1}{2} = 1$ in., and the ratio of the hole radius $r$ to the net width $b$ is $r/b = 0.25/1 = 0.25$. Now, from Fig. 1-18$c$, for $r/b = 0.25$ we find for a circular hole that the stress-concentration factor $K$ is 2.27. Then from Eq. (1-8)

$$\left[ s_{max} = K \frac{P}{A_{min}} \right] \qquad s_{max} = 2.27 \frac{1,500}{(1)(0.25)} = 13,620 \text{ psi} \qquad Ans.$$

## PROBLEMS

**1-32.** Solve Prob. 1-31 when $P = 2,000$ lb and the hole diameter is $\frac{3}{4}$ in.
*Ans.* ($a$) $s_t = 5,333$ psi; ($b$) $s_{max} = 22,700$ psi

**1-33.** A flat steel bar $3\frac{1}{2}$ in. wide and $\frac{3}{8}$ in. thick is reduced in width to $2\frac{1}{2}$ in. by circular fillets of $\frac{1}{2}$-in. radius on each side. If the bar is subjected to an axial pull of 10,000 lb, calculate ($a$) the stress $s_t$ in the wider part of the bar some distance from the change in section, ($b$) the average tensile stress in the narrow part of the bar, and ($c$) the maximum tensile stress adjacent to the circular fillet.
*Ans.* ($a$) $s_t = 7,620$ psi; ($b$) $s_t = 10,670$ psi; ($c$) $s_{max} = 17,700$ psi

**1-34.** A 2-in.-diameter hole is drilled on the center line of a steel bar 4 in. wide by $\frac{1}{2}$ in. thick. What axial pull $P$ can be applied to the bar without exceeding a maximum stress at any point in the bar of 20,000 psi?        *Ans.* $P = 9,380$ lb

**1-35.** A long steel bar is 6 in. wide and 1 in. thick. A 3-ft-long section must be narrowed as much as possible. What is the minimum width that may be used for this section if circular fillets are used at each end of the narrowed section, the maximum permissible local stress is 35,000 psi, and the axial pull on the bar will be 75,000 lb?

## 1-14 Axial Stresses in Members of Two Materials

When a member subjected to axial tensile or compressive stress is made of two materials so joined that both must deform equally under stress, their unit stresses will be proportional to their moduli of elasticity.

Let the subscripts 1 and 2 respectively designate the materials having the lowest and the highest modulus of elasticity. Since $E = s/\epsilon, \epsilon = s/E$, and since the two materials are subjected to equal deformations,

$$\epsilon = \frac{s_1}{E_1} = \frac{s_2}{E_2} \tag{a}$$

thus proving that, for equal unit deformations, the stresses are proportional to the moduli of elasticity of the materials. By cross multiplication we have

$$s_2 = \frac{E_2}{E_1}s_1 \quad \text{or} \quad s_2 = ns_1 \tag{b}$$

where $n = E_2/E_1$. Clearly, the total load $P$ equals the load $s_1A_1$ carried by the one material plus the load $s_2A_2$ carried by the other material. That is,

$$P = s_1A_1 + s_2A_2 \qquad \text{PARALLEL} \tag{c}$$

or, substituting $ns_1$ for $s_2$,

$$P = s_1A_1 + ns_1A_2 = s_1(A_1 + nA_2) \tag{1-9}$$

The quantity $nA_2$ may be considered as an "equivalent area" of the lower-stressed material, replacing the higher-stressed material but carrying the same portion of the total load, since $s_2A_2 = ns_1A_2$.

### ILLUSTRATIVE PROBLEM

**1-36.** A heavy rigid iron casting weighing 1,964 lb is suspended from three $\frac{1}{4}$-in.-diameter wires which are symmetrically spaced and are 24 ft 2 in. long. The two outer wires are of steel having a modulus of elasticity $E$ of 29,000,000 psi and the middle wire is of bronze, for which $E$ is 14,500,000 psi. Calculate (a) the unit stresses in the bronze and the steel wires; (b) the total elongation $\delta$ of each wire. The weight of the wires may be neglected.

FIG. 1-19 Prob. 1-36.

*Solution:* Study of Fig. 1-19 discloses that, because the bar is rigid and the wires symmetrically support the bar, all three wires are subjected to the same total and unit elongations. Let the subscripts $B$ and $S$ denote bronze and steel, respectively.

Since the bronze carries the lowest stress, let $s_B$, $A_B$, and $E_B$ designate the bronze and $s_S$, $A_S$, and $E_S$ the steel; $n = E_S/E_B = 29,000,000/14,500,000 = 2$. Then, since $P = 1,964$ lb, $A_B = 0.0491$ sq in., $n = 2$, and $A_S = 2(0.0491) = 0.0982$ sq in.,

$$[P = s_B(A_B + nA_S)] \qquad 1,964 = s_B[0.0491 + 2(0.0982)]$$

or, the unit stress in the bronze is

$$s_B = \frac{1,964}{0.0491 + 0.1964} = \frac{1,964}{0.2455} = 8,000 \text{ psi} \qquad Ans.$$

and the unit stress in the steel is

$$s_S = ns_B = (2)(8,000) = 16,000 \text{ psi} \qquad Ans.$$

To determine the total elongation $\delta$, we may use the relationships $\epsilon = s/E$ and $\delta = Ls/E$. Either the bronze or the steel may be used. Since $L = 24$ ft 2 in. $= 290$ in., using the bronze,

$$\left[\delta = \frac{Ls}{E}\right] \qquad \delta = \frac{(290)(8,000)}{14,500,000} = 0.16 \text{ in.} \qquad Ans.$$

or, using the steel,

$$\delta = \frac{(290)(16,000)}{29,000,000} = 0.16 \text{ in.} \qquad Ans.$$

## PROBLEMS

**1-37.** A concrete column measuring 10 in. square and reinforced with four 1-in. round steel bars, as shown in Fig. 1-20, supports an axial compressive load $P$ of 91,200 lb. Steel bearing plates at top and bottom of column assure equal deformation of steel and concrete. Calculate the unit stress in each material, assuming that for concrete $E = 2,000,000$ psi and for steel $E = 30,000,000$ psi.

**1-38.** Solve Prob. 1-36 when the casting weighs 1,500 lb, the wires are 10 ft long, the outer wires are aluminum, and the middle wire is of steel.

*Ans.* (a) $s_A = 6,235$ psi; $s_S = 18,090$ psi; (b) $\delta = 0.0748$ in.

**1-39.** In Fig. 1-21 is shown a timber post measuring 8 by 6 in. and reinforced with two $\frac{1}{4}$- by 6-in. steel plates thoroughly fastened to the post. The compressive load

Fig. 1-20 Prob. 1-37.     Fig. 1-21 Probs. 1-39 and 1-40.     Fig. 1-22 Probs. 1-41 and 1-42.

$P$ on the post is 108,000 lb. Calculate the stress in each material. Use $E_W = 1,800,000$ psi and $E_S = 30,000,000$ psi.

**1-40.** The short wood post in Fig. 1-21 is of cypress and measures 2 in. square. It is reinforced on two sides with copper plates, measuring $\frac{1}{8}$ by 2 in., which are thoroughly fastened to the wood. The post supports an axial compressive load of 9,500 lb. Calculate the unit stress in the wood and the copper. See Table 8-1 for the modulus of elasticity of cypress and use 14,500,000 psi for copper.

*Ans.* $s_w = 912.5$ psi (wood) and $s_c = 11,700$ psi (copper)

**1-41.** In Fig. 1-22 is shown a structural-steel rod extending through a bronze tube. The cross-sectional area of the steel rod is 0.6 sq in., and its upper end has 20 threads per inch. The bronze tube is 25 in. long, and its area is 1.8 sq in. Calculate the unit stresses in the unthreaded length of the steel rod and in the bronze tube which would result from a quarter turn of the nut on the bolt. (HINTS: The elongation of the rod plus the shortening of the tube equals the change in position of the nut along the rod. Also, the load $P_S$ on the bolt equals the load $P_B$ on the tube.)

*Ans.* $s_S = 8,820$ psi; $s_B = 2,940$ psi

**1-42.** In the preceding problem, what will be the results if the rod is of steel and the tube is of aluminum?

## 1-15 Temperature Stresses

When a material body is subjected to changes in temperature, it will expand when the temperature increases and contract when it decreases.

Through experiments, it is possible to measure the change in length per unit of length of a bar of material for each degree of change in temperature. This change is called the **linear coefficient of thermal expansion** of the material, and is denoted by the Greek letter $\alpha$ (alpha). Some average values of $\alpha$ for structural materials are given in Table 1-3.

If a straight bar of material is free to change its length with variations in

**Table 1-3  AVERAGE COEFFICIENTS OF THERMAL EXPANSION (PER °F)**

| | | | |
|---|---|---|---|
| Wood | 0.000 0030 | Wrought iron | 0.000 0067 |
| Glass | 0.000 0044 | Copper | 0.000 0093 |
| Concrete | 0.000 0060 | Bronze | 0.000 0100 |
| Cast iron | 0.000 0061 | Brass | 0.000 0104 |
| Steel | 0.000 0065 | Aluminum | 0.000 0128 |

temperature, no change will occur in its internal stress. If, however, the bar is restrained, partially or fully, so as to restrain or prevent change in its length due to temperature variations, internal stresses will be produced in it equal to those required to produce the prevented change in length.

If $\alpha$ is the unit deformation per degree change in temperature, $t$ the number of degrees of change, $L$ the length of a bar, and $\delta$ the total change in length, then the change per unit length is $\alpha t$, and the total change in length $L$ is

$$\delta = \alpha t L \tag{1-10}$$

But change in unit length $\alpha t$ is also denoted by $\epsilon$, in Art. 1-8, and $E = s/\epsilon$, or $s = E\epsilon$. Hence the unit stress produced in a fully restrained bar by a temperature change $t$ is

$$s = E\alpha t \tag{1-11}$$

### ILLUSTRATIVE PROBLEM

**1-43.** The straight bottom chord of a riveted steel bridge is 200 ft long. (*a*) Calculate the change in its length caused by a change in temperature of 70°F (a possible difference between day and night temperatures). (*b*) What unit stress will be produced in this chord if both of its ends are fully restrained against linear expansion of the chord? (*c*) If the lower chord is fully restrained between two virtually "immovable" concrete piers and its average cross-sectional area contains 20 sq in. of steel, what total force $P$ would it exert against its fastenings to the piers?

*Solution:* $\alpha = 0.000\ 0065$, $t = 70°$; $L = 200$ ft $= 2,400$ in. Also $E = 29,000,000$ psi. Hence

| | |
|---|---|
| $[\delta = \alpha t L]$ | $\delta = (0.000\ 0065)(70)(2,400) = 1.092$ in.    *Ans.* |
| $[s = E\alpha t]$ | $s = (29,000,000)(0.000\ 0065)(70) = 13,195$ psi    *Ans.* |
| $[P = sA]$ | $P = (13,195)(20) = 263,900$ lb    *Ans.* |

### PROBLEMS

**1-44.** Calculate the change in length $\delta$ of a copper aerial wire 180 ft long due to a change in temperature of 85°F.          *Ans.* $\delta = 1.708$ in.

**1-45.** A surveyor's steel tape measures exactly 100 ft between end markings at 70°F. What error is made in measuring a distance of 1,000 ft when the temperature of the tape is 32°F?

**1-46.** A continuous straight concrete pavement was laid in 40-ft rectangular sections. The expansion joints between ends of sections measured $\frac{1}{2}$ in. at 70°F. What width $w$ does a joint measure (a) at 30°F and (b) at 110°F?

Ans. (a) $w = 0.6152$ in.; (b) $w = 0.3848$ in.

**1-47.** The steel rails of a continuous, straight railroad track are each 40 ft long and are laid with spaces between their ends of 0.2 in. at 70°F. (a) At what temperature will the rails touch end to end? (b) What unit compressive stress will be produced in the rails, if the temperature rises to 150°F ($E = 29,000,000$ psi)?

**1-48.** Solve Prob. 1-43 when length of chord is 160 ft, change in temperature is 80°F, and the average cross-sectional area of the chord is 6 sq in.

**1-49.** A locomotive tire of hard steel with an inside diameter of 59.972 in. at 70°F is to be expanded by heat to a diameter of 60.050 in. (a) What temperature will produce this expansion? (b) What unit stress will exist in the tire after it has been placed upon a wheel with a diameter of 60 in. and is allowed to cool to 70°F? Let $E = 30,000,000$ psi.                      Ans. (a) 270°F; (b) $s = 14,000$ psi

## SUMMARY

### (By article number)

**1-2. Stress** is resistance to external forces. **Unit stress** is force per unit of area. Total stress is the total force within a member.

**1-3. The two basic stresses** are normal (tensile or compressive) and shearing. **Tensile stresses** pull on the fibers and tend to elongate them. **Compressive stresses** push on the fibers and tend to compress them. A stress directed parallel to the longitudinal axis of a member is called an **axial stress. Shearing stresses** are generally caused by forces acting transversely on a member. **Bearing stresses** are compressive stresses on external surfaces in contact under pressure.

**1-4. Ultimate stress** is the greatest stress a material is capable of resisting. **Allowable stress** is a predetermined stress considered safe for use in design. **Factor of safety** is the ratio of ultimate stress to allowable stress; it is a measure of the probable safety of a design.

**1-5. The direct-stress formula** establishes a relationship by which we may find the unit stress $s$ produced by a total force $P$ on a stressed area $A$. That is,

$$s = \frac{P}{A} \tag{1-1}$$

**1-7. Analysis** relates chiefly to the preliminary investigations which precede design. **Design** is the final determination of size, shape, and strength of a structure and of its component parts.

**1-8. Deformation,** or **strain,** is a measure of the change in form of a body caused by externally applied forces. In a prismatic bar subjected to axial stress,

$$\text{Unit deformation} = \frac{\text{total deformation}}{\text{length}} \quad \text{or} \quad \epsilon = \frac{\delta}{L} \tag{1-2}$$

**1-9. Elasticity** is that property of a material which enables it, within certain limits of stress, to recover its original form after removal of a deforming force. The **elastic limit** is that unit stress beyond which a material will no longer recover its original form. **Permanent set** is that deformation permanently sustained by an elastic material by being stressed beyond the elastic limit.

**1-10.** According to **Hooke's law: below the elastic limit, stress is proportional to strain (deformation).**

The measure of the elasticity of a given material, called the modulus of elasticity and denoted by $E$, is the ratio of its unit stress $s$ to unit strain $\epsilon$. In practical terms, $E$ is actually a measure of the stiffness of an elastic material, or its resistance to deformation under load. That is,

$$\text{Modulus of elasticity} = \frac{\text{unit stress}}{\text{unit deformation}} \quad \text{or} \quad E = \frac{s}{\epsilon} \tag{1-3}$$

Since $\delta = \epsilon L$, $\epsilon = s/E$, and $s = P/A$, the total elongation $\delta$ of a prismatic bar subjected to axial stress is

$$\delta = \epsilon L = \frac{s}{E} \cdot L = \frac{P}{A} \cdot \frac{L}{E} \quad \text{or} \quad \delta = \frac{PL}{AE} \tag{1-4}$$

**1-11. Shearing stress** is related to unit shearing deformation by

$$s_s = G\phi \tag{1-5}$$

The ratio of unrestrained lateral unit deformation to longitudinal unit deformation is called **Poisson's ratio.**

$$\mu = \frac{\delta'/b}{\delta/L} \quad \text{or} \quad \delta' = \mu\epsilon b \tag{1-6}$$

**1-13. Stress concentrations** occur at sudden changes in cross section in members under axial load. The maximum stress at this point is given by

$$s_{\max} = K \frac{P}{A_{\min}} \tag{1-8}$$

**1-14.** In a prismatic member made of two materials so joined that they deform equally under axial stress, the unit stresses in the two materials are proportional to their respective moduli of elasticity. The total load $P$ on such a member is

$$P = s_1 A_1 + n s_1 A_2 = s_1(A_1 + nA_2) \tag{1-9}$$

where $n = E_2/E_1$, $s_1$ is lowest of the two unit stresses, and $E_1$ is the lowest of the two moduli of elasticity.

**1-15.** Material bodies expand or contract with changes in temperature. For any material the change in length per unit of length per degree Fahrenheit change in temperature is called its **coefficient of thermal expansion** and is denoted by the Greek letter $\alpha$. The total linear deformation $\delta$ in an unrestrained bar of length $L$ caused by a change in temperature $t$ is

$$\delta = \alpha t L \tag{1-10}$$

The unit axial stress $s$ produced in a fully restrained bar by a temperature change $t$ is

$$s = E\alpha t \tag{1-11}$$

## REVIEW PROBLEMS

**1-50.** A steel rod 20 ft in length must resist a total axial stress of 60,000 lb in tension. What diameter $d$ is suitable ($a$) if ends are threaded and ($b$) if ends are upset? Allowable stress is 20,000 psi.                    *Ans. ($a$) $d = 2\frac{1}{4}$ in.; ($b$) $d = 2$ in.*

**1-51.** A steel wire 450 ft long is of $\frac{1}{8}$-in. diameter and weighs 0.042 lb per ft. If the wire is suspended vertically from its upper end, calculate ($a$) the maximum tensile stress due to its own weight, and ($b$) the maximum weight $W$ which it can safely support at its lower end if the unit tensile stress is not to exceed 24,000 psi.

**1-52.** If the maximum force $F$ in Fig. 1-23 which must be exerted by the punch $P$ to shear a hole 1 in. in diameter in a steel plate is 30,000 lb, calculate the average unit compressive stress $s_c$ in the punch.                    *Ans. $s_c = 38,200$ psi*

FIG. 1-23 Probs. 1-52 and 1-53.     FIG. 1-24 Probs. 1-54 and 1-55.

**1-53.** In punching a 1-in.-diameter hole in steel plate $\frac{1}{2}$ in. thick, the punch $P$ in Fig. 1-23 causes a cylindrical plug of steel to be sheared out of the plate. Calculate the required force $F$ if the ultimate shearing strength of the steel is 40,000 psi.

**1-54.** The 24-in. pulley shown in Fig. 1-24 is keyed to the 3-in.-diameter shaft with a rectangular steel key which is $\frac{1}{2}$ in. deep, by 2 in. long. The difference in pulls $T_1$ and $T_2$ is 625 lb. Determine the minimum required width $b$ of this key if the allowable average shearing stress is 13,500 psi.                    *Ans. $b = \frac{3}{8}$ in.*

**1-55.** In Prob. 1-54 assume that the pulley diameter is 20 in., the shaft diameter is $2\frac{1}{2}$ in., the given dimensions of the key are unchanged, and the difference in pulls $T_1$ and $T_2$ is 422 lb. Then determine the minimum required width $b$.

*Ans. $b = \frac{1}{4}$ in.*

**1-56.** A structural steel rod 1.5 in. in diameter and 25 ft long in supporting a balcony carries a tensile load of 29,000 lb. ($a$) Find the total elongation $\delta$ of this rod. ($b$) what diameter $d$ is required if the total elongation must not exceed 0.1 in.?

**1-57.** A steel bar 3 in. wide and $\frac{5}{16}$ in. thick must be reduced abruptly to a $1\frac{1}{2}$-in. width. What is the minimum fillet radius $r$ that can be used if the stress next to the fillet must not be greater than $1\frac{1}{2}$ times the average stress in the narrow portion of the bar when under axial load?                    *Ans. $r = 0.54$ in.*

**1-58.** An initially straight aluminum alloy strap $\frac{1}{4}$ in. thick is bent into the form of a circular hoop with a mean diameter of 200 in. The inner fibers are in compression, the outer fibers are in tension, and those midway between the two faces are unchanged in length and therefore unstressed. Compute ($a$) the unit elongation of the outer fibers and ($b$) the unit tensile stress in these fibers.

*Ans. ($a$) $\epsilon = 0.001\ 25$; ($b$) $s_t = 12,500$ psi*

**1-59.** A spanner wrench (see Fig. 1-25) is used to tighten the brass couplings on a fire hose. Compute the minimum diameter of the projecting brass pins if the allowable shearing stress in the pins is 5,000 psi and if the pull on the wrench will not exceed the value shown. Assume a smooth frictionless surface at A.

*Ans. d* = 0.366 in.

FIG. 1-25  Spanner wrench of Prob. 1-59.

**1-60.** The water-pump pliers shown in Fig. 1-26 are gripped with the hand, thus causing the 30-lb forces shown. Find the shearing stress in pin A if its diameter is $\frac{5}{16}$ in. Assume that the crushing force exerted on the pipe is in the direction of line BC.

*Ans. $s_s$* = 2,565 psi

FIG. 1-26  Water-pump pliers of Prob. 1-60.

**1-61.** A structural-steel eyebar $\frac{1}{2}$ in. thick and 20 ft 2 in. long between centers of pins carries a total stress of 36,000 lb in tension. If this bar must not elongate more than 0.15 in. and if E for this steel is 29,000,000 psi, calculate (*a*) the required width *w* of the bar and (*b*) the average unit stress at this maximum elongation.

*Ans.* (*a*) *w* = 4 in.; (*b*) $s_t$ = 18,000 psi

**1-62.** Two steel rods of $\frac{3}{4}$-in. diameter are joined by a turnbuckle (American Standard coarse threads) and are fastened at their outer ends to virtually immovable concrete piers spaced 40 ft apart. If the rods are already stressed, calculate the increase P in the total stress caused by one complete tightening turn of the turnbuckle. Assume that E is 29,000,000 psi and that the cross-sectional area is constant throughout the entire 40 ft.

*Ans. P* = 5,340 lb

**1-63.** To construct a building column of high strength, the inside of a steel cylinder is filled with concrete. The inside diameter of the cylinder is 12 in., and its wall thickness is $\frac{3}{8}$ in. The axial load is applied so as to bear correctly on the entire cross-sectional area of the top of the column. Calculate the stress in each material due to an axial load $P$ of 240,000 lb. Use $E_S = 30,000,000$ psi and $E_C = 2,000,000$ psi.

**1-64.** The cast-brass collar shown in Fig. 1-27 just slides on the 1-in.-diameter steel shaft ($T = 60°F$) when the collar temperature is 100°F. If both the shaft and the collar are at 60°F, what pull $P$ on the steel shaft will allow the collar to slide?

FIG. 1-27  Prob. 1-64.                    FIG. 1-28  Prob. 1-65.

**1-65.** A rigid bar is supported horizontally by two rods of equal diameter, as shown in Fig. 1-28. Disregard the weight of the bar and the rods. (*a*) If rod $A$ is of aluminum and $B$ of hard steel and both are of equal original length, what distance $d$ from $A$ should load $P$ be placed for equal total elongations of rods $A$ and $B$? (*b*) Assume now that rod $A$ is three times as long as rod $B$ but that the bar remains horizontal. If rod $A$ is of structural steel and $B$ of bronze, what distance $d$ from rod $A$ should $P$ be placed for equal total elongations of the rods?

*Ans.* (*a*) $d = 9$ in.; (*b*) $d = 7.2$ in.

**1-66.** A bronze wire is stretched tightly between two "virtually immovable" supports. If the unit tensile stress in the wire is 4,350 psi at 70°F, what is it (*a*) at 50°F and (*b*) at 90°F?        *Ans.* (*a*) $s_{50} = 7,350$ psi; (*b*) $s_{90} = 1,350$ psi

**1-67.** A cylindrical water tank is constructed of vertical wood staves tightly held in place by horizontal encircling $3/4$-in.-diameter steel hoops. The hoops are tightened to a unit stress of 8,000 psi at 70°F, when their diameter is exactly 12 ft. Assuming that the tank does not contract (illogical, but useful for investigative purposes), what will be the unit stress in the hoops at 10°F?        *Ans.* $s_t = 19,310$ psi

**1-68.** A surveyor's tape of hard steel having a cross-sectional area of 0.004 sq in. is exactly 100 ft long when a pull of 16 lb is exerted upon it at 70°F. (*a*) Using the same pull on it, how long will the tape be at 40 and at 100°F? (*b*) What pull $P$ on it will maintain its length of 100.0000 ft at 40 and at 100°F? Use $E = 30,000,000$ psi.

**1-69.** A tension member in a steel bridge truss carries a load of 90 kips. The member (see Fig. 1-29) was fabricated from three eyebars. Each bar was $1/2$ in. thick and 3 in. wide. The length from center to center of eyes was intended to be 30 ft 0 in., but the two outside bars were fabricated $1/8$ in. long and the center bar was fabricated $1/8$ in. short. All eyebars were forcibly assembled, end pins inserted, and the member placed in the truss. Calculate the unit stress in each bar under the design load of 90 kips tension.

FIG. 1-29 Prob. 1-69.                              FIG. 1-30 Prob. 1-70.

**1-70.** The outer cylinder shown in Fig. 1-30 is of steel and has an inside diameter of 4 in., a thickness of $\frac{1}{4}$ in., and a length of 10 in. The inner solid cylinder is of cast brass, has a diameter of 2 in., and is 10.005 in. long. A load of 80 kips is applied as shown. Calculate the stress in each member.

## REVIEW QUESTIONS

**1-1.** What are stress, unit stress, total stress?

**1-2.** Explain the meaning of tensile stress and compressive stress.

**1-3.** What is meant by shearing stress, bearing stress, and axial stress?

**1-4.** In connection with pins, bolts, and rivets, what is meant by single shear and double shear?

**1-5.** Define ultimate stress, allowable stress, and factor of safety.

**1-6.** What are some of the factors that govern the determination of a reasonable factor of safety?

**1-7.** Who determines the allowable stresses which may safely be used in design? Upon what information must such determination necessarily rely?

**1-8.** Give the direct-stress formula and explain each term in it.

**1-9.** What is meant by (a) root area and (b) S4S, and when must they be considered?

**1-10.** Explain the general meaning of the terms analysis and design.

**1-11.** Define strain, elasticity.

**1-12.** What is meant by shearing deformation and how is it related to shearing stress?

**1-13.** Define Poisson's ratio. Give approximate numerical limits.

**1-14.** Sketch two examples of axially loaded members with local stress concentrations.

**1-15.** What is meant by elastic limit, permanent set?

**1-16.** State Hooke's law and define modulus of elasticity.

**1-17.** What is a stress-strain diagram?

**1-18.** What condition in a member will generally produce a concentration of stress? What may be the result of a high concentration of stress in a soft steel and in a hard steel?

**1-19.** When two materials are so combined in a member that they deform equally under load, what is true of their unit stresses?

**1-20.** How do changes in temperature affect material bodies?

**1-21.** Define coefficient of thermal expansion.

# CHAPTER 2

# Engineering Materials
# and Their Properties

## 2-1 Introduction

A considerable number of so-called engineering materials—materials used in the building of structures and machines—are available: the various woods and metals and inert materials such as brick, stone, and concrete. Some of the most commonly used of these materials, the processes by which they are made or obtained, and some of their properties are briefly described in this chapter.

## 2-2 Definitions of Some Properties of Materials

The proper and efficient use of engineering materials requires considerable knowledge of their mechanical properties. Among the most important of these are strength, elasticity, and stiffness. Other properties are ductility, malleability, hardness, resilience, toughness, creep, and machinability.

The methods used for determining these and other properties of materials have been evolved over a long period of years and they have, by now, been carefully standardized. The American Society of Testing Materials (ASTM) publishes complete directions for making such tests together with the limits, agreed upon by engineers, scientists, and manufacturers, within which the test results must fall for any given classification of a material. Numerical values representing the relative properties of strength and elasticity of many engineering materials are given in the several tables included in this text.

**Strength** is the ability of a material to resist stress without failure. Several materials, such as structural steel, wrought iron, aluminum, and copper, have approximately equal strength in tension and compression, while their strength in shear is about two-thirds of that in tension. In other materials, such as gray cast iron, brick, and concrete, the strength in tension and in shear is only a fraction of the strength in compression. The measure of the strength of a material is its ultimate stress, or the greatest force per unit area it can withstand without rupture.

**Elasticity** is that property of a material enabling it to return to its original size and shape after removal of a deforming force. This property is important in all structures subjected to varying loads and is exceedingly important in precision tools and machines.

**Stiffness** is the property by virtue of which a material can resist deformation. This property is desirable in materials used in beams, columns, machines, and machine tools. The measure of the stiffness of a material is its modulus of elasticity, obtained by dividing the unit stress by the unit deformation caused by that stress.

**Ductility** is that property of a material that enables it to be drawn permanently through great change of shape without rupture, as, for example, copper being drawn into wire. Copper, aluminum, and wrought iron are among the ductile metals. The measure of ductility is the percent elongation before rupture of a test specimen. Ductility is desirable in a member or part which may be subjected to sudden and severe loads, since evident excessive deformation would give warning of impending failure. Brittleness is the opposite of ductility. Brittle materials fail suddenly without warning when stressed beyond their strength.

**Malleability** is that property of a material that enables it to undergo great change in shape without rupture under compressive stress, as, for example, copper, aluminum, or wrought iron being hammered into various shapes or steel being rolled into structural shapes or sheets.

**Hardness** is the ability of a material to resist indentation or abrasion. It is most commonly measured by the Brinell test, in which a hardened-steel ball 10 mm in diameter is forced into a flat surface of a test specimen by a force of 3,000 kg. The force divided by the surface area of the spherical indentation is called the **Brinell hardness number.**

**Resilience** is that property of a material which enables it to absorb, without being permanently deformed, the energy produced by the impact of a suddenly applied load or blow. This property is essential in steel used for all types of springs, as in an automobile, on railroad cars, in watches, etc., where energy must be absorbed quickly without causing permanent deformation. Resilience is sometimes referred to as "elastic toughness," since the energy must be absorbed without stressing the material beyond its elastic limit. The measure of resilience is the amount of energy that a unit volume of the material has absorbed after being stressed to its elastic limit.

**Toughness** is that property of a material which enables it to absorb energy at high stress without fracture, usually above the elastic limit. Being above the elastic limit, the stress will cause permanent deformation. Wrought iron, for example, is tough because it can be bent into most any shape without fracture. The measure of toughness is the amount of energy that a unit volume of the material has absorbed after being stressed up to the point of fracture.

**Creep** is that property which causes some materials under constant stress to deform slowly but progressively over a period of time. Creep occurs to a slight extent at room temperatures in timber and concrete subjected to constant stress over a long period of years. In the softer metals, such as lead and zinc, subjected only to moderate, steady stress, creep will occur in a relatively short period of time. Creep occurs also in steel under constant stress, but not in any measurable degree until the temperature of the steel exceeds about 600°F. It assumes importance, however, in the design of steam boilers and turbines, of internal combustion engines, and of other machines which operate at temperatures in excess of 600°F. In such cases, lower allowable stresses must be used. The maximum stress that may be steadily applied to a material at a given temperature and during a given period of time without causing a specified deformation to be exceeded is called the **creep limit** for that material. The creep limit generally used is the maximum unit stress at which the creep will not exceed 1 percent in 100,000 hours of constant stress, equal to a little over 11 years.

**Machinability** is the readiness with which a material yields to shaping with cutting tools.

## 2-3 Methods of Manufacture

Of the engineering materials mentioned earlier in this chapter, only wood and stone are used in the state as found in nature, except for shaping. All the metals are refined and manufactured from raw materials called **ore** found in various forms at or near the earth's surface. Brick and concrete are manufactured from the clays, limestones, and gravels found in great quantities on the earth's surface. The following descriptions of some of the manufacturing processes used serve only as brief introductions to the subject.

## THE WOODS

## 2-4 Wood

Wood is a product of nature and in most cases is used in its natural state. It requires only to be shaped into suitable form for practical use. This process of the sawmill is well known. Among the woods most commonly used for structural purposes are Douglas fir, southern pine, southern cypress, and mountain hemlock. A vast number of experiments have shown, however, that the mechanical properties of these woods, especially strength and stiffness, vary considerably; for example, fir and pine are both stronger and stiffer than cypress and hemlock. Hence for the guidance of engineers and architects who include wood as a structural material in their designs, the lumber industry has adopted standard grading rules for the purpose of classifying structural woods according to their mechanical properties and their recommended uses. The most commonly used of these woods, their classifications, and some properties are listed in Table 8-1.

## 2-5 Preservation of Wood

One of the disadvantages of wood as a structural material is that it is subject to decay. However, if used under proper conditions, the wood in a structure will generally outlive the useful life of the structure itself. Among the causes of decay are the following: (1) alternating wetness and dryness, resulting in wet rot, (2) lack of ventilation, resulting in dry rot, and (3) the destructive action of fungi, worms, and insects. In most structures, the effects of these causes are readily controlled.

When it is known to the designer that the causes of decay cannot be eliminated, two solutions to the problem are available: (1) to allow the decaying process to proceed and to replace decayed timbers when necessary, in which case an average life of about six to ten years can be expected, or (2) to use timber treated with creosote or some other suitable preservative, in which case the average life is ten to fourteen years more, depending upon the degree of exposure to decay. In the creosote treatment, the timber, after having been properly seasoned, is placed in a closed cylindrical chamber, and steam is introduced to soften the wood fibers. Air and moisture in the timber are then removed by a vacuum pump. Finally, creosote is injected into the cylinder under pressure, resulting in almost complete penetration of the wood fibers by the creosote preservative. Other simpler and less expensive preservative treatments are also being used.

## THE METALS

## 2-6 Iron and Steel

Iron ore is found in huge quantities in the earth's crust, at or near the surface. It occurs mostly in the form of iron oxide containing small amounts of silica and other impurities. The iron is separated from the oxide in a **blast furnace,** which consists of a cylindrical steel shell from 20 to 30 ft in diameter and from 150 to 200 ft high, lined inside with firebrick capable of withstanding temperatures higher than any produced in the furnace. Coke is used as a fuel to produce heat. The "charge," consisting of alternate layers of coke and ore into which a "flux," usually limestone, is mixed, is fed into the furnace stack through the top, beginning with a layer of coke from 20 to 25 ft deep at the bottom of the stack. Air under considerable pressure and preheated to a high temperature is blown into the furnace at the bottom through a series of nozzles called tuyeres. There it unites with the carbon of the coke to form carbon monoxide at a temperature exceeding 3000°F. As the hot carbon monoxide rises through the ore in the furnace, melting it, it combines with the oxygen of the iron oxide ore, escaping at the top as carbon dioxide, leaving the iron to trickle down to the bottom through the bed of coke. It is in this bottom high-temperature zone that smelting takes place. As the molten iron trickles down through the bed of coke, it combines with about 3 to 5 percent of carbon from the coke. The limestone flux combines with the silica to form molten slag, which floats above the molten iron near the bottom and is drawn off as a waste. The molten iron which collects and is drawn off at the bottom of the furnace is cast into small bars called "pigs." This pig iron cannot be used directly but requires additional processing and refining before it becomes cast iron, wrought iron, or steel.

## 2-7 Cast Iron

The term **cast iron** covers a series of materials possessing a broad range of mechanical and physical properties which are the result chiefly of the differing compositions of the materials. Pig iron, mixed with scrap cast iron and a limestone flux and, for special purposes, with scrap steel, chromium, copper, nickel, molybdenum, etc., is melted in a cupola furnace fired with coke and air blast. The molten metal is then poured into sand or metal molds, producing castings of desired forms and dimensions. The widely used gray cast iron, recognizable by the color and appearance of its fracture, contains carbon ranging from 2 to 4 percent and silicon from 1 to 3.5 percent. The structure of gray cast iron contains free graphite plates that act to impart properties peculiar to this material such as resistance to wear and corrosion, dampening ability, and machinability. Other types are chilled, or white, cast iron, which is highly resistant to abrasion; malleable cast iron, which has strength and ductility; and nodular cast iron, the structure of which contains spheroidal graphite that tends to develop attributes approaching those of carbon-steel castings.

## 2-8 Wrought Iron

Wrought iron is an iron product so made and refined that it possesses the important properties of ductility, malleability, and toughness. It is, therefore, especially suitable for machine parts to be shaped by forging. Also, its range of usefulness is considerably increased because of the ease with which it can be welded. Its ultimate strength is about three-quarters of that of structural steel.

Several processes are used in the production of wrought iron. Of these, the **puddling process** is perhaps most commonly used. In this process pig iron, together with a quantity of flux, is melted in a reverberatory furnace lined with iron oxide, the oxygen of which reduces the carbon content of the iron to less than 0.15 percent and also removes some sulfur. The mixture is then "puddled" by being stirred with steel rods. Toward the end of the process of purifying the iron, the heat of the furnace is insufficient to keep the purified iron in a molten state, and it therefore forms into balls of pasty iron, which are mechanically mixed with a small amount of slag. These balls are removed from the furnace and are squeezed into "blooms" in a press, thus removing some excess slag. The blooms are finally rolled into smaller bars suitable for commercial use. The presence of slag in wrought iron gives it a fibrous texture. Other methods of producing wrought iron use the "charcoal-hearth" type of furnace, by which a finer grade can be produced.

## 2-9 Steel

Steel is an iron product containing varying amounts of carbon, ranging up to about 1.7 percent. This product is technically referred to as **carbon steel.** An **alloy steel** is a carbon steel to which have been added one or more additional elements in amounts sufficient to give it recognizable properties not originally possessed by the carbon steel. Because of the marked differences in the properties of the carbon and the alloy steels, each group will be discussed separately.

**The Carbon Steels.** Pure iron (ferrite) contains no carbon, of course. It is relatively soft and is both ductile and malleable, but it does not possess great strength. Almost pure iron has an ultimate tensile strength of about 40,000 psi. The addition of carbon to pure iron, in amounts ranging from 0.05 to 1.7 percent, produces what is known as **steel.** The carbon has the effect of increasing strength and hardness, but it also decreases ductility and toughness. A carbon content of about 0.1 percent produces a so-called **soft steel** with an ultimate tensile strength of about 50,000 psi, suitable for rolling into plates. **Structural steel,** commonly used for rolling into structural shapes such as angles, beams, and columns, contains about 0.25 percent carbon, giving a fairly ductile and tough steel with an ultimate strength of about 64,000 to 72,000 psi. An increase to 0.40 percent of carbon produces a **machine steel** with an ultimate tensile strength of about 80,000 psi, while 0.75 percent carbon produces a hard **spring steel** with an ultimate tensile strength of 100,000 psi and 0.90 to 1 percent carbon produces a very hard **tool steel** with an ultimate tensile strength of 120,000 to 130,000 psi.

**The Alloy Steels.** When small amounts of one or more metals are added to a carbon steel in sufficient quantities so as to produce definite recognizable new properties in the steel, the product is called an **alloy steel.** The most commonly used alloying metals are nickel, manganese, chromium, vanadium, and molybdenum. Silicon, an inert material, is also used as an alloy to produce notch toughness. The effect of the addition of these alloys to carbon steel, singly or in combinations such as chromium and nickel, chromium and vanadium, and chromium and molybdenum, is, in general, to increase its strength and hardness, sometimes without sacrificing other desirable properties already possessed by the carbon steel.

## 2-10 The Manufacture of Steel

The making of steel from pig iron is essentially a refining process. It consists in first removing from the molten pig iron the excess carbon together with several impurities, such as silicon, manganese, and sulfur, and then adding carbon and/or any of the alloying elements required to produce a steel having the desired qualities and properties. The three most commonly used processes are (1) the *Bessemer process*, (2) the *open-hearth process*, and (3) the *electric-furnace process*. Of these, the open-hearth process produces the greatest quantity of steel in the United States, and the electric-arc furnace produces the most highly refined steel. The particular process used depends upon the chemical analysis of the pig iron to be refined and also upon the desired quality of the steel to be produced. The finished molten steel is poured into ingots in sizes suitable for use by the rolling mills.

**The Bessemer Process.** A Bessemer furnace, or converter, consists of a steel cylinder lined with firebrick. It is open at the top and is so constructed and supported that it can be revolved horizontally and tipped vertically to receive or discharge its contents. It is charged with 10 to 20 tons of molten pig iron, often directly from a blast furnace. Several thousand cubic feet per minute of unheated air is then blown through the molten iron for from 10 to 15 minutes. This is called the **blow,** and its action is to remove, in the form of burning gases, the silicon, manganese, carbon, and sulfur. After this, the elements required to give the steel the desired chemical properties are added to the molten mass. The Bessemer process is the most rapid of the three, but this rapidity makes control of the chemical properties more difficult.

**The Open-hearth Process.** In this process, preheated air and gas are used to melt the charge consisting generally of scrap iron, pig iron, iron oxide, and a limestone flux. The excess carbon, manganese, and phosphorus are removed by oxidation with the oxygen from the iron oxide, after which the elements required to give the desired chemical properties are added. The capacities of open-hearth furnaces range up to nearly 200 tons. This process is much slower than the Bessemer process, requiring about 10 hours for each charge, but the slowness of the process makes possible a high degree of control of the quality of the steel.

**The Electric-furnace Process.** Two types of electric furnaces are in general use. They are distinguished by the method used to produce heat. In the *electric-arc furnace*, the charge is heated by an electric arc passing between electrodes. In the *electric-induction* furnace, the charge is placed in a crucible encircled by coils and wire through which a high-frequency current is passed. This "induces" secondary eddy currents in the iron and heats it by virtue of its resistance to the passage of these currents. Both methods permit a high degree of control of the steel product, and the *induction* method is especially suitable for the making of high-grade alloy steels.

### 2-11 Aluminum

This metal is very widely used in the manufacture of items in which lightness is an advantage or is required, most notably in aircraft. The ultimate tensile strength of nearly pure aluminum is less than 10,000 psi, but when alloyed with 4 to 5 percent copper and smaller percentages of manganese, magnesium, and/or other metals, various grades of wrought aluminum can be produced whose ultimate tensile strengths vary from about 20,000 to 60,000 psi or better. Aluminum is also extensively used for castings after being alloyed with varying percentages of silicon, copper, iron, zinc, manganese, and magnesium, either singly or in combinations.

Aluminum is obtained principally from an ore called **bauxite,** which, chemically, is hydrated oxide of aluminum, found in great abundance on the earth's surface. When this ore is treated with caustic soda, aluminum oxide is produced, which then is mixed with molten cryolite and is reduced electrolytically to metallic aluminum.

Although aluminum is one of the most chemically active of the commonly used metals, it is comparatively resistant to corrosive influences in its purer forms but is much less resistant in its alloyed forms, especially when alloyed with copper. The destructive effect of corrosion is most serious in thin sheets of the metal, so widely used in aircraft construction. In order to combine the high strength of aluminum alloyed with copper with the corrosion resistance of pure aluminum, sheets and other forms of the high-strength metal are coated with thin surface layers of the pure metal. The trade name "Alclad" is one which designates products so treated.

**Wrought aluminum** is easily drawn into wire, rolled into sheets, or forged into machine parts. Tubing and small structural shapes such as angles and beams are readily formed by extrusion or rolling. Large **aluminum castings** are usually formed in sand molds. Smaller castings are made either in metal molds or by die-casting, a process in which liquid metal is forced into a die under high pressure.

### THE INERT MATERIALS

### 2-12 Stone, Brick, and Concrete

These materials are generally used in large masses in structures where only compressive stresses are to be resisted, such as in foundations, piers, and walls.

A property which is common to all of the materials in this group is their relatively low strength in tension and in shear. Consequently, if any part of such a structure is subjected to tensile stress, as is the case in concrete beams and floor slabs of buildings, such part must be reinforced by steel bars placed in the areas where tensile or shear stresses must be resisted. In geographical areas subjected to earthquake shocks, walls of stone, brick, or concrete must be reinforced with networks of steel bars.

**Stone.** This natural material was still extensively used in heavy foundations and structures until half a century ago. Since then it has gradually been replaced with steel and concrete. Its use at present is confined almost entirely to architectural ornamentation.

**Brick.** Two types of bricks are extensively used: (1) clay bricks and (2) sand-lime bricks. Clay bricks consist of clay mixed to a given consistency and molded to shape by machine. They are then "burned" in a kiln at temperatures below the melting point of the clay until they become semivitrified. Sand-lime bricks are made of a mixture of common sand mixed with lime, which acts as a binder. They are molded by machine under high pressure. Both types of brick are made under controlled conditions, and their strength is purposely made about equal. Depending on grade, the compressive strength of brick varies from about 1,000 psi to 5,000 psi or more. Brick is usually very brittle and has relatively low strength in tension and shear.

**Concrete.** Concrete is a mixture of cement, sand, and gravel with enough water added to bring about complete chemical action of the cement and to make the mixture workable, that is, capable of being poured into forms leaving reasonably smooth outer surfaces after having dried. Because of its strength and fire-resisting qualities and the ease with which it can be mixed and molded into any desired form, concrete has, in the past half-century, become one of the most important of the structural materials. In addition, the necessary ingredients are readily available in huge quantities and at low cost almost everywhere on the face of the earth.

**Cement.** Several types of cement are being manufactured. Of these, *portland cement* has greatest strength and hardens most quickly and is, therefore, most widely used. *Natural cement* is less strong and hardens less quickly.

The main ingredients of cement are calcium carbonate, silica, and alumina. Limestone is the chief source of calcium carbonate. Silica and alumina are obtained from clay, shale, or argillaceous limestone. These raw materials are first ground and mixed in suitable proportions. Next the mixture is burned to a clinker, just short of fusion, after which it is ground to a fine powder.

The aggregates most commonly used are sand and gravel or crushed rock, clean and free of alkali and organic matter. To produce concrete ready for pouring into forms, cement, sand, and gravel are mixed with water in such proportions by volume that the desired strength is obtained. If the proportions are $1:3:5$, a concrete with an ultimate strength of 1,500 psi can be expected; a $1:2:4$ mix should give a 2,500-psi concrete; and a $1:1\frac{1}{2}:2\frac{1}{2}$ should give a 3,500-psi concrete. Low-strength concrete is used in heavy masses, such as foundations, piers, dams, while higher-strength concrete is generally used in bridges and in the reinforced-concrete frames of buildings.

The amount of water used per sack of cement varies from 5 to 8 gal. A low water-cement ratio gives high strength but poor workability, while a high ratio gives low strength and good workability. The concrete will *set* in a few hours after pouring and the expected strength is usually reached in 28 days, during which period it "cures." During hot, dry weather, the concrete must be kept damp to permit proper curing. Too quick drying interferes with the chemical action of the cement and results in loss of strength. Freezing of concrete before it has set will materially destroy its strength.

## REVIEW QUESTIONS

2-1. What is meant by the term "properties of materials"?

2-2. Define in general the properties of strength and elasticity.

2-3. What is meant by ductility and malleability?

2-4. Explain the property of hardness and how it is measured.

2-5. What is meant by resilience? How is it measured?

2-6. What is meant by toughness and how is it measured?

2-7. Explain the terms "creep" and "machinability."

2-8. Name at least four kinds of wood commonly used as structural materials. Discuss some of their properties.

2-9. How may wood be preserved against decay?

2-10. From what ore is iron most generally obtained?

2-11. Describe the common process of smelting iron ore. What is the final product of this process called?

2-12. Explain the process of producing iron castings from pig iron. What are some of the usual properties of cast iron?

2-13. What are the three most important properties of wrought iron? Name some uses for wrought iron.

2-14. Describe what is known as the "puddling process" of making wrought iron.

2-15. Describe in general terms the product steel, and explain the distinction between carbon steel and alloy steel.

2-16. Name at least four elements with which carbon steel is commonly alloyed.

2-17. Name three processes by which steel is commonly produced.

2-18. From what ore is metallic aluminum usually obtained and how is it generally produced?

2-19. What are some of the metals with which aluminum is commonly alloyed?

2-20. Discuss some differences in properties of pure aluminum and aluminum alloyed with copper.

2-21. Describe one process by which a copper-aluminum alloy is commonly made more resistant to corrosion.

2-22. Name some commercial shapes into which wrought aluminum is generally formed.

2-23. Discuss processes by which bricks are made.

2-24. What are some of the strength characteristics of brick?

2-25. Describe concrete and how it is produced.

2-26. Discuss some strength properties of concrete.

2-27. Describe in general the manufacture of cement.

# CHAPTER 3

# Riveted and Welded Joints
# Thin-walled Pressure Vessels

### 3-1 Riveted Joints

Steel structures, such as building frames composed of beams and columns, roof and bridge trusses, cranes, railroad cars, boilers, water tanks, and many others, are built up of a number of separate pieces, which then must be effectively joined together. A common method of joining is that of **riveting.**

A **rivet** is a short bar of metal with a half-spherical *head* already formed on one end. Matching holes are punched or drilled in the two pieces to be joined. After the pieces have been placed in their relative final positions, rivets are inserted into the holes and are then *driven,* usually with a pneumatic hammer. That is, the head end of a rivet is held tightly against the joined pieces while the opposite end is hammered until another similar head is formed there, thus gripping the pieces tightly together. The force to be resisted is then transmitted from the one piece to the rivets and by the rivets to the other piece. Rivets may be driven hot or cold.

Steel rivets, as used in most structural and boiler work, when heated to a bright red heat, "soften" greatly and are then more easily driven. During cooling a hot-driven rivet will shrink, thus producing tension in the rivet and consequently gripping the joined pieces even more tightly together. Copper and aluminum rivets, which are frequently used in aircraft work, are generally driven cold. Such rivets are usually small and the metal is somewhat softer than steel.

When rivets thus grip plates tightly together, a considerable amount of frictional resistance is developed between the plates, which, of course, increases the safety of a joint. In practical design, however, this frictional resistance is often disregarded.

## 3-2 Types of Riveted Joints

Two types of riveted joints in general use are the **structural joint** and the **boiler joint.** The structural joint, as the term implies, is generally used for connecting members of structures such as building frames, trusses, cranes, etc. The boiler joint is used in all cases requiring an absolutely tight joint, such as in boilers, water tanks, compressed air tanks, and vessels of all types subjected to pressure. When several rivets are used at a joint, each rivet is presumed to resist its proportional part of the total load on the joint. Certain differences which exist in the methods of design of structural and boiler joints are explained in later paragraphs.

* **Structural Joints.** In Fig. 3-1 are shown two typical riveted structural joints. Such joints are generally referred to as **connections.** In the beam-to-girder connections, two beams frame into a large supporting girder from opposite sides. The connection is made possible by riveting two short opposite angles to the web of each beam, using four rivets $B$ in each. The two sets of angles are then fastened to the web of the girder by eight common rivets $G$. Note here that all these rivets are in double shear.

The beam-to-column connection in Fig. 3-1b is very similar. Here, a single beam frames into the flange of a column. Two short opposite angles are riveted to the web of the beam with three rivets $B$. These angles are then fastened to the column by means of six rivets $C$ through its flange. Note here that, while the three rivets $B$ through the web of the beam are in double shear, the six rivets $C$ through the column flange are in single shear. If the rivets all have the same diameter, we see that the three in double shear have the same shear strength as the six in single shear. A "seat" angle is often used, thus giving the joint greater strength and rigidity and greatly simplifying the task of erection.

FIG. 3-1 Riveted structural joints. (*a*) Beam-to-girder connection. (*b*) Beam-to-column connection.

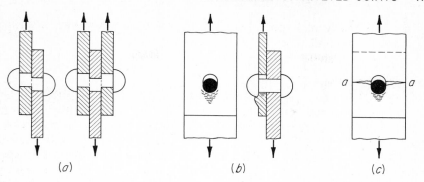

FIG. 3-2 Failures of riveted joints. (*a*) Shearing failure. (*b*) Bearing failure. (*c*) Tearing.

## 3-3 Failures of Riveted Joints

When a riveted joint or connection is subjected to a load which is beyond its capacity to resist, it may fail in one of several ways, as indicated in Fig. 3-2. While failure may occur entirely in shear, in bearing, or in tension, it may also occur simultaneously or progressively in several combinations of these.

**Shearing Failure.** As shown in Fig. 3-2*a*, a rivet may fail in shear, either in *single* shear on *one* cross-sectional area of the rivet or in *double* shear on *two* cross-sectional areas, depending on the type of joint.

**Compression or Bearing Failure.** The material of the plate in front of the rivet where it bears against the plate may crush, as indicated in Fig. 3-2*b*. The rivet itself, of course, is subjected to the same bearing stresses, and the rivet metal may fail, or failure may be partly in plate and partly in rivet.

**Tension Failure.** Failure in tension occurs when a plate tears apart along some line of least resistance as, for example, along line *a-a* in Fig. 3-2*c*.

**Other Types of Failure.** A riveted joint subjected to a load beyond its capacity to withstand will very likely fail either in shear, in bearing, or in tension, as described in preceding paragraphs. Certain other types of failures are possible, however, but are usually preventable if correct design procedures are followed.

In Fig. 3-3 are illustrated two other possible types of failure, **end shearing** and **end tearing.** Such failures result only when the *end distance e in line of stress* is insufficient. Failure of this type is most often a combination of crushing and shearing or of crushing and tearing. Because of this the computed shear stress along the failure planes in Fig. 3-3*a* is an unreliable indicator of adequate design. Hence, in considering the safety of a design, it is best to follow the empirical rule that the minimum value of *e* should not be less than the shearing area of the rivet, single or double, divided by the thickness of the plate.

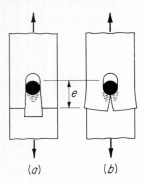

FIG. 3-3 End failures. (a) End shearing. (b) End tearing.

(a)        (b)

## 3-4 Stresses in Riveted Joints

The discussion of possible failures in the preceding article has indicated the stresses, shearing, bearing, and tensile, existing in riveted joints. However, rivets and plates are also subjected to other stresses which are not so apparent. For example, in the simple lap joint shown in Fig. 3-4a, it is clear that the forces $P$ in the plates are transmitted to the rivet at approximately the mid-points of the plates, the center lines of which are a distance $e$ apart.

These forces tend to bend both the rivet and the plates. If a substantial rivet is used to fasten together two rather light plates and if a sufficiently great force is then applied, the plates will bend somewhat as illustrated in Fig. 3-4b, until the plate forces $P$, at short distances from the rivet, are nearly in line. In Fig. 3-4c the forces presumably acting on the rivet are shown. The two forces $P$ form a counterclockwise couple $Pb$, which must be and is counteracted by an opposed clockwise couple $P'a$. Clearly, the forces $P$ produce shearing stresses in the rivet and also bend it; the forces $P'$ likewise bend the rivet and, in addition, produce tensile stresses in it. Naturally, the actual forces acting on the rivet are nonuniformly distributed, and $P$ and $P'$ are merely resultants.

Despite their known existence, these bending and tensile stresses are not

FIG. 3-4 Bending in lap joints.

directly considered in the design of riveted joints. Their intensities would be very difficult to determine. However, allowable stresses permitted in design are sufficiently conservative to compensate for disregarding them.

**Shearing Stresses.** The shearing stresses on a cross-sectional area of a rivet are not uniformly distributed. Consequently, the use of the direct-stress formula $s = P/A$ gives only the *average unit shearing stress*. The stressed area in single shear is the cross-sectional area and in double shear is twice the cross-sectional area.

**Bearing Stresses.** The area in bearing between a plate and a rivet is semi-cylindrical. The actual distribution of stress over this area is rather complex. Therefore, as in the case of shear, the use of the direct-stress formula gives only an *average unit bearing stress*. The area in bearing commonly used for the purpose of determining this average stress is the **projected area,** which is the diameter of the rivet times the thickness of the plate against which it bears.

In some joints, either structural or boiler, two bearing stresses of different intensities exist and must be considered. For example, in all pinned, bolted, or riveted connections where a main or inside piece or plate is *enclosed* and is held in place by two outside pieces or plates, the pressure between the enclosed or inside piece and the pin, bolt, or rivet is called **inside bearing.** The pressure between the two outside pieces and the pin, bolt, or rivet is called **outside bearing.** These conditions are clearly illustrated in Fig. 3-5. When only two pieces are used, as, for example, to form a lap joint, as in Fig. 3-2a, b, and c, only outside bearing exists. In some literature on the subject, inside bearing is called **enclosed bearing** and outside bearing is called **plain bearing.** These terms, however, especially the latter, seem inadequately explanatory.

**Tensile Stresses.** The tensile stresses here dealt with are those in the plates or in the connected structural members that are subjected to direct tensile stress. Obviously, to determine the unit tensile stress, the *net area* in tension must be used. That is, only the area of material actually resisting stress may be included.

FIG. 3-5 Bolted structural connection. Inside and outside bearing.

In **structural joints**, rivet holes are generally laid out and punched separately in the parts to be joined. To avoid difficulty in assembly of the parts, the holes are punched with a diameter $\frac{1}{16}$ in. greater than the rivet diameter. The punching operation damages the material of the wall of the cylindrical hole to some slight depth, perhaps $\frac{1}{32}$ in., thus making that much of the material of the plate or part incapable of resisting tensile stress. Therefore, *for the purpose of calculating the net area in tension of punched plates or parts, the diameter of the hole is considered to be $\frac{1}{8}$ in. greater than the diameter of the rivet to be used.* This specification applies also when *unfinished bolts* are used instead of rivets. *In structural work, areas in shear and bearing are based on the original nominal diameter of the rivet.* Steel rivets are available in diameters which are multiples of $\frac{1}{8}$ in. from $\frac{3}{8}$ to $1\frac{1}{4}$ in.

**Allowable Stresses.** Safe allowable stresses for use in design are generally based upon the results of a large number of tests in which bolts, rivets, plates, and made-up joints are tested to failure to determine their ultimate strength. The allowable stress is then arrived at by dividing the ultimate strength by a reasonable and predetermined factor of safety, generally varying between 3 and 5. The allowable stresses recommended for use in design of riveted joints are specified by various codes and vary somewhat according to the type of joint. The most generally accepted allowable stresses for structural joints are listed in Table 3-1.

**Table 3-1 ALLOWABLE STRESSES FOR STRUCTURAL RIVETS, BOLTS, AND PINS. AISC SPECIFICATIONS. RECOMMENDED FOR STRUCTURAL JOINTS**

| | Tension, psi | * Shear, psi | | Bearing, psi, ASTM A36 steel |
|---|---|---|---|---|
| | | Friction-type connection | Bearing-type connection | |
| A141 hot-driven rivets | 20,000 | | 15,000 | 48,500 |
| Unfinished bolts, ASTM A307 | 14,000 | | 10,000 | 45,000[a] |
| A325 bolts when threading is excluded from shear planes | 40,000 | 15,000 | 22,000 | 48,500 |
| A325 bolts when threading is *not* excluded from shear planes | 40,000 | 15,000 | 15,000 | 48,500 |

[a] ASTM A7 steel.

Commercially available rivet and bolt diameters:

rivets: $\frac{3}{8}$, $\frac{1}{2}$, $\frac{5}{8}$, $\frac{3}{4}$, $\frac{7}{8}$, 1, $1\frac{1}{8}$, $1\frac{1}{4}$

unfinished bolts: $\frac{1}{4}$, $\frac{5}{16}$, $\frac{3}{8}$, $\frac{7}{16}$, $\frac{1}{2}$, $\frac{5}{8}$, $\frac{3}{4}$, $\frac{7}{8}$, 1, $1\frac{1}{8}$, $1\frac{1}{4}$, then in multiples of $\frac{1}{4}$ in. up to and including 4 in.

**Allowable Stresses for Structural Joints.** According to the American Institute of Steel Construction, *the safe allowable stress in tension in* ASTM A36 structural steel is 22,000 psi. It also recommends the allowable stresses given in Table 3-1 for riveted and bolted joints.

✳ When slip between the connected parts cannot be tolerated and must therefore be prevented by high clamping force, the allowable shear values are those given for *friction-type connections*. When a small amount of initial slip is permissible (to allow the rivets or bolts to seat on the bearing surfaces), the allowable stresses are those for *bearing-type connections*.

## 3-5 Analysis of Riveted Structural Joints

An analysis of a given joint is made to determine one of two things: (1) the average unit stresses, $s_s$, $s_c$, and $s_t$, in shear, bearing, and tension produced in a connection of given dimensions by a given load $P$ or (2) the maximum load $P$ which can safely be resisted by a connection of given dimensions. Both are illustrated in Probs. 3-1 and 3-2.

**Allowable Loads on Bolts and Rivets.** In design practice, it is customary to work with the *allowable load* on a rivet in shear and bearing, inside and outside, rather than the allowable unit stress. This allowable load is the product of the allowable stress and the area under stress. Thus the allowable load on a $\frac{3}{4}$-in.-diameter A325 bolt in a friction-type connection is $(15,000)(0.442) = 6,630$ lb and in double shear is $(15,000)(0.442)(2) = 13,260$ lb. In the structural joint of Fig. 3-5, let the inside plate be $\frac{3}{8}$ in. thick and each side plate be $\frac{1}{4}$ in. thick. All plates are of ASTM A36 steel. The allowable load on an A325 bolt in a friction-type connection is the least of the following calculated values: $(15,000)(0.442)(2) = 13,260$ lb or $(48,500)(\frac{3}{8})(\frac{3}{4}) = 13,620$ lb or $(48,500)(2)(\frac{1}{4})(\frac{3}{4}) = 18,200$ lb. Since the least, or controlling, value is that calculated for shear, the allowable load for this bolt is 13,260 lb. In any connection, the *least allowable load on a bolt or rivet* is called its value.

### ILLUSTRATIVE PROBLEMS

**3-1.** In Fig. 3-6 is shown a simple structural connection composed of two steel straps held together by a bolt 1 in. in diameter. The upper strap is $\frac{3}{8}$ in. thick, and the lower strap is $\frac{1}{2}$ in. thick. Both straps are $2\frac{3}{8}$ in. wide. Calculate the maximum unit stresses in shear, bearing, and tension caused by a pull $P$ of 7,500 lb.

*Solution:* In each case we shall need to calculate the size of the area resisting the pull. By dividing this area into the pull $P$, we obtain the average unit stress.

*Shear.* The bolt is in single shear. Hence, $A_s = 0.7854$ sq in. and

$$\left[ s = \frac{P}{A} \right] \qquad s = \frac{7,500}{0.7854} = 9,550 \text{ psi} \qquad Ans.$$

*Bearing.* The upper strap bears against the right side of the bolt, and the area in bearing is presumed to be the bolt diameter times the thickness of the strap. That

FIG. 3-6 Bolted connection.

area is $(1)(0.375) = 0.375$ sq in. The lower strap bears against the left side of the bolt, and that bearing area is $(1)(0.5) = 0.5$ sq in. Clearly, since each strap is subjected to a pull of 7,500 lb, the maximum stress will be exerted on the smallest of these two areas and is

$$\left[ s = \frac{P}{A} \right] \qquad s_c = \frac{7,500}{0.375} = 20,000 \text{ psi} \qquad Ans.$$

*Tension.* When either rivets or unfinished bolts are used in *structural* joints, the hole is considered to be of $\frac{1}{8}$-in. larger diameter than the rivet or bolt. The maximum tensile stress will be found in the upper strap, where the net resisting area is smaller than in the lower strap. This area is $(2.375 - 1.125)(0.375) = 0.469$ sq in. Hence

$$\left[ s = \frac{P}{A} \right] \qquad s_t = \frac{7,500}{0.469} = 16,000 \text{ psi} \qquad Ans.$$

**3-2.** An ASTM A36 steel beam, W 21 × 62, is connected to a steel column, W 14 × 95, with $\frac{3}{4}$-in.-diameter rivets, as illustrated in Fig. 3-7. All connecting angles are $\frac{7}{16}$ in. thick. Calculate the allowable load $R$ that may safely be transmitted from the beam to the column through the bearing-type connection.

*Solution:* In order to determine the allowable load on this connection, we must first find the allowable load on one rivet in shear, single or double, and in bearing.

FIG. 3-7 Beam-to-column connection.

In this connection, however, there are two groups of rivets which must be considered separately: group $A$ of five rivets through the web of the beam and group $B$ of four rivets which connect the seat angle to the column flange. The two rivets fastening the beam flange to the seat angle do not resist any vertical load and must be disregarded.

From Table E-6 we find that the web of the W 21 × 62 beam is 0.40 in. thick. The side angles are $\frac{7}{16}$ in. (0.4375 in.) thick. The rivets through the beam web are in double shear, in bearing against 0.40-in. plate and two $\frac{7}{16}$-in. angles.

Allowable load on one $\frac{3}{4}$-in. rivet in group $A$ (through beam web):

In double shear $(15,000)(2)(0.442)$                     $= 13.26$ kips ← Governs
In inside bearing on 0.40-in. plate $(48,500)(0.4)(\frac{3}{4})$ $= 14.57$ kips
In outside bearing on $(2)(0.4375) = 0.8750$-in. plate $= 31.80$ kips

$3/8 \times 48500$

In group $B$, the four rivets are in single shear, in bearing against one $\frac{7}{16}$-in. angle and also in bearing against the 0.748-in.-thick column flange.

Allowable load on one $\frac{3}{4}$-in. rivet in group $B$ (through seat angle):

In single shear $(15,000)(0.442)$                     $= 6.63$ kips ← Governs
In bearing on $\frac{7}{16}$-in. plate $(48,500)(\frac{7}{16})(\frac{3}{4})$ $= 15.90$ kips
In bearing on 0.748-in. plate                     $= 27.20$ kips

The total allowable load on the connection then is

Allowable load on 5 group $A$ rivets $= (5)(13.26) = 66.30$ kips
Allowable load on 4 group $B$ rivets $= (4)(6.63)  = 26.52$ kips
Total allowable load $R$ on connection           $= 92.82$ kips    *Ans.*

## ANALYSIS PROBLEMS

NOTE: In the following problems assume ASTM A36 steel and A325 bolts in bearing-type connections unless otherwise stated. Furthermore, assume care will be taken to exclude threads from the shear planes.

3-3. In the bolted structural joint shown in Fig. 3-6, assume that each strap is $\frac{1}{4}$ in. thick and $1\frac{3}{8}$ in. wide and that they are held together by a $\frac{5}{8}$-in.-diameter bolt. Calculate the allowable pulls $P_s$, $P_c$, and $P_t$ in shear, bearing, and tension on this connection.

3-4. Assume in Fig. 3-7 that the beam is a W 21 × 96, the column is a W 12 × 53, all angles are $\frac{7}{16}$ in. thick, and 1-in.-diameter rivets are used. Then calculate the allowable load $R$ on the connection.

3-5. In Fig. 3-1a assume that the beam is a W 18 × 70, the girder is a W 24 × 110, the angles are $\frac{3}{8}$ in. thick, and $\frac{3}{4}$-in.-diameter rivets are used. Then calculate the allowable load $R$ that may safely be transmitted from the beam to the girder.

*Ans.* 53 kips

3-6. Assume in Fig. 3-1b that the beam is a W 14 × 61, the column is a W 10 × 33, all rivets are of $\frac{7}{8}$ in. diameter, and all angles are $\frac{1}{4}$ in. thick. Calculate the value of the maximum load $R$ (end reaction) that may safely be transmitted from the beam to the column.

**3-7.** A typical chain link is shown in Fig. 3-8. The side straps are $\frac{3}{8}$ in. thick by $1\frac{1}{2}$ in. wide between the pins and $2\frac{1}{4}$ in. wide across the pins. The finished pins are $\frac{1}{2}$ in. in diameter and are close fitting. Bushings are used to space the inner straps. Calculate the unit stresses in shear, bearing, and tension in this link when it is subjected to a pull of 3,000 lb.

*Ans.* $s_s = 7{,}650$ psi; $s_c = 8{,}000$ psi; $s_t = 2{,}667$ psi

FIG. 3-8 Probs. 3-7 and 3-8.　　　　FIG. 3-9 Probs. 3-9 and 3-10.

**3-8.** Calculate the value of the maximum load $P$ to which the chain link in the preceding problem may be subjected. The side straps are of structural steel having an allowable bearing stress of 32,000 psi and an allowable tensile stress of 20,000 psi. The allowable shear stress in the pin is 15,000 psi.

**3-9.** The bell crank shown in Fig. 3-9 is held in the position shown by the two forces $H$ and $T$. The pin $A$ is $\frac{1}{2}$ in. in diameter and fits snugly in a reamed hole. (*a*) Calculate the unit shearing stress in pin $A$ when $H = 2{,}400$ lb, $a = 4$ in., and $b = 3$ in. (*b*) What force $T$ will produce a shearing stress in pin $A$ equal to 15,000 psi?

**3-10.** In Fig. 3-9, the pin $A$ is $\frac{3}{4}$ in. in diameter and fits snugly into the reamed hole. Calculate the thickness $t$ of the bell crank which will be required to give the pin equal strength in shear and bearing under the conditions shown. Use allowable stresses of 15,000 psi for shear and 42,000 psi for bearing.

## 3-6 Design of Riveted Structural Joints

The problem of designing a riveted joint to resist a given load $P$ involves the use of predetermined allowable stresses $s$ in order to evaluate the required area $A$ in shear, bearing, and, in some connections, tension. In design, then, the basic equation $s = P/A$ is used in the form $A = P/s$.

A complete design would involve also consideration of the shearing and tensile stresses in the gusset plate. If, however, the usual recommended rivet spacings and minimum edge distances are used, the gusset-plate stresses will invariably be less than those allowed, thus making design of gusset plates unnecessary except in special cases.

## ILLUSTRATIVE PROBLEMS

**3-11.** A typical structural joint of a light roof truss is shown in Fig. 3-10. Each member is made up of two $\frac{5}{16}$-in.-thick angles placed back to back and riveted to a gusset plate. (*a*) Calculate the minimum thickness *t* of gusset plate for each rivet to be approximately as strong in bearing as it is in shear. (*b*) If force $A = 26,100$ lb, $B = 43,500$ lb, and $C = 34,800$ lb, how many $\frac{3}{4}$-in.-diameter rivets are required in each member?

FIG. 3-10 Riveted structural joint.

*Solution:* We note that all rivets are in double shear and inside bearing. From Table 3-1 the allowable stresses in rivets are shear, 15,000 psi; bearing, 48,500 psi. We may now calculate the "value" of a $\frac{3}{4}$-in.-diameter rivet in double shear, which is

$$(2)(0.4418)(15,000) = 13,250 \text{ lb}$$

The area in bearing on a $\frac{3}{4}$-in. rivet is 0.75*t* sq in., and its value in bearing then is $(0.75t)(48,500) = 36,400t$. Then for equal strength in shear and bearing

$$36,400t = 13,250 \qquad \text{or} \qquad t = 0.365 \text{ in.}$$

$$\text{Use } t = \tfrac{3}{8} \text{ in. (0.375 in.).} \qquad Ans.$$

If we use a $\frac{3}{8}$-in. gusset plate, the strength of a $\frac{3}{4}$-in. rivet in bearing is $(0.75)(0.375)(48,500) = 13,600$ lb, which is approximately the same as its strength in double shear. Now, for equal or greater strength in bearing, each angle must be at least half as thick as the gusset plate. Since each angle is $\frac{5}{16}$ in. thick, this requirement is satisfied. Consequently, we may disregard bearing on the angles.

The value in shear, being less than the value in bearing, will then determine the number of rivets required. That is,

Member *A*: $\qquad N = \dfrac{26,100}{13,250} = 1.97 \qquad$ Use 2 rivets. $\qquad Ans.$

Member *B*: $\qquad N = \dfrac{43,500}{13,250} = 3.28 \qquad$ Use 4 rivets. $\qquad Ans.$

Member *C*: $\qquad N = \dfrac{34,800}{13,250} = 2.62 \qquad$ Use 3 rivets. $\qquad Ans.$

FIG. 3-11 Bolted structural connection.

**3-12.** Design a bolted structural connection of the type shown in Fig. 3-11 to withstand a pull $P = 20,000$ lb. Use high-strength ASTM A325 bolts. The steel straps are of ASTM A36 steel. No slippage between plates is to be permitted, and care will be taken to exclude threads from the shear planes.

*Solution:* The design requires the determination of (*a*) the bolt diameter *d*, (*b*) the thickness *t* and the width *w* of the main straps, and (*c*) the thickness and width of the splice plates. Item (*c*) is easily determined after the thickness and width of the main strap are computed. Each splice plate will be made the same width and at least half as thick as the main strap.

The required size of the main strap cannot be determined until the size of the bolt hole is known. The bolt diameter is obtained from the required shear area. From Table 3-1, for this friction-type joint the allowable stresses are $s_s = 15,000$ psi and $s_c = 48,500$ psi. Since each bolt must resist half the total load, or 10,000 lb, the required shear area is

$$\left[ A_s = \frac{P}{s_s} \right] \qquad A_s = \frac{10,000}{15,000} = 0.667 \text{ sq in.}$$

Since each bolt has two shear planes, the required area per shear plane is 0.333 sq in. and the diameter is $\frac{3}{4}$ in. ($A = 0.442$ sq in.)                     *Ans.*

We may now determine the thickness *t* from the required bearing area, which is *td*. That is,

$$\left[ A_c = \frac{P}{s_c} \right] \quad A_c = \frac{10,000}{48,500} = 0.206 \text{ sq in.} \quad \text{and} \quad t = \frac{0.206}{0.75} = 0.275 \text{ in.}$$

Use $t = \frac{3}{8}$ in. (0.375 in.).     *Ans.*

The width *w* may now be determined by the required net area in tension. The critical section is through the exterior bolt. The tensile force at this section is 20,000 lb. The required area is

$$\left[ A_t = \frac{P}{s_t} \right] \qquad A_t = \frac{20,000}{22,000} = 0.909 \text{ sq in.}$$

and the net width then is

$$\frac{A_t}{t} = \frac{0.909}{0.375} = 2.42 \text{ in.}$$

To obtain the gross width *w*, we must add the bolt diameter *d* plus $\frac{1}{8}$ in. The required width *w* is then

$$w = 2.42 + 0.75 + 0.125 = 3.295 \text{ in.}$$

Use $w = 3\frac{1}{2}$ in.     *Ans.*

As previously stated, the thickness of each side plate should be at least half that of the main strap.

Use $\frac{3}{16}$- $\times$ $3\frac{1}{2}$-in. splice plates.     *Ans.*

**3-13.** In Fig. 3-12, the load to be transmitted from each of the opposite W 16 $\times$ 88 steel beams to the W 24 $\times$ 110 girder is 96 kips. All connecting angles are $\frac{3}{8}$ in. thick, and all rivets are of $\frac{7}{8}$-in. diameter. If only four rivets can be used through the web of each beam, how many rivets are required to fasten the two opposite seat angles to the web of the girder?

FIG. 3-12 Structural beam-to-girder connection. (*a*) Section through girder. (*b*) Section through beam.

*Solution:* We must first determine the amount of load carried by the four rivets through the web of the beam. The remainder then must be carried by the seat-angle rivets. The rivets connecting the beam flange to the seat angle do not resist any of the vertical load. From Table E-6, we find that the web thickness of each beam is 0.504 in. and of the girder is 0.510 in.

Allowable load on one $\frac{7}{8}$-in. rivet through beam web:

In double shear $(2)(0.601)(15,000)$     $= 18.04$ kips $\leftarrow$ Governs
In bearing on 0.504-in. plate $(\frac{7}{8})(0.504)(48,500) = 21.40$ kips
In bearing on $(2)(0.375) = 0.75$-in. plate     $= 31.80$ kips

Then load carried by the four beam-web rivets $= (4)(18.04) = 72.16$ kips, and the load to be carried by the seat-angle rivets $= 96 - 72.16 = 23.84$ kips from each side or $(2)(23.84) = 47.68$ kips from both sides.

Allowable load on one $\frac{7}{8}$-in. rivet through seat angle and girder web:

In double shear $(2)(0.601)(15,000)$     $= 18.04$ kips $\leftarrow$ Governs
In bearing on web $(0.510)(\frac{7}{8})(48,500)$     $= 21.65$ kips
In bearing on angle $(2)(0.375)(\frac{7}{8})(48,500) = 31.85$ kips

The number of rivets through the seat angle and girder web is

$$N = \frac{47.68}{18.04} = 2.64 \qquad \text{Use 3 rivets.} \qquad \textit{Ans.}$$

## DESIGN PROBLEMS. STRUCTURAL JOINTS

NOTE: The following problems are all of the structural type. Assume ASTM A36 steel and A325 bolts in bearing-type connections unless otherwise stated. Also assume that threads will be excluded from the shear planes.

**3-14.** In the structural joint in Fig. 3-10, three double-angle members are riveted to a gusset plate. All angles are $\frac{3}{8}$ in. thick. Rivets of $\frac{7}{8}$-in. diameter are to be used. (a) Calculate the required thickness $t$ of gusset plate which will give the $\frac{7}{8}$-in. rivets approximately equal strength in shear and bearing. (Give $t$ as a multiple of $\frac{1}{16}$ in.) (b) If force $A = 70$ kips, $B = 118$ kips, and $C = 95$ kips, how many rivets are needed in each member?

**3-15.** Design a bolted structural connection of the type shown in Fig. 3-11 capable of safely resisting a tensile load $T$ of 50,000 lb.

**3-16.** The steel bar in Fig. 3-13 is held in place by two unfinished bolts. The bar and bolts must be designed to support a tensile load $T$ of 20,000 lb. (a) Based on their strength in shear, what diameter of bolts should be used? (b) If a $\frac{1}{2}$-in.-thick ASTM A36 bar is used, what width $w$ should it be?

*Ans.* (a) $\frac{7}{8}$ in.; (b) $w = 3$ in.

Seat

Beam-to-column connection

Splice plate

FIG. 3-13 Probs. 3-16 and 3-17.     FIG. 3-14 Probs. 3-19 and 3-20.     FIG. 3-15 Probs. 3-21 and 3-22.

**3-17.** The vertical steel bar in Fig. 3-13 is $\frac{3}{8}$ in. thick and must be designed to withstand a tensile load $T$ of 26,400 lb. Two high-strength A325 bolts will be used. Calculate (a) the required diameter $d$ of the bolts and (b) the required width $w$ of the bar.         *Ans.* (a) $d = \frac{5}{8}$ in.; (b) $w = 4$ in.

**3-18.** Assume, in Fig. 3-12, that the beams are W 14 × 61, the girder is a W 21 × 82, three rivets are used through each beam web, all angles are $\frac{1}{2}$ in. thick, and all rivets are 1 in. in diameter. How many rivets will be required to connect the seat angle to the girder web if the load to be transmitted from each beam to the girder is 90 kips?

**3-19.** A typical structural beam-to-column connection is shown in Fig. 3-14. The beam is a W 16 × 64 and the column is a W 10 × 49. All angles are $\frac{3}{8}$ in. thick. If the connection must be capable of transmitting a load of 63 kips from the beam to the column, what diameter of rivets should be used?         *Ans.* $\frac{3}{4}$-in. rivets

**3-20.**  A structural connection similar to the one shown in Fig. 3-14 must be capable of transmitting 90 kips from the beam to the column. The beam is a W 18 × 55, and the column is a W 12 × 53. All angles are $\frac{3}{8}$ in. thick, and all rivets are $\frac{7}{8}$ in. in diameter. If two rivets connect the seat angle to the column flange, how many rivets are required through the beam web?

**3-21.**  In Fig. 3-15 is shown a typical splice often used in the lower chord of double-angle trusses. A splice plate is used to connect the outstanding legs, thus enabling stress in them to be transmitted from one side of the splice to the other. Thicknesses: angles, $\frac{5}{16}$ in.; gusset plate, $\frac{7}{16}$ in.; splice plate, $\frac{3}{8}$ in. If $T = 90$ kips and the splice plate is fastened as shown with $\frac{3}{4}$-in. rivets, how many $\frac{3}{4}$-in. rivets should be used through the connected legs on each side of the center?

Ans. Five rivets

**3-22.**  The type of connection shown in Fig. 3-15 is used to join two sections of the lower chord of a double-angle truss. The angles are $\frac{3}{8}$ in. thick, the gusset plate $\frac{1}{2}$ in., and the splice plates $\frac{3}{8}$ in. The rivets are $\frac{3}{4}$ in. in diameter. If four rivets are used through the connected legs on each side of where they join, how many rivets will be required on each side through the splice plate to resist a pull of 86 kips?

## 3-7 Riveted Boiler Joints

Generally speaking, a boiler joint is one in which two plates are fastened together in order to make a joint that will remain airtight under pressure, such as normally found in steam boilers and in compressed-air tanks. The simplest way to accomplish this is to lap the two plates and to fasten them together with rivets fitted into matched holes. Such a joint is called a **lap joint,** three examples of which are shown in Fig. 3-16a, b, and c. When greater strength or efficiency is required, the main plates to be fastened together are butted, and cover plates, or splice plates, riveted to both main plates then fasten them together. Such a joint is called a **butt joint,** of which Fig. 3-16d, e, and f are examples.

If absolute tightness of a joint is required, as perhaps in the case of joints in water tanks and steam boilers, the edge of the outer plate is beveled and the joint is then calked. That is, with the aid of a blunt-edged chisel-shaped calking tool, the lower edge of the outer plate is hammered into the seam, thus sealing the joint tightly. Occasionally, inner seams are calked also.

The single-riveted lap joint is commonly used in low-pressure tanks. It is most suitable with plates less than $\frac{1}{2}$ in. in thickness. In the double-riveted lap joint, greater tightness is secured by staggering the rivets in the two rows. In the triple-riveted lap joint, both the strength and the "efficiency" of the joint are increased by making the pitch $p$ of the two outer rows of rivets twice as great as the pitch of the middle row. This can be seen when we consider the strength in tension of these joints. Given the same plate thicknesses, rivet diameters, and pitch $p$ for three such lap joints and assuming that all are weakest in tension, failure in tension of the main plates would occur along

FIG. 3-16 Riveted boiler joints. (*a*) Single-riveted lap joint. (*b*) Double-riveted lap joint. (*c*) Triple-riveted lap joint. (*d*) Single-riveted butt joint. (*e*) Double-riveted butt joint. (*f*) Triple-riveted butt joint.

a line between the rivet holes in the outer row. The joint with the smallest pitch in the outer row has the least net resisting area and therefore is the weakest.

A disadvantage of increased rivet spacing in the outer rows is that the effectiveness of calking decreases as the rivet pitch increases, since the most effective calking naturally is secured when rivets are closely spaced.

For boilers and tanks subjected to higher loads and pressures, and when plates $\frac{1}{2}$ in. or more in thickness are used, the butt joint illustrated in Fig. 3-16*d*, *e*, and *f* is most suitable. The main plates are butted together and are held in place by thinner cover plates, with most of the rivets passing through all three plates. When only one cover plate is used, its thickness must be equal to or greater than the thickness of the main plates. When two cover plates are used, the thickness of each should be not less than one-half of the thickness of the main plates in the case of boiler joints, and not less than five-eighths in the case of similar structural joints. When these specifications are adhered to, the factor of safety of the cover plates will be as high as or higher than that of the main plates and the cover plates need not be separately designed or investigated for stresses.

**Definitions.** We shall need here to define certain terms that are commonly used in connection with riveted joints:

**Pitch** is the distance $p$ between centers of adjacent rivets in a row. In joints using two or more rows of rivets, each row may use a different pitch. The longer pitch $p_1$, as in Fig. 3-16c and f, is generally twice as great as the shorter pitch $p$. The minimum recommended pitch is generally three times the rivet diameter.

**Gage** is the distance $g$ between center lines of parallel rows or rivets.

**Repeating Section.** In continuous riveted joints the rivets are generally so spaced as to form a pattern which repeats itself along the rows of rivets. The length of joint required by one complete pattern is called a **repeating section.** In Fig. 3-16a, b, and d, the length of a repeating section equals the pitch $p$. In Fig. 3-16c, e, and f the length of a repeating section equals the longer pitch $p_1$, which in both of these instances equals $2p$.

**Cover plate** is the longitudinal plate, as in a butt joint, that is the element connecting the two main plates. A single cover plate may be sufficient at a butt joint, but two are generally used.

**Edge distance** is the distance from the center of a rivet to the edge of the plate. This is distance $e$ in Fig. 3-16, generally a minimum of $1\frac{1}{2}$ times the rivet diameter.

**Tension Failures of Boiler Joints.** Failure in tension occurs when a plate tears apart along some line of least resistance as, for example, along line $a$-$a$ in Fig. 3-17.

In a continuous joint with two or more rows of staggered rivets, a possibility

FIG. 3-17 Tension failures of riveted boiler joints. ($a$) Tearing. ($b$) Tearing of plate between rivet holes.

exists of failure along the diagonal lines between the closest adjacent rivets. This type of failure is easily prevented by making the gage $g$ not less than $1\frac{3}{4}$ times the rivet diameter.

### 3-8 Strength and Efficiency of Riveted Boiler Joints

The strength of a riveted joint is the maximum load it can resist without failure. Since a joint may fail in several ways, its strength in each must be calculated. The strength having the least value, then, is the strength of the joint.

The efficiency of a riveted boiler joint is the ratio of the strength of the joint to the strength in tension of the unpunched plate. That is,

$$\text{Efficiency} = \frac{\text{strength of joint}}{\text{strength of unpunched plate}} \qquad (3\text{-}1)$$

To determine the strength of a continuous joint, a section equal in length to a repeating section is generally used.

### 3-9 Stresses in Thin-walled Pressure Vessels

The most common form of pressure vessel in which riveted (or welded) joints are used is the cylinder, as used for steam boilers, air-compressor tanks, and water tanks. Spheres are also used to some extent for containing gas under pressure. Liquids and gases causing internal pressure in a closed vessel are both referred to as **fluids.** When the fluid is a gas, the pressure is constant in all parts of the vessel. When it is liquid, the pressure is lowest at the top and increases roughly $\frac{1}{2}$ psi per foot of depth of liquid. Because of its comparative insignificance, this increase is generally disregarded except when the depth exceeds, say, 10 ft. When pressure is caused by water, as in open water tanks, it is equal at any point to the weight of a column of water extending from the point vertically to the level of the free water surface. This vertical distance is called the **head** $h$. Assuming water to weigh 62.5 pcf (pounds per cubic foot) the pressure $p$ at any point would be $62.5h$ psf, or $62.5h/144 = 0.434h$ psi. That is, a head of 10 ft of water would cause a pressure of $(0.434)(10) = 4.34$ psi.

In order that joints or seams in these vessels, longitudinal or circumferential, may be designed, the force to be resisted per linear inch of seam must first

FIG. 3-18 Riveted cylindrical pressure vessel.

(a)    (b)    (c)

FIG. 3-19 Stress on longitudinal seam of cylindrical pressure tank. (a) Internal pressure due to gas or liquid. (b) Steam and water under pressure. (c) Stress in wall of tank shell.

be determined. We shall consider here only cylinders and spheres in which the walls are thin relative to their diameters. For example, a wall thickness not exceeding 10 percent of the diameter of the vessel may safely be considered to be thin. In such vessels, the intensity of stress between inner and outer surfaces is approximately constant. In thick-walled vessels, the stress variation becomes more complex, being highest at the outer surface.

**Stress on Longitudinal Seam of Cylinder.** Consider a thin-walled cylinder, shown in cross section in Fig. 3-19, to be subjected to internal fluid pressure. A law of fluid mechanics is that *fluid pressure at any point is equal in all directions and is always directed perpendicular to the resisting surface,* as indicated in Fig. 3-19a.

In Fig. 3-19b, a steam boiler is half filled with water with steam under pressure in the upper half. The steam pressure is exerted against the semi-cylindrical surface of the upper half of the boiler and also against the level water surface in the lower half, perpendicular to those resisting surfaces. The water acts as a solid surface resisting the downward steam pressure.

If now we cut this vessel at the water line and isolate the lower half as a free body, it becomes apparent that the total downward pressure $P$ must be balanced by the two forces $T$.

A section of length $L$ of the lower half has been isolated in Fig. 3-20. Since the intensity of pressure $p$ is constant over the area $dL$, $P = pdL$. Similarly, $T = s_t tL$. Then

$$pdL = 2s_t tL$$

and

$$\textbf{Unit wall tensile stress } s_t = \frac{pd}{2t} \qquad (3\text{-}2)$$

where $p$ = internal pressure, psi
$d$ = inside diameter of cylinder, in.
$t$ = thickness of wall material, in.

FIG. 3-20 Stress on longitudinal seam of cylinder.

In the illustration above, water and steam were used for explanatory purposes. Identical results will be obtained, however, with any medium properly classified as a fluid.

We should note here that the stress on a **longitudinal** seam is a **circumferential** stress, since it is directed around the circumference. Similarly, the stress on a **circumferential** seam is a **longitudinal** stress, since it is directed parallel to the longitudinal axis of the cylinder. See Fig. 3-20.

When designing riveted or welded seams in pressure vessels, we must know the force $T_L$ to be resisted per linear inch of seam. If the force on a longitudinal section $L$ inches long is $s_t tL$, as previously found, the force $T_L$ per inch of length is $s_t tL/L$, or $T_L = s_t t$. But $s_t = pd/2t$. Hence

$$\text{Force per inch of longitudinal seam } T_L = \frac{pd}{2} \tag{3-3}$$

To obtain the minimum required wall thickness of a cylinder to resist a given pressure $p$, Eq. (3-2) may be rewritten thus:

$$\text{Required wall thickness of cylinder } t = \frac{pd}{2s_t} \tag{3-4}$$

Equation (3-4) may be used to determine the required wall thickness of tanks where welded seams with an efficiency of 100 percent can be obtained. Since the efficiencies of riveted boiler joints usually vary between 65 and 85 percent, somewhat thicker plate must be used in riveted cylinders.

**Stress on Circumferential Seam of Cylinder or Sphere.** We shall show now that the force $T_C$ per linear inch of a circumferential seam of a cylinder is only half of that in a longitudinal seam.

Consider the end portion of a cylinder as shown in Fig. 3-21a. Let this portion be filled with liquid under pressure $p$ per square inch. Since the

FIG. 3-21 Force on circumferential seams of cylinders and spheres. (*a*) Pressure on end section of cylinder. (*b*) Cross section of cylinder or sphere. (*c*) Pressure on half of sphere.

cross-sectional area of the cylinder is $\pi d^2/4$, the total force against the end is $p\pi d^2/4$. This force is equaled by the total resistance offered by the material of the cylinder wall around the entire circumferential seam. If $T_C$ is the resisting force per linear inch of the circumference, which is $\pi d$, then the total resisting force is $T_C\pi d$. But $T_C\pi d = p\pi d^2/4$, or

$$\text{Force per inch of circumferential seam } T_C = \frac{pd}{4} \tag{3-5}$$

which is seen to be just half of the force exerted on the longitudinal seam. The stress on the circumferential seam is then

$$s_t = \frac{T_C}{t} = \frac{pd}{4t} \tag{3-6}$$

which is seen to be one-half that on the longitudinal seam as given by Eq. (3-2).

## 3-10 Allowable Stresses for Boiler Joints
The Boiler Code of the American Society of Mechanical Engineers (see Table 3-2) gives the *ultimate* strengths of boiler steel to be used for boilers and tanks and also recommends a factor of safety of 5.

Table 3-2 ULTIMATE AND ALLOWABLE STRESSES. ASME BOILER CODE. RECOMMENDED FOR BOILERS AND TANKS

| Kind of stress | Symbol | Ultimate strength, psi | Factor of safety | ✳ Allowable stress, psi |
|---|---|---|---|---|
| Tension | $s_t$ | 55,000 | 5 | 11,000 |
| Shear | $s_s$ | 44,000 | 5 | 8,800 |
| Bearing | $s_c$ | 95,000 | 5 | 19,000 |

In the case of boiler joints, no distinction is made between inside and outside bearing. Areas in shear and bearing are based on the rivet-hole diameters rather than on the smaller rivet diameters, as in structural joints.

## 3-11 Analysis and Design of Boiler Joints

The design of riveted boiler joints so as to obtain the greatest efficiency consistent with the practical aspects of the design is a somewhat laborious problem and is beyond the scope of this text. Various handbooks list a great many standard designs of both lap and butt joints whose efficiencies have been calculated and range from about 50 percent for some single-riveted lap joints to as high as 94 percent for a butt joint of great strength. In most cases, these standard designs are used. The following examples, however, indicate how strength and efficiency of a given boiler joint are calculated.

### ILLUSTRATIVE PROBLEMS

**3-23.** A cylindrical air-compressor tank with an inside diameter of 22 in. is to be designed to withstand safely a working pressure of 125 psi. Boiler plate with an allowable tensile stress of 11,000 psi will be used (see Table 3-2). All seams will be butt welds with an expected efficiency of 100 percent. Determine (a) the force $T_L$ per linear inch of longitudinal seam; (b) the required thickness $t$ of wall; (c) the unit tensile stress in the circumferential seam (see Fig. 3-22).

FIG. 3-22 Design of welded air-compressor tank.

*Solution:* The force per linear inch is obtained from Eq. (3-3). That is,

$$\left[ T_L = \frac{pd}{2} \right] \qquad T_L = \frac{(125)(22)}{2} = 1{,}375 \text{ lb} \qquad Ans.$$

Equation (3-4) gives the required thickness of wall. That is,

$$\left[ t = \frac{pd}{2s_t} \right] \qquad t = \frac{(125)(22)}{(2)(11{,}000)} = 0.125 \text{ in.} \qquad \text{or} \qquad \frac{1}{8} \text{ in.} \qquad Ans.$$

To obtain the unit wall stress caused by internal pressure of 125 psi, we first find from Eq. (3-5) the force $T_C$ per inch of circumference. That is,

$$\left[ T_C = \frac{pd}{4} \right] \qquad T_C = \frac{(125)(22)}{4} = 687.5 \text{ lb per in.}$$

and

$$\left[ s_t = \frac{T_C}{t} \right] \qquad s_t = \frac{687.5}{0.125} = 5{,}500 \text{ psi} \qquad Ans.$$

This last result was to be expected, since the wall thickness was designed to resist 11,000 psi and the longitudinal stress is half of the circumferential stress.

**3-24.** The longitudinal seam of a boiler having an inside diameter of 5 ft is a double-riveted butt joint similar to that shown in Fig. 3-23. The main boiler plate is $\frac{3}{4}$ in. thick, the cover plates (butt straps) are $\frac{1}{2}$ in. thick, and the rivets are 1 in. in diameter in $1\frac{1}{16}$-in. holes. The pitch $p$ is $3\frac{1}{4}$ in., and $p_1$ is $6\frac{1}{2}$ in. Calculate (a) the strength and efficiency of a repeating section and (b) the maximum allowable internal steam pressure, if a factor of safety of 5 is required.

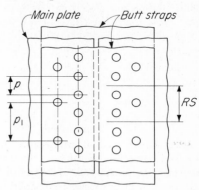

FIG. 3-23 Strength and efficiency of riveted boiler joint.

*Solution:* The strength of a repeating section $6\frac{1}{2}$ in. in length and containing three rivets, two in the inner row and one in the outer row, is its least ultimate strength in shear, bearing, or tension, or in whatever combination of these is least. In shear and bearing we find that

$$P_S = (3)(2)(0.8866)(44,000) = 234,000 \text{ lb}$$

$$P_B = (3)(0.75)(1.0625)(95,000) = 227,000 \text{ lb}$$

In tension, the main plate may fail by tearing the material between the rivets in the outer row. But it may also fail by tearing the lesser area of material between the rivets in the inner row. This latter can happen only if simultaneous therewith one rivet in the outer row fails either in shear or in bearing. The calculated values of $P_S$ and $P_B$ indicate that failure will occur in bearing. Then

$$\text{Outer row: } P_T = (6.5 - 1.0625)(0.75)(55,000) = 224,000 \text{ lb}$$
$$\text{Inner row: } P_T = (6.5 - 2.125)(0.75)(55,000) + (\tfrac{1}{3})(227,000)$$
$$= 180,500 + 75,700 = 256,200 \text{ lb}$$

Clearly, now, tension failure would occur between rivets in the outer row. Consequently, this being least,

$$\text{Strength of repeating section: } P = 224,000 \text{ lb} \qquad Ans.$$

The efficiency of the joint is the ratio of its strength to the strength of the solid unpunched plate. That is,

$$\text{Efficiency} = \frac{224,000}{(6.5)(0.75)(55,000)} = 0.835 \qquad \text{or} \qquad 83.5 \text{ percent} \qquad Ans.$$

To determine the maximum allowable internal steam pressure $p$, we must first find the allowable force $T_L$ per linear inch of longitudinal seam. With a factor of safety of 5

$$\text{Allowable } T_L = \frac{224,000}{(6.5)(5)} = 6,880 \text{ lb}$$

Then from Eq. (3-3)

$$\left[ p = \frac{2T_L}{d} \right] \qquad p = \frac{(2)(6,880)}{60} = 230 \text{ psi} \qquad Ans.$$

## PROBLEMS

**3-25.** Assume that the tank in Prob. 3-23 must have an inside diameter of 29.5 in. and that it must withstand a pressure of 165 psi. Then solve.

**3-26.** In Prob. 3-24 assume the following: boiler diameter, 4 ft; main boiler plate, 1 in. thick; cover plates, $\frac{5}{8}$ in. thick; rivets, $1\frac{1}{8}$ in. diameter in $1\frac{3}{16}$-in. holes; $p = 4$ in. and $p_1 = 8$ in. Then solve.

**3-27.** A seamless-tube steam pipe with an inside diameter $d$ of 8.071 in. and a wall thickness $t$ of 0.277 in. is subjected to internal steam pressure $p$ of 200 psi. (a) Calculate the unit tensile stress $s_t$ in the wall material. (b) What maximum internal pressure $p$ can this pipe safely stand if the allowable tensile stress is 11,000 psi?
    Ans. (a) $s_t = 2,910$ psi; (b) $p = 755$ psi

**3-28.** A 16-in. seamless-tube water pipe must be capable of withstanding a pressure head of 900 ft. What minimum wall thickness is required if allowable tensile stress is 12,000 psi?

**3-29.** A lubricating hoist is supported by a hollow cylindrical shaft whose inside diameter is 9 in. The pressure of compressed air against its closed upper end raises and lowers the shaft. What minimum air pressure $p$ must be supplied to raise a load of 8,000 lb?
    Ans. $p = 126$ psi

**3-30.** A 20-in.-diameter air-compressor tank must safely withstand a working pressure of 145 psi. Boiler plate $\frac{1}{4}$ in. thick will be used. The longitudinal seam will be a single-riveted lap joint. What diameter of rivet and what pitch should be used? Minimum rivet spacing is 3 diameters.

**3-31.** A spherical tank 50 ft in diameter is used to contain gas under pressure. The plate is $\frac{1}{2}$ in. thick and is butt-welded to give a joint efficiency of 100 percent. (a) What maximum interior pressure can the tank safely stand if allowable tensile plate stress is 11,000 psi? (b) If the tank must stand a pressure of 55 psi, what should the plate thickness be?

**3-32.** The longitudinal seam in a boiler whose inside diameter is 66 in. is a double-riveted butt joint with an efficiency of 80 percent. The boiler plate is $\frac{3}{4}$ in. thick. Calculate the maximum safe internal pressure $p$ if allowable tensile plate stress is 11,000 psi.
    Ans. $p = 200$ psi

**3-33.** A seamless-tube water pipe with an inside diameter of 10.192 in. runs from a reservoir to a small nearby town. (a) Calculate the pressure $p$ and the force $T_L$ per linear inch of (imaginary) longitudinal seam in the pipe at a point where it is 300 ft vertically below the open (free) water surface of the reservoir. (b) If the wall thickness is 0.279 in. and allowable tensile stress is 11,000 psi, what maximum head $h$ may this pipe safely withstand?

**3-34.** A large water pipe has an inside diameter of 4 ft. It is made of $\frac{3}{8}$-in. steel plate using double-riveted lap joints and $\frac{7}{8}$-in.-diameter rivets (see Fig. 3-16b), with a pitch of 4 in. in both rows. (a) Calculate the strength and efficiency of the joint. (b) Calculate the maximum head $h$ which this pipe may safely withstand if a factor of safety of 5 is used.        *Ans.* (a) 60,700 lb, 73.6 percent; (b) 290 ft

**3-35.** Assume in Prob. 3-34 that the pipe diameter is 6 ft, the steel plate is $\frac{1}{2}$ in. thick, $1\frac{1}{4}$-in.-diameter rivets are used in $1\frac{5}{16}$-in. holes, and the pitch is 5 in. Then solve.

## 3-12 Eccentrically Loaded Riveted Joints

As in the case of riveted boiler joints, most riveted structural connections have been standardized so as to provide the maximum possible strength and efficiency along with the simplicity of standardized fabrication.

Maximum strength and efficiency are obtained when the line of action of the load to be resisted by the connection passes through the centroid $C$ of the cross-sectional resisting shear areas of the rivets in the group, as illustrated in Figs. 3-24a and b and 3-26a and c. All such connections are said to be **concentrically loaded,** and, although perhaps not strictly true, because of imperfections of material and in fabrication, the very reasonable assumption is made that each rivet in the group carries its proportional part of the total load, as illustrated in Fig. 3-26a and c.

When the line of action of the load on the connection fails to pass through the centroid $C$, as it does in Figs. 3-25 and 3-26b and d, the connection is said to be **eccentrically loaded.** This condition results in additional forces $Q$ on the rivets due to the twisting action on the connection caused by the eccentric *moment Pe* of the load with respect to the centroid, as discussed in Prob. 3-36 and illustrated in Fig. 3-28. Consequently, designers carefully avoid eccentricity in connections when possible to do so. There are, however,

FIG. 3-24 Concentrically loaded connections.

FIG. 3-25 Eccentrically loaded connections.

FIG. 3·26 Rivet loads in concentrically and eccentrically loaded connections.

many special-purpose types of connections in which eccentricity cannot be avoided. A typical case is illustrated in Prob. 3-36.

The **location of the centroid of the rivet areas** in riveted connections therefore becomes a matter of importance. Since in most structural connections all rivets of a group have the same diameter and are generally arranged in a symmetrical pattern, the location of the centroid $C$ of the group can usually be seen at a glance, as in Figs. 3-24 to 3-27. For rivet groups not symmetrically spaced, the centroid may be located by the method outlined in Chap. 5.

Consider the eccentrically loaded connection shown in Fig. 3-26$b$. Two equal but oppositely directed forces may be added at the centroid $C$ without altering the equilibrium condition. Let each of these forces have the magnitude of the applied load $P$ (see Fig. 3-27$a$). The three forces $P$ may be grouped as shown, that is, as a couple and a single concentric force. The rivet forces due to the concentric force $P$ are obviously $P/2$, as shown in Fig. 3-27$b$. The

FIG. 3-27 Rivet loads in eccentric connection by superposition. ($a$) Equal and opposite loads $P$ added at $C$. ($b$) Rivet forces due to direct load. ($c$) Rivet forces due to couple $Pe$. ($d$) Rivet forces by superposition.

couple, or moment, $Pa$ must be balanced by another couple. This couple is formed by the rivet forces shown in Fig. 3-27$c$. Equating the couples gives $(F)(4a) = Pa$, or $F = P/4$. The values shown in Fig. 3-26$d$ could well be checked by the technique suggested here.

In Fig. 3-28$a$ is shown a typical case of a plate supporting a crane rail and riveted to the flange of a column. Clearly, the eccentric moment $Pe$ will twist, or rotate, the plate about some center. We shall now show that *the center of rotation is the centroid of the group of rivet areas.*

The plate is shown isolated in Fig. 3-28$b$. The centroid of the rivet areas is $C$, and $O$ is the center of rotation whose position is unknown. Let $c$ be the distance from $O$ to the rivet farthest away and $r$ the distance to any other rivet. If $s$ is the unit shearing stress in the rivet distant $c$ from $O$, then the unit shearing stress in any rivet distant $r$ from $O$ is $(r/c)s$, since *the intensity of the shearing stress in any rivet is proportional to its distance $r$ from $O$.* Proof of this statement may be found in Chap. 4, Torsion, Art. 4-2 and Fig. 4-3 (see also Observation and Conclusion 2 in Art. 7-1). The resisting force $Q$ of *any* rivet due to the moment $Pe$ then is $(r/c)sA$, where $A$ is the cross-sectional area of the rivet. The line of action of $Q$ is always perpendicular

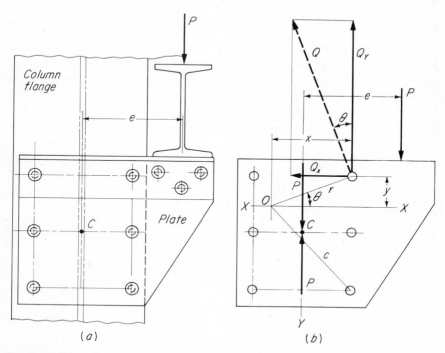

FIG. 3-28 Proof that center of rotation $O$ coincides with centroid $C$ of rivet group under rotational couple $Pe$. (*a*) Column bracket. (*b*) Bracket plate as free body.

to the radial distance $r$. The horizontal component of $Q$ is $(r/c)sA \sin \theta$, or $Q_X = (s/c)Ay$, since $r \sin \theta = y$. But, since the force system acting on the plate is necessarily in equilibrium, the sum of all horizontal components is zero. That is, $\Sigma Q_X = 0$, or $(s/c)\Sigma Ay = 0$. Since $s/c$ is not zero, it follows that $\Sigma Ay$ must be zero, which it can be only when the $X$ axis passes through the centroid $C$. We may similarly show that the $Y$ axis must also pass through $C$, thus proving that *the center of rotation is the centroid C.*

✳ The remaining problem now is to evaluate these moment-resisting forces $Q$. It has already been pointed out that (a) the rivet forces developed in resisting the applied twisting moment will each be perpendicular to a line joining the rivet and the center of rotation $C$, and (b) the rivet forces will be proportional to their distances from $C$. We can state this second fact mathematically by saying $Q = kr$, where $Q$ is the rivet force, $r$ is the distance from the centroid $C$, and $k$ is a proportionality constant.

If we take moments about the centroid $C$, we have

$$Pe = Q_1 r_1 + Q_2 r_2 + Q_3 r_3 + \cdots$$

$Q_1$, $Q_2$, and $Q_3$ are the forces acting on rivets 1, 2, and 3, respectively. The distance from each rivet to the centroid of the rivet group is denoted by $r_1$, $r_2$, $r_3$, etc. Since each rivet force is equal to $kr$,

$$Pe = kr_1^2 + kr_2^2 + kr_3^2 + \cdots = k(r_1^2 + r_2^2 + r_3^2 + \cdots)$$

This can be abbreviated $Pe = k\Sigma r^2$. The $\Sigma$ sign indicates "the sum of." Now, since $x$ and $y$ are the space components of $r$, and since $r^2 = x^2 + y^2$,

$$Pe = k\Sigma(x^2 + y^2) = k(\Sigma x^2 + \Sigma y^2)$$

or

$$k = \frac{Pe}{\Sigma x^2 + \Sigma y^2} \tag{3-7}$$

When $k$ has thus been evaluated, the twisting force $Q$ on any rivet in the group is easily calculated. That is,

$$Q = kr \tag{3-8}$$

in which $r$ is the distance to the rivet from the centroid of the group.

When the *twisting* force $Q$ on any given rivet has thus been evaluated, it must be combined vectorially with the direct force $P/N$ on that rivet. To do so algebraically, we shall find it convenient to use the $x$ and $y$ components of $Q$, which are easily obtained by simply multiplying $Q_1$ by the appropriate $y$ or $x$ distance from $O$. That is,

$$Q_X = Q \sin \theta = kr\frac{y}{r} \quad \text{or} \quad Q_X = ky \tag{3-9}$$

and

$$Q_Y = Q \cos \theta = kr\frac{x}{r} \quad \text{or} \quad Q_Y = kx \qquad (3\text{-}10)$$

Note carefully here that the force components $Q$ are proportional to the *opposite* space components ($x$ or $y$), not to the corresponding space components.

### ILLUSTRATIVE PROBLEM

**3-36.** The bracket plate shown in Fig. 3-29a supports a crane-rail load of 12 kips at a distance of 9 in. from a vertical axis passing through the centroid of the group of rivets which fasten the plate to a column flange. Calculate the total force $F$ on each of the six rivets. All rivets will have same diameter. (NOTE: In solving this type of problem, the plate, and not the rivets, is isolated as a free body. Consequently, instead of solving for the force exerted by the plate on each rivet, we solve for the equal but opposite force exerted by each rivet on the plate.)

(a)                                      (b)

FIG. 3-29 Eccentrically loaded riveted connection. (a) Bracket plate. (b) Forces exerted by rivets on plate.

*Solution:* As a first step in solving this problem, we locate at a glance the centroid $C$ of the rivet group. Next, we add two equal and opposite 12-kip forces at $C$, acting parallel to $P$. This we may do without changing the final results. Now, we have a downward-acting 12-kip force at the centroid resisted equally by all six rivets ($P/N = 12/6 = 2$ kips), and a clockwise couple or twisting moment, which is equal to $Pe = (12)(9) = 108$ kip-in. and is resisted by the twisting forces $Q$. We then evaluate $\Sigma x^2$ and $\Sigma y^2$ and calculate $k$. That is, measuring from the centroid,

$$\Sigma x^2 = (-3)^2 + (-3)^2 + (-3)^2 + 3^2 + 3^2 + 3^2 = 54$$
$$\Sigma y^2 = 4^2 + 4^2 + 0 + 0 + (-4)^2 + (-4)^2 \qquad = 64$$

and

$$k = \frac{Pe}{\Sigma x^2 + \Sigma y^2} = \frac{(12)(9)}{54 + 64} = 0.915 \text{ kip/in.}$$

Each of the four corner rivets is 5 in. from the centroid $C$, and the two middle rivets are 3 in. from $C$. Hence

$$[Q = kr] \qquad\qquad Q_5 = (0.915)(5) = 4.58 \text{ kips}$$

and

$$Q_3 = (0.915)(3) = 2.75 \text{ kips}$$

Since $P$ is downward acting, the six 2-kip resisting forces, referred to as *direct* forces, are upward acting. Also, since the moment $Pe$ is clockwise with respect to the centroid $C$, the moments of the resisting twisting forces are counterclockwise. This must be clearly understood in order for these forces to be properly directed, as in Fig. 3-29$b$. If this diagram is laid out to scale, the total force $F$ on each rivet can readily be determined graphically. Analytically we find for the four corner rivets that the vertical and horizontal components of $Q$ are

$$[Q_X = ky] \qquad\qquad Q_X = (0.915)(4) = 3.66 \text{ kips}$$

and

$$[Q_Y = kx] \qquad\qquad Q_Y = (0.915)(3) = 2.75 \text{ kips}$$

When these components are now added vectorially to the direct forces, we obtain the force $F$ on each of the rivets. That is,

$$F_A = F_E = \sqrt{(3.66)^2 + (2 - 2.75)^2} = 3.74 \text{ kips}$$
$$F_B = F_F = \sqrt{(3.66)^2 + (2 + 2.75)^2} = 5.99 \text{ kips}$$
$$F_G = 2.75 - 2 \qquad\qquad\qquad = 0.75 \text{ kip}$$
$$F_H = 2.75 + 2 \qquad\qquad\qquad = 4.75 \text{ kips} \qquad Ans.$$

## PROBLEMS

**3-37.** Calculate the resultant forces exerted by each of the two rivets on the angle clip shown in Fig. 3-30 when $P = 6,625$ lb. Check $F$ against each rivet by moments.
$$Ans. \ F = 12,600 \text{ lb}; \ F = 5,960 \text{ lb}$$

**3-38.** In Fig. 3-31, calculate the maximum force $F$ exerted by any one of the four rivets on the angle clip when $P = 13,250$ lb. Check the $Q$ force against each rivet by moments.

**3-39.** The angle clip shown in Fig. 3-32 is acted upon by a force $P$ of 15,000 lb. Calculate the maximum force $F_{max}$ exerted on the angle by any one of the three rivets. Check the $Q$ force against each rivet by moments.

**3-40.** Isolate the crane-rail-bracket plate in Fig. 3-33 and calculate the force $F$ exerted on the plate by the most highly stressed rivet in the group; $a = 3$ in., $b = 4$

Fig. 3-30 Prob. 3-37.    Fig. 3-31 Prob. 3-38.    Fig. 3-32 Prob. 3-39.

in., $e = 10$ in., and $P = 22,000$ lb. What diameter rivets should be used assuming bearing will not control?

**3-41.** In the riveted connection shown in Fig. 3-25, $P = 9,000$ lb, $e = 2$ in., the rivet spacing $a = 3$ in., and $\theta = 30°$. Calculate the forces $F$ exerted on the gusset plate by each of the three rivets.

**3-42.** Each of the two plates in Fig. 3-34 is $\frac{3}{8}$ in. thick. (a) If $a = 4$ in., $e = 5$ in., and $P = 28$ kips, what diameter of rivets should be used if they are in single shear? (b) Calculate the maximum shearing and bearing stresses on the most highly stressed rivet.

**3-43.** If, in Fig. 3-34, both plates are $\frac{5}{16}$ in. thick, $a = 4$ in., $e = 5$ in., and $\frac{3}{4}$-in.-diameter rivets are used (in single shear), what maximum load $P$ may be applied as shown?                   *Ans.* $P = 20,700$ lb

**3-44.** In Fig. 3-35, let $a = 4$ in., $e = 6$ in., $\theta = 30°$, and $P = 16,000$ lb. (a) What diameter of rivets should be used? (b) Calculate the maximum shearing and bearing stresses on the most highly stressed rivet if each plate is $\frac{5}{16}$ in. thick.

Fig. 3-33 Probs. 3-40 and 3-68.

Fig. 3-34 Probs. 3-42 and 3-43.

Fig. 3-35 Prob. 3-44.

### 3-13 Welded Joints

Welding is a process of joining two pieces of metal by fusion. That is, the two parts to be welded are securely held in place and are heated by an electric arc or an oxyacetylene torch. Fused metal from a slender welding rod is then deposited between them and allowed to cool.

While the strength of the welding-rod metal is usually higher than that of the parts to be welded, the strength of the finished welded joint depends largely upon the skill of the welder. An improperly done weld may be dangerous, although shop welding done under carefully supervised conditions is very dependable. The quality of a weld is difficult to determine. Where small factory-made welded parts are produced in large quantities, it is usually feasible and practical to x-ray the welded joint. By this means, flaws in a weld can generally be detected. For most large parts and for welds made in the field, the x-ray method is seldom feasible.

Welding is especially suitable for repairing broken parts and for joining many of the steel parts of shop-fabricated roof trusses, derricks, and frames for tractors, railroad cars, and many other types of heavy equipment and machinery.

### 3-14 Types of Welds

Various types of welds are illustrated in Fig. 3-36 and also, in more detail, in Figs. 3-37 to 3-42.

The two types of welds most frequently used are **fillet welds** and **butt welds.** Detailed illustrations and explanations of terminology are shown in Figs. 3-37 to 3-39.

The critical section in a fillet weld is the **throat.** The size of a fillet weld is the leg length of the largest inscribed right isosceles triangle, as is illustrated

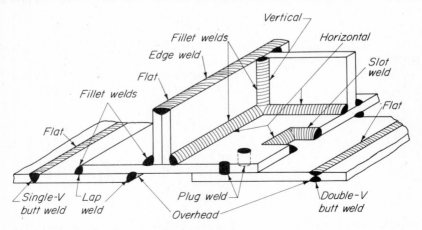

FIG. 3-36 Types of welds. Examples of locations.

FIG. 3-37 Fillet welds.

FIG. 3-38 Fillet-weld joints. (*a*) Single-fillet lap joint. (*b*) Double-fillet lap joint.

in Fig. 3-37, and the throat is 0.707 times that leg length. Fillet welds are usually made with equal legs and with a somewhat convex rather than a flat surface, since such a weld is most efficient and is also easiest to make. It is customary to add about ¼ in. to the length of each weld for starting and stopping.

Three types of butt welds are shown in Fig. 3-39. The square butt weld is generally used with plates not exceeding ¼ in. in thickness, the single V with plates from ¼ to ½ in., and the double V with plates over ½ in. thick. Butt welds are usually subjected mainly to tension and compression, not to shear. The **throat** of a butt weld joining two plates is the thickness of the thinnest of the plates.

Three other types of welds are illustrated in Figs. 3-40 to 3-42. In **plug welds,** punched holes in one of the two plates are filled with weld metal, which fuses partly into the other plate. **Slot welds** are sometimes used when other types of welds are not suitable and also at times to provide additional strength to a fillet-welded joint. **Spot welding** is used extensively in the fabrication of sheet-metal parts, merely as a simple and quick method of fastening light pieces together at intervals along a seam. Opposite metal rods clamp the pieces together tightly for an instant, during which a high electric current fuses the metal of both pieces at a small spot, as illustrated in Fig. 3-42.

FIG. 3-39 Butt-weld joints. (*a*) Square butt weld. (*b*) Single-V butt weld. (*c*) Double-V butt weld.

FIG. 3-40  Plug weld.          FIG. 3-41  Slot          FIG. 3-42  Spot
                               weld.                    weld.

### 3-15 Strength of Welded Joints

The Code of the American Welding Society (AWS) recommends the following allowable unit stresses in welds:

Butt welds: 20,000 psi in tension on section through throat
Fillet welds: 13,600 psi in shear on throat plane

Since the throat length of a fillet weld is 0.707 times its leg length (size), the allowable load $F$ on a $\frac{1}{2}$-in. fillet weld 1 in. long would be $F = (\frac{1}{2})(0.707)(1)(13,600) = 4,800$ lb.

In the case of a fillet weld along the square edge of a plate or the rounded edge of an angle or flange, it may be uncertain whether the fillet weld will be built up to equal the thickness of the plate, angle, or flange. Therefore, the American Institute of Steel Construction recommends the following maximum effective sizes of fillet welds:

At square edge: $\frac{1}{16}$ in. less than nominal thickness of plate
At rounded edge: $\frac{3}{4}$ of nominal thickness of angle or flange
Maximum effective size of any fillet weld: $\frac{5}{8}$ in.

Table 3-3 lists the recommended sizes of fillet welds at square and rounded edges, together with the allowable load per linear inch on each.

However, when required by design conditions and when specifically designated on drawings, fillet welds equal in size to the thickness of a plate or rolled section may be used. Nevertheless, the effective sizes listed in Table 3-3 are generally used in good practice.

The net required length of a fillet weld should be specified on drawings. The welder will put down the length of weld specified and will then taper it off at the ends, usually for an additional length equal to the size of the weld.

In practice, it is advisable to maintain the same fillet-weld size throughout

Table 3-3 RECOMMENDED MAXIMUM SIZES OF EDGE FILLET WELDS WITH ALLOWABLE LOADS $F$ IN POUNDS PER LINEAR INCH

| Thickness of plate | Square edge | | Round edge | |
|---|---|---|---|---|
| | Size of fillet, in. | Allowable load per in. | Size of fillet, in. | Allowable load per in. |
| $1^{3}/_{16}$ and over | $^{5}/_{8}$ | 6,000 | $^{5}/_{8}$ | 6,000 |
| $^{3}/_{4}$ | $^{5}/_{8}$ | 6,000 | $^{9}/_{16}$ | 5,400 |
| $1^{1}/_{16}$ | $^{5}/_{8}$ | 6,000 | $^{1}/_{2}$ | 4,800 |
| $^{5}/_{8}$ | $^{9}/_{16}$ | 5,400 | $^{7}/_{16}$ | 4,200 |
| $^{9}/_{16}$ | $^{1}/_{2}$ | 4,800 | $^{7}/_{16}$ | 4,200 |
| $^{1}/_{2}$ | $^{7}/_{16}$ | 4,200 | $^{3}/_{8}$ | 3,600 |
| $^{7}/_{16}$ | $^{3}/_{8}$ | 3,600 | $^{5}/_{16}$ | 3,000 |
| $^{3}/_{8}$ | $^{5}/_{16}$ | 3,000 | $^{9}/_{32}$ | 2,700 |
| $^{5}/_{16}$ | $^{1}/_{4}$ | 2,400 | $^{7}/_{32}$ | 2,100 |
| $^{1}/_{4}$ | $^{3}/_{16}$ | 1,800 | $^{3}/_{16}$ | 1,800 |
| $^{3}/_{16}$ | $^{1}/_{8}$ | 1,200 | $^{1}/_{8}$ | 1,200 |
| $^{1}/_{8}$ | $^{3}/_{32}$ | 900 | $^{3}/_{32}$ | 900 |

the connection, since a change of fillet size necessitates a change of welding rods and therefore slows the work and may cause errors.

Experiments have shown that the ends of a fillet weld, lying parallel to the line of action of the load, carry higher unit stresses than the mid-portion of the weld, as illustrated in Fig. 3-43. Also, when an end weld is combined with side welds, the unit stress in the end weld will be somewhat higher than those in the side welds. For this reason, designers prefer not to combine end and side welds unless lack of space makes it necessary.

FIG. 3-43 Ordinates show variations in stress in fillet welds.

## ILLUSTRATIVE PROBLEM

**3-45.** A $\frac{3}{8}$- by 4-in. square-edged structural-steel strap is to be welded to a heavy steel plate in the manner illustrated in Fig. 3-44. Calculate the length $L$ of fillet weld required to withstand safely a tensile pull $P$ of 30,000 lb.

FIG. 3-44 Required length of fillet weld.

*Solution:* From Table 3-3, the recommended size of fillet weld along the square edge of the $\frac{3}{8}$-in.-thick strap is $\frac{5}{16}$ in., on which the allowable load is 3,000 lb per linear inch. Since the total length of fillet weld is $2L$,

$$2L = \frac{30,000}{3,000} = 10 \text{ in.} \qquad \text{and} \qquad L = 5 \text{ in.} \qquad Ans.$$

## 3-16 Eccentricity in Welded Joints

One of the most common examples of eccentrically loaded welded joints is that of a structural angle welded to a gusset plate, as shown in Fig. 3-45. The total load $P$ in the angle is presumed to act along its gravity axis, the location of which may be obtained from a handbook. Consequently, a condition of eccentric loading exists, since $d_1$ is smaller than $d_2$.

In riveted joints, because of rather rigid adherence to standard patterns of placing rivets, the rivets are often overstressed because of eccentricity of loading, which is not always taken into account. In welded joints, however, we often find it possible to place the welds so that these excess stresses are eliminated. The following problem illustrates a common case.

FIG. 3-45 Balancing effect of eccentricity in welded connection.

## ILLUSTRATIVE PROBLEM

**3-46.** The structural-steel angle shown in Fig. 3-45 is 5 by 3 by $\frac{1}{2}$ in. Its long leg is to be welded to a heavy plate as indicated. (*a*) If a full end weld is used, calculate the additional lengths $L_1$ and $L_2$ of side welds to resist a tensile load of 60,000 lb, proportioned to eliminate the effect of eccentricity. (*b*) If $L_1$ is limited to approximately 5 in. in length and a full-length end weld is used, what size fillet weld should be used along the heel of the angle? (NOTE: Unless otherwise specified, the same size of weld is to be used all around.)

*Solution:* From Table E-11 we find that $d = 5$ in., $d_1 = 1.75$ in., and $d_2 = 3.25$ in. $R_1$, $R_2$, and $R_3$ are the resistances which must be offered, respectively, by fillet welds 1, 2, and 3. From Table 3-3 we find that a $\frac{3}{8}$-in. fillet weld is recommended along the rounded edge of a $\frac{1}{2}$-in.-thick angle, with an allowable load of 3,600 lb per linear inch.

(*a*) Since the length of the end weld is 5 in., $R_3$ may be evaluated. That is,

$$R_3 = (5)(3,600) = 18,000 \text{ lb}$$

Since the forces $P$, $R_1$, $R_2$, and $R_3$ are necessarily in equilibrium, $R_1$ and $R_2$ may be determined by moments; thus:

$[\Sigma M_2 = 0]$      $5R_1 + (18,000)(2.5) = (60,000)(3.25)$
                        $5R_1 = 195,000 - 45,000 = 150,000$      and      $R_1 = 30,000 \text{ lb}$

$[\Sigma M_1 = 0]$      $5R_2 + (18,000)(2.5) = (60,000)(1.75)$
                        $5R_2 = 105,000 - 45,000 = 60,000$      and      $R_2 = 12,000 \text{ lb}$
                        $18,000 + 30,000 + 12,000 = 60,000$      *Check.*

The required lengths $L_1$ and $L_2$ then are

$$\left[ L = \frac{R}{F} \right] \qquad\qquad L_1 = \frac{30,000}{3,600} = 8.33 \text{ in.} \qquad Ans.$$

$$L_2 = \frac{12,000}{3,600} = 3.33 \text{ in.} \qquad Ans.$$

$$\frac{60,000}{3,600} = 16.66 \text{ in.} = 5 + 8.33 + 3.33 = 16.66 \text{ in.} \qquad Check.$$

(*b*) If $L_1$ is limited to 5 in., the strength per linear inch must be

$$\frac{30,000}{5} = 6,000 \text{ lb per linear inch}$$

Use $\frac{5}{8}$-in. fillet weld.      *Ans.*

## 3-17 Design of Welded Joints

The four preceding articles together with the problems in them cover only the most elementary phases of welding. For a more complete treatment of the subject, the various handbooks on welding[1] should be referred to.

[1] "Procedure Handbook of Arc Welding Design and Practice," The Lincoln Electric Co., Cleveland; "Manual of Design for Arc Welded Steel Structures," Air Reduction Sales Company, New York.

## PROBLEMS

**3-47.** A steel strap $\frac{5}{16}$ in. thick by 2 in. wide is welded to a steel plate by means of $\frac{1}{4}$-in. fillet welds, as indicated in Fig. 3-46. (*a*) Determine required length $L$ of fillet, if $P = 10{,}800$ lb. (*b*) Determine $L$ required to develop the full tensile strength of the strap at 20,000 psi.

**3-48.** Solve Prob. 3-46 when the angle is 4 by $3\frac{1}{2}$ by $\frac{1}{2}$ in. and $\frac{3}{8}$-in. fillet welds are used except in (*b*), where $L_1$ is limited to 4 in. and the full-length end weld is increased to $\frac{7}{16}$ in. $P = 36{,}000$ lb.

**3-49.** In Fig. 3-47, $A$ is a double-angle structural-steel member welded on both sides to the gusset plate. The angles are 4 by 3 by $\frac{5}{16}$ in. with their long legs back to back. The connection must resist a total stress $P$ of 40,000 lb. Using only side fillet welds, determine $L_1$ and $L_2$. *Ans.* 6 by $\frac{7}{32}$ in.; 9.15 by $\frac{5}{16}$ in.

Fig. 3-46 Prob. 3-47.    Fig. 3-47 Probs. 3-49 and 3-50.    Fig. 3-48 Prob. 3-51.

**3-50.** The structural member $B$ in Fig. 3-47 is made of two steel angles $3\frac{1}{2}$ by 3 by $\frac{5}{16}$ in. with their long legs welded back to back to the gusset plate. The total stress $T$ in the member is 32,000 lb. The size of the gusset limits the length $L_1$ of the heel fillet to 4.5 in. A full $3\frac{1}{2}$-in.-long end weld should be used. Determine $L_1$ and $L_2$ and the fillet sizes, properly proportioned to counteract eccentricity.

**3-51.** In Fig. 3-48 is shown a cross-sectional view of the upper chord of a roof truss, in which the chords are T-sections (WT 6 × 18) and all web members are of standard pipe sections. The slotted pipe ends fit over the web of the T-section and are welded to the web all around. If the pipe shown fits at right angles with the T-section, $h = 5.58$ in., and $D = 4.0$ in., what tensile pull $T$ can be developed by a $\frac{1}{4}$-in. fillet weld?

## SUMMARY

### (*By article number*)

**3-1.** Whenever two or more pieces are fastened together by tightly driven rivets, the connection is called a **riveted joint.**

**3-3. Failures of riveted joints** are generally in *shear* or *bearing* in the rivets or in *tension* in the connected plates or parts. Failure in *end shearing* or *tearing* is prevented when proper end distances are maintained.

**3-4.** In **structural joints,** rivet holes are generally punched $\frac{1}{16}$ in. greater in diameter than the rivets. Rivet *shearing and bearing stresses* are based on the original diameter of the undriven rivet. *Tensile stresses* in the connected parts are based on an assumed rivet-hole diameter $\frac{1}{8}$ in. greater than the original rivet diameter.

The unit stresses in riveted joints, shearing, bearing, and tensile, are not of uniform intensity, but may vary considerably. Hence, the equation $s = P/A$ gives only an *average stress*. Commonly used allowable stresses have been adjusted to compensate for this nonuniformity.

When, in a riveted joint, two *outside* members enclose a third *inside* member, the rivets are in *double shear*. Bearing stress between the rivets and the outside members is called *outside bearing*, or plain bearing, while bearing stress between the rivets and the inside member is called *inside bearing*, or enclosed bearing.

**3-5.** An **analysis of a structural joint** is made to determine (1) the stresses in shear, bearing, or tension or (2) the allowable load on the joint.

**3-6.** The **design of a riveted structural joint** involves determination of the number of rivets required and their diameters and the required thicknesses of plates to resist a given load.

**3-7.** In **riveted boiler joints,** rivet holes are usually drilled, or are punched and reamed to a diameter $\frac{1}{16}$ in. larger than the rivet. All stresses, shearing, bearing, and tensile, are based on the drilled or reamed size of the hole.

**3-8.** The **strength of a riveted joint** is the maximum load it can resist without failure. The **efficiency of a riveted joint** is the ratio of its strength to the strength in tension of the unpunched plate.

**3-9.** **Thin-walled pressure vessels** are generally cylindrical with riveted or welded seams. When internal pressure is caused by a fluid, gas or liquid, the unit stress in the uniformly thick wall material is

$$\text{Unit wall tensile stress } s_t = \frac{pd}{2t} \tag{3-2}$$

and

$$\text{Force per inch of longitudinal seam } T_L = \frac{pd}{2} \tag{3-3}$$

and

$$\text{Force per inch of circumferential seam } T_C = \frac{pd}{4} \tag{3-5}$$

**3-12.** A **concentrically loaded joint** is one in which the line of action of the load passes through the centroid of the rivet group. When so loaded, all rivets are presumed to be equally loaded and the joint is efficient.

An **eccentrically loaded joint** results when the line of action bypasses the centroid. The product of the load $P$ and the eccentric distance $e$ is a twisting moment $Pe$ whose center of rotation is the centroid of the rivet group. As a result, twisting forces act on the rivets in addition to the direct force $P/N$ on each rivet.

**3-13.** **Welding** is a process of joining two or more parts by depositing fused metal from a slender rod along the joint between them and allowing it to cool, the heat of fusion being applied either by an electric arc or by an oxyacetylene torch.

## REVIEW PROBLEMS

**3-52.** A typical structural joint of a light roof truss is shown in Fig. 3-49. Each member is made up of two angles placed back to back and riveted to a gusset plate $\frac{5}{16}$ in. thick. The sloping member is made up of two angles $3\frac{1}{2}$ by 3 by $\frac{1}{4}$ in., which are fastened to the gusset plate with four $\frac{5}{8}$-in.-diameter rivets. Calculate the unit shearing and bearing stresses in these rivets caused by a tensile load $T$ of 32,000 lb.

**3-53.** Calculate the maximum compressive force $C$ which the vertical double-angle member in Fig. 3-49 may safely transmit to the joint if it is fastened with three $\frac{5}{8}$-in.-diameter rivets to a $\frac{3}{8}$-in.-thick gusset plate. The thickness of each angle is $\frac{5}{16}$ in.  *Ans.* $C_{max} = 27,600$ lb

FIG. 3-49  Probs. 3-52 and 3-53.    FIG. 3-50  Probs. 3-57 and 3-58.

**3-54.** Assume that the beam in Fig. 3-1$a$ is a W 16 × 36, the girder is a W 21 × 62, all rivets are of $\frac{3}{4}$-in. diameter, and the side angles are $\frac{3}{8}$ in. thick. Calculate the value of the maximum load $R$ (beam-end reaction) that may safely be transmitted from the beam to the girder.

**3-55.** In a beam-to-column connection similar to that shown in Fig. 3-14, the beam is a W 24 × 76 and the column is a W 12 × 65. All connecting angles are $\frac{7}{16}$ in. thick. If $\frac{3}{4}$-in.-diameter rivets are used and four rivets connect the seat angle to the column flange, how many additional rivets are required through the web of the beam to transmit a load of 106 kips?

**3-56.** Assume that the beam in Fig. 3-12 is a W 16 × 40, the girder is a W 27 × 94, 1-in.-diameter rivets are used, and the load to be transmitted from each beam to the girder is 105 kips. How many rivets must be used to connect the seat angle to the girder? All angles are $\frac{1}{2}$ in. thick.

**3-57.** A connection of the type shown in Fig. 3-50 must be designed to support a tensile load $T$ of 36,000 lb. Calculate ($a$) the required diameter $D$ of the unfinished bolt $B$; ($b$) the required thickness $t$ of bar $A$ and $t'$ of each of the two side plates $C$ (based on bearing); ($c$) the required width $w$ of bar $A$ if allowable tensile stress on net section is 16,000 psi; ($d$) the minimum distances $a$ and $c$ to preclude end shearing in bar $A$ and in plates $C$, if allowable plate shearing stress is 12,000 psi; ($e$) the required number of $\frac{3}{4}$-in. rivets to fasten plates $C$ to plate $D$. Where necessary, round off calculated values to nearest higher multiple of $\frac{1}{16}$ in.

**3-58.** The structural connection shown in Fig. 3-50 is to be designed to support a tensile load $T$ of 18,000 lb. Calculate (a) the required diameter $D$ of the unfinished bolt $B$; (b) the required thickness $t$ of bar $A$ and $t'$ of each of the two side plates $C$; (c) the required width $w$ of bar $A$ if allowable tensile stress on net area is 15,000 psi; (d) the minimum distances $a$ and $c$ to preclude end shearing in bar $A$ and in plates $C$ if allowable plate shearing stress is 12,000 psi; (e) the required number of $\frac{5}{8}$-in. rivets to fasten side plates $C$ to the main plate $D$. Round off calculated dimensions to nearest higher multiple of $\frac{1}{16}$ in. where necessary.

**3-59.** The pin $A$ of Fig. 3-51 is $1\frac{1}{2}$ in. in diameter and is in double shear. Calculate the shearing stress in the pin if $W = 21$ kips.                 *Ans. $s_s = 10.3$ ksi*

FIG. 3-51  Prob. 3-59.

FIG. 3-52  Prob. 3-60.

**3-60.** The tubing cutter shown in Fig. 3-52 is used to cut $1\frac{1}{4}$-in.-outside-diameter pipe. If the axle of wheel $A$ is $\frac{1}{4}$ in. in diameter and is supported at both ends, calculate the shearing stress $s_s$ in the axle. The movable wheel $B$ can exert a force of 500 lb on the pipe.

**3-61.** In the continuous triple-riveted lap boiler joint shown in Fig. 3-53, the short pitch $p$ is $2\frac{13}{16}$ in., the long pitch $p_1$ is $5\frac{5}{8}$ in., the rivet holes are $\frac{15}{16}$ in. in diameter, and the thickness of the plates is $\frac{1}{2}$ in. Calculate the unit stresses in this joint in shear, bearing, and tension due to a load $P$ of 24,000 lb on a repeating section.

*Ans. $s_s = 8,690$ psi; $s_c = 12,800$ psi; $s_t = 10,250$ psi*

FIG. 3-53  Probs. 3-61 and 3-62.

FIG. 3-54  Prob. 3-67.

**3-62.** Using the data given in Prob. 3-61, calculate (a) the strength of a repeating section of the boiler joint shown in Fig. 3-53 and (b) the efficiency of the joint.

**3-63.** At a small hydroelectric power plant, a water pipe with an inside diameter of 30 in. and a wall thickness of 0.375 in. operates under a maximum head h of 800 ft. If the ultimate strength of the steel is 50,000 psi, what is the factor of safety?

**3-64.** A spherical gas tank 24 ft in diameter is constructed of boiler plate to withstand safely internal gas pressure of 67 psi. Allowable tensile plate stress is 11,000 psi. What thickness t of plate should be used (a) if seams are butt-welded with 100 percent efficiency and (b) if seams are riveted and are 70 percent efficient?
*Ans.* (a) $t = \frac{7}{16}$ in.; (b) $t = \frac{5}{8}$ in.

**3-65.** An open-top water surge tank is 20 ft in diameter and must withstand the pressure caused by a static depth of 98 ft of water. (a) What thickness t of plate is required if 100 percent efficient butt-welded seams are used? (b) Calculate the length of a repeating section and the required plate thickness t if 1-in. rivets are used in a triple-riveted lap joint.

**3-66.** Using the eccentrically loaded riveted connection shown in Fig. 3-26d, show that the forces exerted on the plate by the three rivets are $P/12$, $4P/12$, and $7P/12$.

**3-67.** The gusset plate shown in Fig. 3-54 will be riveted to a column flange with six equally spaced rivets whose centers lie on a 10-in.-diameter circle. The load P is 15,000 lb. (a) Calculate the maximum force $F_{max}$ exerted on the plate by the most highly stressed rivet. (b) If the rivets are of $\frac{5}{8}$-in. diameter, by what maximum percentage may the load P be increased without stressing any rivet beyond its capacity in single shear?
*Ans.* (a) $F_{max} = 3,870$ lb; (b) 19 percent

**3-68.** In the eccentrically loaded riveted connection shown in Fig. 3-33, let $a = 4$ in., $b = 5$ in., and $e = 10$ in. and assume that ten $\frac{7}{8}$-in.-diameter rivets are used in single shear. Thickness of the column flange is 0.688 in. and of the plate is 0.375 in. Calculate the maximum safe load P.

**3-69.** In Fig. 3-55, two $\frac{3}{4}$-in. main straps are fastened together by welding $\frac{3}{8}$-in. splice plates 3 in. wide on both sides. If $\frac{5}{16}$-in. fillet welds and full-width end welds are used, calculate required length L to resist a pull P of 72,000 lb.

FIG. 3-55  Prob. 3-69.              FIG. 3-56  Prob. 3-70.

**3-70.** A $\frac{1}{2}$-in.-thick circular plate 6 in. in diameter is welded to a heavier plate by means of a $\frac{1}{2}$-in. fillet weld around its perimeter, as shown in Fig. 3-56. Calculate the maximum moment Pe that this weld may safely resist. The forces P are exerted against a lug projecting from the plate.
*Ans. Pe = 271,400 lb-in.*

## REVIEW QUESTIONS

**3-1.** Name some structures in which the separate members are usually riveted together.

**3-2.** Describe an undriven rivet and the manner in which it is usually driven.

**3-3.** What are some of the standard practices in the design of riveted structural joints? Name some examples of structural joints.

**3-4.** Name some of the standard practices in the design of riveted boiler joints. Give some examples and types of boiler joints.

**3-5.** Describe several possible types of failures of riveted joints.

**3-6.** Discuss shearing stresses in riveted joints. Give examples of single shear and double shear.

**3-7.** Explain bearing stresses and give examples of inside bearing and outside bearing.

**3-8.** Discuss tensile plate stresses in a riveted joint.

**3-9.** Are unit stresses in riveted joints fairly uniform or rather nonuniform?

**3-10.** When the equation $s = P/A$ is used in connection with riveted joints, what stress $s$ is thus obtained?

**3-11.** On what bases are allowable rivet unit stresses usually determined? Approximately what factor of safety is already included in the allowable stresses for steel rivets?

**3-12.** What is regarded as the strength of a riveted joint?

**3-13.** Define the term "efficiency of a riveted joint." Estimate a common efficiency of a boiler lap joint.

**3-14.** What is meant by a concentrically loaded riveted joint?

**3-15.** Explain what is meant by an eccentrically loaded riveted joint.

**3-16.** By what method may the centroid of an unsymmetrical group of rivets be located?

**3-17.** By what quantities is an eccentric moment usually measured?

**3-18.** What portion of the total load $P$ is each rivet of a concentrically loaded group of $n$ rivets presumed to carry?

**3-19.** What is the effect of an eccentric moment on the rivets of a group? Describe some characteristics of the force $Q$ and how $Q$ is determined.

**3-20.** Against which rivet of a group of equal-diameter rivets is the maximum force $F$ generally exerted and how is $F$ determined?

**3-21.** Describe generally the process of welding.

**3-22.** For what type of work is welding especially useful?

**3-23.** What are the best assurances of good-quality welding?

**3-24.** Describe briefly fillet welds and butt welds.

**3-25.** What determines the size of a fillet weld?

**3-26.** Where is the critical section in a fillet weld and how are its dimensions determined?

**3-27.** How may eccentric loading of welded joints sometimes be compensated for?

# CHAPTER 4

# Torsion

## 4-1 Twisting Moments. Torque

Structural members are subjected to various kinds of forces according to the intended function of each member. For example, the wire ropes supporting an elevator cage are acted upon by *tensile forces,* a timber column supporting beams is subjected to *compressive forces,* a beam supporting a footbridge across a stream is subjected to *bending forces,* and a steel shaft transmitting power is subjected to *twisting forces* which produce a **twisting moment,** commonly called a **torque.** A torque is normally applied in a plane perpendicular to the longitudinal axis of the member. When so applied, the member is said to be in **torsion.**

In Fig. 4-1, for example, a piece of shafting is welded to a baseplate $B$, which in turn is fastened to a support by three bolts at equal distances from the shaft center. When the shaft is twisted by the application of the couple $Fa$ to the circular wheel $A$, a resisting torque is formed by the forces $P$ exerted against the plate by the three equispaced bolts. The applied torque is $Fa$, and if $N$ designates the number of bolts, the resisting torque is $NPr$. The axis of rotation is the centroidal axis of the shaft. When two or more bolts are symmetrically spaced about the axis of rotation, each bolt is presumed to resist its proportional part of the total torque.

### ILLUSTRATIVE PROBLEM

**4-1.** If in Fig. 4-1, $F = 1,500$ lb, $a = 20$ in., and $r = 5$ in., (*a*) what force $P$ must be resisted by each bolt, and (*b*) what diameter bolts should be used if the allowable shearing stress is 12,000 psi?

FIG. 4-1 Twisting moment. Torque.

*Solution:* The applied torque on $A$ is the moment of the couple shown, or *Fa*. This torque is resisted by the moment of the three forces acting on $B$. Equating the acting and resisting torques gives

$[Fa = NPr]$    $(1,500)(20) = (3)(P)(5)$      or      $P = 2,000$ lb      *Ans.*

The required bolt diameter can now be found:

$$\left[A = \frac{F}{s_s}\right] \qquad \frac{\pi D^2}{4} = \frac{2,000}{12,000} \qquad \text{or} \qquad D = 0.461 \text{ in. Use } \tfrac{1}{2}\text{-in. bolts.} \qquad \textit{Ans.}$$

## PROBLEMS

**4-2.** Solve Prob. 4-1 when $F = 3,000$ lb, $a = 24$ in., $r = 3$ in., and four equally spaced bolts are used.

**4-3.** Assume in Fig. 4-1 that two $\tfrac{1}{4}$-in.-diameter bolts fasten the baseplate $B$ to its support and that $a = 10$ in. and $r = 4$ in. If the ultimate shearing strength of the steel in the bolts is 42,000 psi, what value of $F$ is required to shear the bolts off?                                                                         *Ans.* $F = 1,650$ lb

**4-4.** If, in Fig. 4-1, $F = 2,300$ lb, $a = 24$ in., $r = 3$ in., and allowable bolt shearing stress is 10,000 psi, what number $N$ of $\tfrac{5}{8}$-in. bolts would be required, equally spaced on a circle of radius $r$?

## 4-2 Stresses in Circular Shafts. The Torsion Formula

If we assume a circular shaft to be made up of a series of thin disks clamped tightly together, it becomes apparent that when the shaft is twisted, a system of forces or stresses is produced on each set of contacting surfaces of adjacent disks which will tend to prevent slippage of any disk with respect to both adjacent disks. The resultants of these stress systems form equal and opposite couples on opposite faces of each disk, one couple being the *torque* and the other the *resisting torque*. Similar stress conditions exist on every pair of cross-sectional planes of a solid shaft.

For the purpose of designing circular shafts to withstand a given torque, we shall need to develop an equation giving the relationship between the

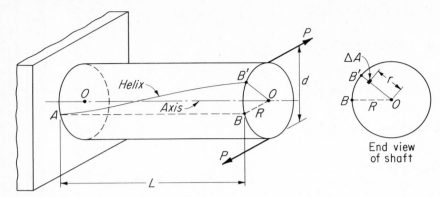

Fɪɢ. 4-2 Development of torsion formula.

twisting moment or torque, the maximum stress produced, and a quantity representing the size and shape of the cross-sectional area of the shaft.

Let the shaft shown in Fig. 4-2 be held completely stationary at its left end. When a torque $Pd$ is applied at its right end, that end is rotated until the line $ABO$ assumes the position $AB'O$, of which $AB'$ is a helix and $B'O$ is assumed to be straight. The deformation on a cross-sectional plane (Fig. 4-3) is seen to increase at a uniform rate from zero at the center $O$ to a maximum $BB'$ at the outer surface of the shaft. Consequently the shearing stresses on any cross-sectional area will vary in the same manner, since stress is proportional to deformation.

In Fig. 4-2, consider a small area $\Delta A$ on any cross-sectional plane of the shaft. The following relationships may now be established:

Fɪɢ. 4-3 Variation of deformations, and corresponding shearing stresses, on cross section of circular shaft.

If maximum unit stress at distance $R$ from $O$ is $s_s$,

unit stress at unit distance from $O$ is $\dfrac{s_s}{R}$    (a)

unit stress at $r$ distance from $O$ is $\dfrac{s_s r}{R}$    (b)

shearing force on area $\Delta A$ is $\dfrac{s_s r \, \Delta A}{R}$    (c)

moment about $O$ of shearing force on $\Delta A$ is $\dfrac{s_s r^2 \, \Delta A}{R}$    (d)

moment about $O$ of shearing forces on all $\Delta A$ areas is

$$\frac{s_s}{R} \Sigma r^2 \, \Delta A = \frac{s_s}{R} \cdot J \tag{e}$$

The quantity $\Sigma r^2 \, \Delta A$ is called the **polar moment of inertia** of the cross-sectional area of the shaft with respect to its centroidal axis $OO$. It is denoted by the symbol $J$.

Now, the twisting moment producing these shearing stresses is the torque $T$, and the moment about $O$ of the shearing forces on all $\Delta A$ areas of the cross section is the resisting torque and equals $s_s J/R$. Since torque and resisting torque constitute action and reaction, they are equal. That is, $T = s_s J/R$ or, since $R = D/2$,

**Twisting moment or torque** $T = \dfrac{2 s_s J}{D}$    (4-1)

from which

**Maximum unit shearing stress** $s_s = \dfrac{TD}{2J}$    (4-2)

For a circular area, $J = \frac{1}{2}\pi R^4$, as illustrated in Prob. B-5, Appendix B, or $J = \pi D^4/32$, since $R = D/2$.

For design purposes we use an expression giving the required diameter $D$ of a circular shaft to resist a given torque $T$, using a predetermined allowable shearing stress $s_s$. Since $J = \pi D^4/32$, Eq. (4-1) becomes $T = 2 s_s \pi D^4/32D = s_s \pi D^3/16$, and

**Required diameter of solid shaft** $D = \sqrt[3]{\dfrac{16T}{\pi s_s}} = 1.72 \sqrt[3]{\dfrac{T}{s_s}}$    (4-3)

where $T$ = twisting moment or torque, lb-in.
    $s_s$ = allowable unit shearing stress, psi

**Hollow Shafts.** Owing to the fact that the material of a shaft near its center is of comparatively little value in resisting torque, many large-diameter shafts are made hollow. For example, the strength of a hollow shaft whose inside diameter is one-half of its outside diameter is about 93 percent of that of a solid shaft.

Hollow shafts are often used in large ships, where a single shaft may transmit more than 10,000 hp and may exceed 30 in. in diameter. As is more completely stated in Art. 5-6, "the moment of inertia of an area from which some part has been removed equals the moment of inertia of the total area less the moment of inertia of the part removed." Hence the polar moment of inertia $J$ of a hollow shaft of outside diameter $D$ and inside diameter $d$ is

**Polar moment of inertia of hollow shaft** $J = \dfrac{\pi}{32}(D^4 - d^4)$

$$= 0.0982\,(D^4 - d^4) \qquad (f)$$

Equation (4-3) for the design of solid shafts cannot be used for hollow shafts. Equation (4-1) should be used. In order to eliminate one of the two unknown diameters in $J$, the inside diameter $d$ is generally expressed in terms of the outside diameter $D$, such as $d = 0.7D$ or $0.8D$, etc.

**Noncircular Shafts.** The design of square, rectangular, and other noncircular shafts is a complex problem beyond the scope of this text. By way of comparison, however, it has been found experimentally that the strength of a square shaft of side dimensions $d$ exceeds the strength of a circular shaft of diameter $D$ by a relatively small percentage. Consequently a square shaft may safely be designed by the circular-shaft formula.

## 4-3 Analysis and Design of Circular Shafts

The following problems make use of the basic torsion formula, Eq. (4-1), and of the derived Eqs. (4-2) and (4-3).

### ILLUSTRATIVE PROBLEMS

**4-5.** A torque $T$ is applied to a shaft in which the shearing stress must not exceed 10,000 psi. Calculate the maximum allowable $T$ (a) if the shaft is solid and is 4 in. in diameter and (b) if the shaft is hollow and its outside diameter $D = 4$ in. and inside diameter $d = 2$ in.

*Solution:* The polar moment of inertia $J$ of a solid shaft is $\pi D^4/32$ and of a hollow shaft is $\pi(D^4 - d^4)/32$. Hence for the solid shaft when $D = 4$ in.,

$$\left[ J = \frac{\pi D^4}{32} \right] \qquad J = \frac{\pi(4)^4}{32} = \frac{256\pi}{32} = 25.1 \text{ in.}^4$$

$$\left[ T = \frac{2s_s J}{D} \right] \qquad T = \frac{(2)(10,000)(25.1)}{4} = 125,500 \text{ lb-in.} \qquad Ans.$$

For the hollow shaft when $D = 4$ in. and $d = 2$ in.,

$$\left[ J = \frac{\pi(D^4 - d^4)}{32} \right] \qquad J = \frac{\pi(256 - 16)}{32} = 23.55 \text{ in.}^4$$

$$\left[ T = \frac{2s_s J}{D} \right] \qquad T = \frac{(2)(10,000)(23.55)}{4} = 117,700 \text{ lb-in.} \qquad Ans.$$

**4-6.** A shaft of 3-in. diameter is subjected to a torque $T$ of 40,000 lb-in. Calculate the maximum unit shearing stress in the shaft (a) when it is solid and (b) when it is hollow with an inside diameter $d$ of 2 in. (c) What is the unit stress at the inner surface of the shaft?

*Solution:* For the solid shaft when $D = 3$ in.,

$$\left[ J = \frac{\pi D^4}{32} \right] \qquad J = \frac{\pi(3)^4}{32} = 7.95 \text{ in.}^4$$

$$\left[ s_s = \frac{TD}{2J} \right] \qquad s_s = \frac{(40,000)(3)}{(2)(7.95)} = 7,550 \text{ psi} \qquad Ans.$$

For the hollow shaft when $D = 3$ in. and $d = 2$ in.,

$$\left[ J = \frac{\pi(D^4 - d^4)}{32} \right] \qquad J = \frac{\pi(81 - 16)}{32} = 6.38 \text{ in.}^4$$

$$\left[ s_s = \frac{TD}{2J} \right] \qquad s_s = \frac{(40,000)(3)}{(2)(6.38)} = 9,400 \text{ psi} \qquad Ans.$$

The intensity of stress varies from zero at the center to 9,400 psi at the outer surface. Hence at the inner surface

$$s_s = 9,400 \frac{1}{1.5} = 6,270 \text{ psi} \qquad Ans.$$

**4-7.** Design a solid steel shaft to resist a torque of 6,630 lb-in. Allowable shearing stress is 10,000 psi.

*Solution:* Equation (4-3) will give directly the required diameter of a *solid* steel shaft. That is,

$$\left[ D = \sqrt[3]{\frac{16T}{\pi s_s}} \right] \qquad D = \sqrt[3]{\frac{(16)(6,630)}{\pi(10,000)}} = \sqrt[3]{3.38} = 1.5 \text{ in.} \qquad Ans.$$

**4-8.** A hollow shaft must be designed to withstand an estimated torque of 1,708,000 lb-in. It has been predetermined that an inside diameter $d$ equal to 0.6$D$ is satisfactory. Maximum allowable unit shearing stress is 10,000 psi.

*Solution:* In terms of $D$, the polar moment of inertia $J$ is

$$\left[ J = \frac{\pi(D^4 - d^4)}{32} \right] \qquad J = \frac{\pi}{32}[D^4 - (0.6D)^4] = (0.0982)(0.8704D^4) = 0.0854D^4$$

From Eq. (4-1)

$$\left[ T = \frac{2s_s J}{D} \right] \qquad 1,708,000 = \frac{(2)(10,000)(0.0854D^4)}{D} = 1,708D^3$$

from which

$$D^3 = \frac{1,708,000}{1,708} = 1,000 \qquad D = 10 \text{ in.} \qquad \text{and} \qquad d = 6 \text{ in.} \qquad Ans.$$

## PROBLEMS

**4-9.** Calculate the polar moment of inertia $J$ of a 3-in.-diameter solid circular shaft.
                                          Ans. $J = 7.95$ in.$^4$

**4-10.** A hollow circular shaft has an outside diameter $D$ of 4 in. and an inside diameter $d$ of 2 in. (*a*) Calculate its polar moment of inertia $J$. (*b*) Show by compara-tive equations that the polar moment of inertia of this hollow circular shaft is $^{15}/_{16}$ that of a solid shaft having the same outside diameter.
                    Ans. (*a*) $J = 23.6$ in.$^4$; (*b*) $J = (\pi/32)(^{15}/_{16})D^4$

**4-11.** What maximum torque in pound-inches may safely be applied to a shaft in which the unit shearing stress must not exceed 12,000 psi (*a*) if the shaft is solid and $D = 5$ in. and (*b*) if the shaft is hollow, $D = 5$ in., and $d = 4$ in.?

**4-12.** The propeller shaft on a large ship is hollow with an outside diameter $D$ of 24 in. and an inside diameter $d$ of 16 in. If the shearing stress must not exceed 8,000 psi, what maximum torque $T$ can the shaft safely withstand?
                                          Ans. $T = 17,400,000$ lb-in.

**4-13.** Assume in Fig. 4-1 that the left end of the shaft is held stationary and that $T = 1,800$ lb-in. and $D = 1$ in. Then calculate the maximum unit shearing stress in the shaft. Forces $P$ and $F$ lie in planes perpendicular to the shaft.

**4-14.** Calculate the unit shearing stresses at the outer and inner surfaces of a hollow shaft in which $D = 6$ in. and $d = 4$ in. due to a torque $T$ of 300,000 lb-in.
                              Ans. $s_s = 8,810$ psi; $s_s = 5,870$ psi

**4-15.** Determine the required diameter $D$ of a solid shaft to resist an estimated torque $T$ of 125,000 lb-in. Allowable unit shearing stress is 9,000 psi.

**4-16.** Calculate the required diameter $D$ of a solid shaft to withstand a torque $T$ of 510,000 lb-in. if allowable shearing stress is 12,000 psi.

**4-17.** Design a hollow shaft in which $d = 0.7D$ to withstand a torque $T$ of 9,550,000 lb-in. Allowable shearing stress is 8,000 psi.        Ans. $D = 20$ in.; $d = 14$ in.

**4-18.** Calculate the required inside and outside diameters, $d$ and $D$, of a hollow shaft to withstand a torque $T$ of 500,000 lb-in. if $d = ^3/_4D$ and allowable stress is 12,500 psi.

**4-19.** A threaded machine bolt of $^3/_4$-in. diameter is screwed up tightly, fastening together two steel parts. To avoid failure of bolt in shear, the unit shearing stress must not exceed the elastic-limit stress of 36,000 psi. What maximum pull may safely be exerted on a wrench with a 20-in. moment arm to avoid failure on the root area if its root diameter is 0.6688 in. and if we assume that friction under the bolt head nullifies 25 percent of the pull?

## 4-4 Shaft Couplings

Two methods are available for coupling together two sections of a shaft: (1) flanges may be forged on the ends to be connected, which are then bolted

FIG. 4-4 Flange coupling. Prob. 4-20.

together as illustrated in Fig. 4-4, or (2) a separate two-piece coupling may be used, each piece cast with flange and hub. The flanges are then bolted together and the hubs are keyed to the shafts. Since a keyway weakens a shaft, the allowable torque on a keyed shaft is 25 percent below that on an unkeyed shaft.

When the flange is forged on the shaft, its thickness $t$ must be sufficient to prevent the shaft from shearing out of the flange on the cylindrical surface whose area is the circumference of the shaft $\pi D$ times the thickness $t$, or $\pi D t$. If the unit shearing stress on this surface is $s_s$, the total resisting force developed on that area is $s_s \pi D t$ and the resisting moment or resisting torque of this force about the center of the shaft is $s_s \pi D t R$. But any applied torque $(T = 2s_s J/D)$ equals the resisting torque $(s_s \pi D t R)$. Hence, for equal strength in shear on the cross-sectional area of the shaft and the cylindrical area in the flange, $2s_s J/D = s_s \pi D t R$, from which the required $t$ of flange may be calculated. This reduces to $t = D/8$ for solid shafts. The two allowable stresses $s_s$ are usually, but not always, equal. The resistance offered by the fillet between shaft and flange may be disregarded.

As previously stated, when several bolts in a coupling resist a given torque $T$, each bolt is presumed to resist its proportional part. Also, while the unit shearing stress in a flange bolt is greatest in the portion farthest from the shaft center, it is, for practical purposes, considered to be uniform over the bolt cross section when the bolt diameter is comparatively small in relation to the bolt-circle diameter.

### ILLUSTRATIVE PROBLEM

**4-20.** Two sections of a shaft, connected by a flange coupling similar to that shown in Fig. 4-4, must be capable of transmitting a torque $T$ of 424,000 lb-in. Calculate (a) the required diameter $D$ of shaft; (b) the minimum thickness $t$ of flange; (c) the required diameter of bolts if six are equally spaced on a 9-in.-diameter bolt-circle. Allowable $s_s$ in shaft and flange is 10,000 psi, in bolts is 15,000 psi.

*Solution:* The diameter is determined by Eq. (4-3). That is,

$$\left[ D = \sqrt[3]{\frac{16T}{\pi s_s}} \right] \qquad D = \sqrt[3]{\frac{(16)(424,000)}{\pi(10,000)}} = \sqrt[3]{216} = 6 \text{ in.} \qquad Ans.$$

Since the torque equals $2s_s J/D$, the resisting torque equals $s_s \pi DtR$, and $J = \pi D^4/32$,

$$\frac{2s_s J}{D} = s_s \pi DtR \qquad \text{or} \qquad t = \frac{2J}{\pi D^2 R}$$

from which

$$t = \frac{2}{\pi D^2 R} \frac{\pi D^4}{32} = \frac{D^2}{16R} = \frac{36}{(16)(3)} = 0.75 \text{ in.} \qquad Ans.$$

Then since $T = NPr$, the force on each bolt is

$$\left[ P = \frac{T}{Nr} \right] \qquad\qquad P = \frac{424,000}{(6)(4.5)} = 15,700 \text{ lb}$$

The required bolt diameter may now be found.

$$\left[ A = \frac{P}{s_s} \right] \qquad \frac{\pi D^2}{4} = \frac{15,700}{15,000} \qquad \text{or} \qquad D = 1.155 \text{ in.} \qquad \text{Use } 1\frac{1}{4}\text{-in. bolts} \qquad Ans.$$

## PROBLEMS

**4-21.** Calculate the thickness $t$ of flange required to develop the full strength of a 4-in.-diameter shaft if allowable shearing stresses $s_s$ in flange and shaft are equal.
Ans. $t = 0.5$ in.

**4-22.** Two 3-in.-diameter shafts have flanged ends coupled together with eight bolts equally spaced on a $7\frac{1}{2}$-in.-diameter circle. What diameter bolts (to nearest higher $\frac{1}{8}$ in.) should be used for coupling and shaft to have same strength in torque? Allowable shearing stress in shaft and bolts is 12,000 psi.

**4-23.** In Prob. 4-20, let $T = 674,000$ lb-in. and let the bolt-circle be 12 in. Solve.

## 4-5 Angle of Twist

When a shaft of length $L$ is subjected to a torque $T$, it will twist about its centroidal axis. If a shaft connects parts whose motions must be closely synchronized, the angle $\theta$ through which the shaft will twist must often be determined. We shall now develop an expression giving the angle of twist $\theta$ in terms of torque $T$, length $L$, modulus of elasticity in shear $G$, and polar moment of inertia $J$.

If, in Fig. 4-5, the **total twist** or deformation in length $L$ of an extreme fiber at $B$ is $BB'$, then the **unit twist**, or deformation $\phi$ per inch of length of shaft, is $BB'/L$. But arc $BB' = R\theta$, when $\theta$ is expressed in radians ($360° = 2\pi$

FIG. 4-5 Portion of shaft. Angle of twist.

radians). Hence $\phi = R\theta/L$. The modulus of elasticity in shear $G$ is the ratio of unit stress to unit deformation, or $G = s_s/\phi$, from which $\phi = s_s/G$ [see Eq. (1-5)]. Consequently

$$\phi = \frac{R\theta}{L} = \frac{s_s}{G} \tag{a}$$

from which

$$\theta = \frac{s_s L}{RG} \tag{4-4}$$

But from Eq. (4-2)

$$s_s = \frac{TD}{2J} = \frac{TR}{J} \tag{b}$$

Hence, substituting Eq. (b) in Eq. (4-4), we obtain

$$\text{Angle of twist } \theta = \frac{TL}{GJ} \tag{4-5}$$

where $\theta$ = angle of twist, radians (1 radian = 57.3°)
    $T$ = torque, lb-in.
    $L$ = length, in.
    $G$ = modulus of elasticity in shear, psi
    $J$ = polar moment of inertia, in.$^4$

## ILLUSTRATIVE PROBLEM

4-24. Calculate (a) the maximum permissible angle of twist $\theta$ in a 1-in.-diameter shaft 30 in. long if allowable stress is 10,000 psi; (b) the angle of twist $\theta$ in a 3-in.-diameter shaft 20 ft long due to a torque of 50,000 lb-in.; (c) the required diameter of shaft 10 ft long to withstand a torque of 20,100 lb-in. if $\theta$ must not exceed 3° ($G$ = 12,000,000 psi).

Solution: Part (a) is solved directly by Eq. (4-4) and parts (b) and (c) are solved by Eq. (4-5). That is,

$$\left[\theta = \frac{s_s L}{RG}\right] \qquad \theta = \frac{(10,000)(30)}{(0.5)(12,000,000)} = 0.05 \text{ radian} = 2.87° \qquad Ans.$$

$$\left[\theta = \frac{TL}{GJ}\right] \qquad \theta = \frac{(50,000)(240)(32)}{(12,000,000)(\pi)(3)^4} = 0.1258 \text{ radian} = 7.22° \qquad Ans.$$

$$\left[J = \frac{TL}{\theta G}\right] \qquad J = \frac{(20,100)(120)(57.3)}{(3)(12,000,000)} = 3.87 \text{ in.}^4$$

$$\left[D^4 = \frac{32J}{\pi}\right] \qquad D^4 = \frac{(32)(3.87)}{\pi} = 39.06 \text{ in.}^4 \qquad D = 2\tfrac{1}{2} \text{ in.} \qquad Ans.$$

## PROBLEMS

**4-25.** Calculate the angle of twist $\theta$ of a 2-in.-diameter shaft if its length is 10 ft and the shearing stress is not to exceed 10,000 psi. $G = 12,000,000$ psi.

**4-26.** What unit stress $s_s$ is produced in a 1-in.-diameter shaft 60 in. long when it is twisted through angle $\theta$ of 5°? $G = 12,000,000$ psi.

**4-27.** A solid steel shaft 10 ft long is subjected to a torque $T$ of 2,740 lb-in. Because this shaft must closely synchronize two operations, its angle of twist $\theta$ must not exceed 1° in 10 ft. (a) Calculate the minimum required diameter $D$ of this shaft. (b) What unit stress is produced at the maximum angle of twist? $G = 12,000,000$ psi.

Ans. (a) $D = 2$ in.; (b) $s_s = 1,745$ psi

**4-28.** Calculate the angle of twist $\theta$ of a hollow steel ship propeller shaft ($D = 24$ in., $d = 16$ in., $J = 26,100$ in.⁴) 100 ft long when subjected to a torque of 1,740,000 lb-in.

**4-29.** In order to determine the modulus of elasticity $G$ in shear of a certain type of steel, a 1-in.-diameter shaft 60 in. long was tested in a torsion machine. A torque of 1,715 lb-in. produced an angle of twist of 5°. (a) Calculate the value of $G$. (b) What maximum twist can be given this shaft without exceeding elastic limit of 36,000 psi in shear?

Ans. (a) $G = 12,000,000$ psi; (b) $\theta = 20.6°$

## 4-6 Power Transmission by Shafts

Power is defined as the time rate of doing work, that is, the amount of work done in a unit of time as, for example, a minute. Work is done when a force displaces a body. The measure of work done by a force is the product of the force and the displacement of its point of application. If the force is in pounds and the displacement in inches, the unit of work done is the inch-pound (in.-lb.). Power, then, being work done per unit of time, may be expressed in inch-pounds per minute (in.-lb per min). But this is seen to be the product of a force $P$ in pounds and the velocity $v$, or rate of displacement in inches per minute, of its point of application. That is, power is $Pv$, the product of force and velocity. The following illustration will show the relationship between power, torque $T$, and speed $N$ of pulley or shaft in revolutions per minute (rpm).

In Fig. 4-6 is illustrated the cable drum of a materials hoist. The pull $P$

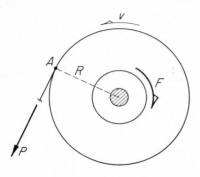

FIG. 4-6 Power transmission by shaft.

in the cable turns the drum counterclockwise, the motion being resisted by a friction force $F$.

Let $A$ be a point on the drum where the cable takes off in a straight line. In one revolution of the drum, point $A$ and force $P$ have been displaced a distance equal to the circumference $2\pi R$ of the drum. If the velocity $v$ of the drum is $N$ rpm, then $v = 2\pi RN$ and the power is $Pv = P2\pi RN$. But $PR = T$, the torque. Hence

$$\textbf{Power} = 2\pi NT \qquad (a)$$

**Horsepower.** The most commonly used unit of mechanical power is the *horsepower*, equal to 33,000 ft-lb of work per minute or

$$(33,000)(12) = 396,000 \text{ in.-lb}$$

of work per minute. Consequently in terms of torque $T$ and shaft speed $N$ in rpm

$$\textbf{Horsepower } H = \frac{2\pi NT}{396,000} \qquad (4\text{-}6)$$

from which

$$\textbf{Torque } T = \frac{63,025H}{N} \qquad (4\text{-}7)$$

where $H$ = horsepower
$N$ = number of revolutions per minute
$T$ = torque, in.-lb

Equation (4-6) is used to determine the power-transmitting capacity of a given shaft and Eq. (4-7) to determine the diameter of shaft required to transmit a given horsepower.

When Eq. (4-7) is substituted in Eq. (4-3) and reduced to its simplest form, we obtain

$$D = \sqrt[3]{\frac{16T}{\pi s_s}} = \sqrt[3]{\frac{(16)(63,025H)}{\pi s_s N}} \qquad (b)$$

from which, for a solid shaft,

$$D = 68.5 \sqrt[3]{\frac{H}{s_s N}} \qquad (4\text{-}8)$$

Equation ($b$) may be put into another useful form, thus:

$$D^3 = \frac{(16)(63,025H)}{\pi s_s N}$$

from which, for a solid shaft,

$$H = \frac{s_s N D^3}{321,000}$$  (4-9)

For hollow shafts, the basic Eqs. (4-1) and (4-7) should be used.

**Keyed Shafts.** A torque is most commonly transmitted to a shaft through a pulley fastened to the shaft by a key fitted into opposite keyways cut into both pulley and shaft. When the torque is relatively small, the pulley may be fastened to the shaft by a set screw. For higher torques, the pulley may be "shrunk" onto the shaft.

A keyway cut into a shaft naturally weakens it. Customarily, a keyed shaft is permitted to transmit only three-fourths as much torque (or horsepower) as an unkeyed shaft. Conversely, a keyed shaft must be designed to withstand $\frac{4}{3}T$ (or $\frac{4}{3}H$). In the following problems, a shaft shall not be considered as keyed unless it is specifically so stated.

## ILLUSTRATIVE PROBLEM

**4-30.** A motor delivers 60 hp at 200 shaft rpm to pulley $C$ in Fig. 4-7. Of this, 10 hp is taken off by pulley $A$, 30 hp by pulley $B$, and 20 hp by pulley $D$. (*a*) Determine the required diameter $D$ of the shaft. (*b*) If $a = 200$ in., $b = 100$ in., and $c = 120$ in., what is the angle of twist between pulleys $A$ and $C$? (*c*) By how many degrees will pulley $A$ lag behind pulley $D$? Allowable $s_s = 8,000$ psi; $G = 12,000,000$ psi. Appropriately spaced bearings prevent bending of shaft.

FIG. 4-7 Power transmission by shaft.

*Solution:* Close study of Fig. 4-7 will show that 40 hp tends to shear the shaft on some cross-sectional plane between pulleys $B$ and $C$, while only 20 hp tends to shear it on a different cross-sectional plane between pulleys $C$ and $D$. Consequently the shaft need be designed to transmit only 40 hp. Hence from Eq. (4-8)

$$\left[ D = 68.5 \sqrt[3]{\frac{H}{s_s N}} \right] \qquad D = 68.5 \sqrt[3]{\frac{40}{(8,000)(200)}} = 2 \text{ in.} \qquad Ans.$$

The total twist of the shaft between pulleys $A$ and $C$ is the twist between $A$ and $B$ caused by 10 hp plus that between $B$ and $C$ caused by 40 hp. We shall obtain

$\theta_{AB}$ from Eq. (4-5) after substituting in it the value of $T$ given by Eq. (4-7). That is, since $L = 200$ in., $N = 200$ rpm, $G = 12,000,000$ psi, and $J = \pi D^4/32$,

$$\left[\theta_{AB} = \frac{TL}{GJ}\right] \qquad \theta_{AB} = \frac{(63,025)(10)(200)(32)}{(200)(12,000,000)(\pi)(2)^4} = 0.0336 \text{ radian} = 1.924°$$

The twist between pulleys $B$ and $C$ may be obtained in the same manner, although, since the stress in this section of the shaft is known to be 8,000 psi, it may be obtained more easily from Eq. (4-4). That is, since $L = 100$ in., $R = 1$ in., and $G = 12,000,000$ psi,

$$\left[\theta_{BC} = \frac{s_s L}{RG}\right] \qquad \theta_{BC} = \frac{(8,000)(100)}{(1)(12,000,000)} = 0.0667 \text{ radian} = 3.825°$$

Then

$$\text{Angle of twist } \theta_{AC} = 1.924 + 3.825° = 5.749° \qquad Ans.$$

The angle of twist between $C$ and $D$ is

$$\left[\theta_{CD} = \frac{TL}{GJ}\right] \qquad \theta_{CD} = \frac{(63,025)(20)(120)(32)}{(200)(12,000,000)(\pi)(2)^4} = 0.0402 \text{ radian} = 2.320° \qquad Ans.$$

and

$$\text{Lag of } A \text{ behind } D = 5.749 - 2.320° = 3.429° \qquad Ans.$$

## PROBLEMS

NOTE: In the following problems, no shaft shall be considered to be keyed to its pulley unless specifically so stated.

**4-31.** The required diameter of shaft in Prob. 4-30 was found to be 2 in. What diameter will be required if pulleys $C$ and $D$ in Fig. 4-7 are interchanged, all other conditions remaining unchanged?

**4-32.** Calculate the maximum unit shearing stress produced in a 3-in.-diameter shaft transmitting 27 hp at 32 rpm.

**4-33.** What maximum horsepower may be transmitted by a 2-in.-diameter shaft at 320 rpm if allowable shearing stress must not exceed 10,000 psi? What horsepower can safely be transmitted by this shaft if it is keyed to its pulleys?

*Ans.* 79.8 hp; 59.8 hp

**4-34.** Design a solid shaft with keyed pulleys capable of transmitting 60 hp at a shaft speed of 100 rpm, using an allowable stress of 8,000 psi. What diameter would be required if no keyways were cut in the shaft?

*Ans.* $D = 3.18$ in.; $D = 2.89$ in.

**4-35.** Pulleys $B$, $C$, and $D$ shown in Fig. 4-8 have diameters of 6, 8, and 10 inches, respectively. The shaft $AE$ revolves at 1,000 rpm. What horsepower is taken off the shaft at $B$ and $D$? What horsepower is added at $C$?

**4-36.** In Fig. 4-9 a motor transmits 40 hp to gear $B$ and turns it at 600 rpm. Ten horsepower is taken off at gear $C$. All pulleys are adequately attached to the shaft. Calculate the required shaft diameter if the allowable shearing stress is 12,000 psi.

FIG. 4-8  Probs. 4-35 and 4-49.                    FIG. 4-9  Probs. 4-36 and 4-50.

By what angle does the gear $C$ lag the driving gear $B$? Assume a constant diameter from $B$ to $C$ and $G = 10,000,000$ psi.

**4-37.** An automobile engine develops a maximum of 96 hp at an engine speed of 3,400 rpm, 90 hp of which reaches the rear axle at 240 rpm in low gear and at 725 rpm in high gear. The axle is made of high-strength heat-treated nickel steel having a proportional limit of 96,000 psi. Its minimum diameter is $1\frac{1}{4}$ in. Calculate the maximum shearing stresses in the shaft in low gear and in high gear.

**4-38.** A 25-hp electric motor drives a centrifugal pump through two gears whose ratio is 5:1. The motor shaft turns at 1,750 rpm and the pump at 350 rpm. Shaft bending and gear friction are negligible. One gear is keyed to each shaft. Calculate the minimum required diameters of these shafts, using an allowable stress of 8,000 psi.                    *Ans.* Motor $D = 0.915$ in.; pump $D = 157$ in.

**4-39.** The engine of a ship delivers 5,000 hp to the propeller shaft at 120 rpm. (*a*) Design a hollow shaft in which $d = \frac{2}{3}D$ and allowable shearing stress is 12,000 psi. (*b*) If the shaft is 40 ft long from engine to propeller, through what angle $\theta$ will it twist under full power?     *Ans.* (*a*) $D = 11.8$ in., $d = 7.86$ in.; (*b*) $\theta = 4.85°$

**4-40.** A certain shaft must transmit 500 hp. The allowable shearing stress is 10,000 psi. Calculate the required diameter $D$ when the shaft speed $N$ equals, respectively, 125, 512, 1,000, 1,728, 3,375, and 8,000 rpm. Plot these values and draw a curve in which the vertical scale is the required diameter $D$ and the horizontal scale is speed $N$. [HINT: Use Eq. (4-8). The plot could best be made on log-log paper.]

## SUMMARY

*(By article number)*

**4-1.** When a member is twisted in a plane perpendicular to its longitudinal axis, it is said to be in **torsion**. The **twisting** moment is called a **torque**. Shafts are the most common examples of torsion members.

**4-2. Torsional stresses in circular shafts** caused by a torque are shearing stresses induced on every transverse plane of the shaft. These shearing stresses increase uniformly from zero at the centroidal axis to a maximum at the surface of the shaft. If $T$ = torque, $s_s$ = maximum shearing stress, $J$ = polar moment of inertia, and $D$ = diameter of shaft, the relationship of these quantities is

$$\text{Twisting moment or torque } T = \frac{2s_s J}{D} \tag{4-1}$$

The polar moment of inertia $J$ is

**For solid shaft:**
$$J = \frac{\pi D^4}{32}$$

**For hollow shaft:**
$$J = \frac{\pi(D^4 - d^4)}{32}$$

The diameter of a solid shaft is obtained from the equation

$$D = \sqrt[3]{\frac{16T}{\pi s_s}} \tag{4-3}$$

Hollow shafts are most easily designed by solving for the required $J$, using Eq. (4-1).

**4-4.** Twisting sections of a shaft may be coupled together end to end (1) by forging flanges on the shaft ends, which are then bolted together, or (2) by use of a two-piece coupling in which the two pieces are bolted together and one piece is keyed to each section of the shaft.

**4-5.** The **angle of twist** is a function of the torque $T$, the length $L$, the shear modulus of elasticity $G$, and the polar moment of inertia $J$. The relationship is

$$\text{Angle of twist } \theta = \frac{TL}{GJ} \tag{4-5}$$

**4-6. Transmission of power** is the usual function of shafts. In terms of horsepower $H$ and speed $N$ in rpm, the torque $T$ exerted on a shaft is

$$\text{Torque } T = \frac{63,025H}{N} \tag{4-7}$$

When this value of $T$ is substituted in Eq. (4-3), the required

$$\text{Diameter of solid shaft } D = 68.5 \sqrt[3]{\frac{H}{s_s N}} \tag{4-8}$$

Another convenient form of this equation is

$$H = \frac{s_s N D^3}{321,000} \tag{4-9}$$

**Keyways** weaken shafts. A shaft having one or more keyways is permitted to transmit only three-fourths as much torque (or horsepower) as an unkeyed shaft. Therefore, keyed shafts must be designed to withstand $\frac{4}{3}T$ (or $\frac{4}{3}H$).

## REVIEW PROBLEMS

**4-41.** A threaded bolt of $\frac{1}{2}$-in. diameter is being screwed up tightly, fastening together two steel parts. To avoid failure of the bolt in shear, the unit shearing stress in it must not exceed the elastic limit of 36,000 psi. What maximum pull $P$ on a wrench with a 12-in. moment arm will cause this torsional stress (*a*) if the gross cross-sectional area resists the stress and (*b*) if the root area resists the stress?

*Ans.* (*a*) $P = 73.6$ lb; (*b*) $P = 37.6$ lb

**4-42.** Calculate the maximum permissible torque $T$ on a shaft if the maximum allowable shearing stress is 12,000 psi and if (*a*) the shaft is solid and $D = 2\frac{1}{2}$ in. and (*b*) the shaft is hollow, $D = 2\frac{1}{2}$ in., and $d = 2$ in.

**4-43.** The elastic limit in shear $s_s$ of the steel in a shaft is determined by twisting the shaft in a torsion machine. This limit was reached when the angle of twist $\theta$ measured 14.8°. Calculate (*a*) the unit shearing stress at the elastic limit and (*b*) the modulus of elasticity $G$. The shaft is 48 in. long and of 1-in. diameter. Torque $T = 6,285$ lb-in.           *Ans.* (*a*) $s_s = 32,000$ psi; (*b*) $G = 11,900,000$ psi

**4-44.** What maximum horsepower may be transmitted by a 2-in.-diameter shaft at 100 rpm if maximum shearing stress is not to exceed 8,000 psi and if (*a*) the shaft is not keyed and (*b*) the shaft is keyed?

**4-45.** The engine of an automobile develops a maximum of 100 hp at an engine speed of 3,200 rpm. The velocity ratio in low gear between engine and rear axle is 13.68. Calculate the minimum required diameter $D$ of axle if allowable stress in the heat-treated nickel steel is 70,000 psi.

**4-46.** A 3-in.-diameter shaft is to transmit 120 hp. (*a*) Calculate the minimum speed $N$ (rpm) in order that the maximum shearing stress shall not exceed 10,000 psi. What maximum stress will be induced in the shaft when the same horsepower is delivered to the shaft (*b*) at 120 rpm and (*c*) at 160 rpm?

*Ans.* (*a*) $N = 142$ rpm; (*b*) $s_s = 11,890$ psi; (*c*) $s_s = 8,930$ psi

**4-47.** The engines of a large ship deliver 15,000 hp to each of four hollow propeller shafts at 150 rpm. (*a*) Design one shaft with an inside diameter $d$ equal to $\frac{3}{4}$ of the outside diameter $D$. (*b*) If the shaft is 100 ft long, what will be the angle of twist $\theta$? Allowable shearing stress is 10,000 psi.

*Ans.* (*a*) $D = 16.8$ in., $d = 12.6$ in.; (*b*) $\theta = 6.82°$

**4-48.** Assume that the shaft in Prob. 4-47 is in two 50-ft sections connected by means of a flange coupling. (*a*) How many $1\frac{3}{4}$-in.-diameter turned bolts are required if they are equally spaced on a 22-in.-diameter bolt-circle? (*b*) Calculate the minimum required thickness $t$ of flange to prevent shaft from shearing out of flange. Use $s_s = 15,000$ psi for bolts.

**4-49.** Pulleys $B$ and $D$ are driven by a motor connected to the shaft $AE$ by means of pulley $C$, as shown in Fig. 4-8. Pulley diameters $B$, $C$, and $D$ are 6, 8, and 10 inches, respectively. (*a*) Calculate the belt tension $T_3$. (*b*) Calculate the torque in section $CD$. (*c*) If the shaft $CD$ is 2 in. in diameter, calculate the maximum shearing stress in this section.

**4-50.** Gear $C$, shown in Fig. 4-9, is driven by an electric motor. Five horsepower is taken off the $1\frac{1}{2}$-in.-diameter steel shaft at $B$ and 15 hp is taken off at $A$. Calculate the maximum shearing stress in the shaft and the angle by which the gear $C$ leads gear $B$. The shaft rotates at 120 rpm.

## REVIEW QUESTIONS

**4-1.** What is meant by a torque? In what manner is it usually applied to a member with respect to its longitudinal axis?

**4-2.** What type of stress is induced in a member subjected to a torque?

**4-3.** Describe the variation of deformations and torsional stresses in a circular shaft subjected to a torque.

**4-4.** Give the expression for the polar moment of inertia (*a*) of a solid circular shaft and (*b*) of a hollow circular shaft.

**4-5.** In the design of hollow shafts, is it necessary to predetermine the ratio of *d* to *D*?

**4-6.** What methods are generally used to connect two sections of a shaft end to end?

**4-7.** What determines the minimum thickness *t* of a flange forged at the end of a shaft?

**4-8.** What is meant by the term angle of twist? In what units must the angle of twist be expressed?

**4-9.** How may the shear modulus of elasticity of the steel in a circular shaft be determined?

**4-10.** If 100 hp is to be transmitted by each of two separate shafts, one turning at 200 rpm and the other at 300 rpm, which shaft must have the greater diameter?

**4-11.** How does a keyway affect the strength of a shaft? What is the permissible torque on a keyed shaft compared with that on an unkeyed shaft?

**4-12.** In the design of a keyed shaft, by what fraction must the given torque be increased?

# CHAPTER 5

# Centroids and Moments of Inertia of Areas

## 5-1 Centroids

In the analysis and design of beams and columns, covered in Chaps. 7, 8, and 12, we shall find it necessary (a) to locate the centroids of their cross-sectional areas, (b) to calculate the first, or static, moment of such areas with respect to some axis, and (c) to calculate the second moment, or moment of inertia, of such areas with respect to some axis. This necessity accounts for the insertion here of this chapter containing material[1] which may, at first glance, appear to the student to be unrelated to the subject matter of this text.

The concept of the centroid of an area is most easily understood if first we consider the concept of center of gravity of a thin plate of uniform thickness and of homogeneous material. To determine *mathematically* the center of gravity of a flat plate of irregular shape but of uniform thickness and of homogeneous material, the plate may be divided into a number of small elements of equal size, as shown in Fig. 5-1a. If the weight of each element is $w$, concentrated at its center, a parallel force system is formed whose resultant $W$, the weight of the entire plate, passes through the center of gravity of the plate.

Let the weights of these elements be designated $w_1$, $w_2$, etc., let their coordinates be $x_1$ and $y_1$, $x_2$ and $y_2$, etc., and let the coordinates of the resultant weight $W$ be $\bar{x}$ and $\bar{y}$ (bar $x$ and bar $y$). Now, about any point, the moment

---

[1] The material in this chapter is taken from the authors' "Applied Engineering Mechanics," 3d ed., McGraw-Hill Book Company, New York, 1972.

FIG. 5-1   Center of gravity of homogeneous flat plate. (*a*) Plate divided into equal-size elements. (*b*) Forces projected into *YZ* plane. (*c*) Forces projected into *XZ* plane.

of the resultant *W* equals the algebraic sum of the moments of the separate weights. Consequently, by moments about *O* in the *XZ* plane,

$$W\bar{x} = w_1x_1 + w_2x_2 + \cdots = \Sigma wx \tag{a}$$

and by moments about *O* in the *YZ* plane,

$$W\bar{y} = w_1y_1 + w_2y_2 + \cdots = \Sigma wy \tag{b}$$

or

$$\bar{x} = \frac{\Sigma wx}{W} \quad \text{and} \quad \bar{y} = \frac{\Sigma wy}{W} \tag{5-1}$$

thus locating the center of gravity.

Let us imagine now that the flat plate in Fig. 5-1 gradually reduces in thickness. When its thickness finally becomes zero, it no longer has weight and only a surface or an area remains. The point in the plate which formerly was the center of gravity of the body now is the **centroid** of the area. If *a* is the surface area of one small element in Fig. 5-1 and *t* is the thickness of the plate, its volume is *ta* and its weight *w* is $\delta ta$, where $\delta$ is the density. If $\delta ta$ is substituted for *w* in Eq. (*a*) above and *A* is the total surface area, $W = \delta tA$, and

$$\delta tA\bar{x} = \delta ta_1x_1 + \delta ta_2x_2 + \cdots = \delta t\Sigma ax \tag{c}$$

When the constants $\delta$ and $t$ are canceled from all terms of this equation, we find that $A\bar{x} = \Sigma ax$. Similarly from Eq. (b) we would find that $A\bar{y} = \Sigma ay$. Consequently Eq. (5-1) would become

$$\bar{x} = \frac{\Sigma ax}{A} \qquad \text{and} \qquad \bar{y} = \frac{\Sigma ay}{A} \tag{5-2}$$

Apparently, then, *about any axis* **the moment of an area equals the algebraic sum of the moments of its component areas,** wherein *the moment of an area is defined as the product of the area multiplied by the perpendicular distance from the moment axis to the centroid of the area.* By means of this principle, we may locate the centroid of any area.

The terms *centroid* and *center of gravity* are often used interchangeably. Also, a **gravity axis** passing through the center of gravity of a body is often called a **centroidal axis.**

## 5-2 Centroids of Simple Geometric Areas

By geometry and by calculus (see Appendix A), the centroids of some simple geometric areas are found to be located as illustrated in Fig. 5-2.

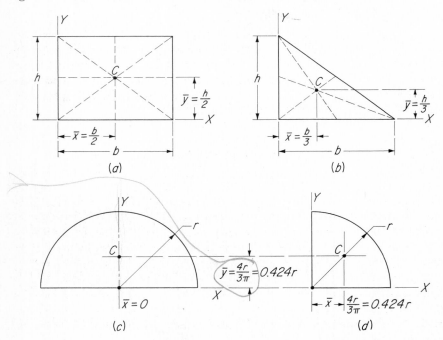

FIG. 5-2 Centroids of simple geometric areas. (*a*) Rectangle. (*b*) Triangle. (*c*) Semicircle. (*d*) Quarter-circle.

The centroid of a rectangle or of any parallelogram lies at the intersection $C$ of its two diagonals. This point is also the intersection of the two lines bisecting the two pairs of opposite sides. The truth of this statement can be deduced as follows: Let the rectangle be divided into narrow strips running parallel to the two opposite sides. The centroid of each strip lies at its mid-point, the centroids of all the parallel strips will lie on a line bisecting the two opposite sides, and the centroid $C$ of the entire area will lie at the intersection of the two bisecting lines, which is at their mid-points.

The centroid of a triangle lies at the intersection of its medians. If the triangle is divided into infinitesimally narrow strips running parallel to any base, the centroid of each strip is at its mid-point, the centroids of all the parallel strips will lie on the median of that base, and the centroid $C$ of the entire area lies at the intersection of the three medians, at a distance of one-third of the length of the median from its base.

The centroid of a semicircle is most easily determined by integration, as is shown in Prob. A-3, Appendix A. By symmetry, the centroids of a quarter-circle and of a semicircle, shown in Fig. 5-2, are seen to lie at equal distances from the $X$ and $Y$ reference axes.

## 5-3 Centroids of Composite Areas

A composite area is one made up of a number of simple areas. To determine its centroid, a composite area is generally divided into two or more simple component areas. The centroid of the composite area is then found by the previously stated principle: about any axis, *the moment of an area equals the algebraic sum of the moments of its component areas.* The following example illustrates the application of this principle.

### ILLUSTRATIVE PROBLEMS

**5-1.** Locate the centroid of the composite area shown in Fig. 5-3 with respect to the $X$ and $Y$ axes.

*Solution:* Let the $X$ and $Y$ axes intersect at the lower left corner of the area. Then all moments are positive. This composite area divides naturally into three simple geometric areas, a rectangle, a triangle, and a semicircle, whose centroids we obtain from Fig. 5-2. To aid in computing $\bar{x}$ and $\bar{y}$, we represent these areas as forces concentrated at their respective centroids and shown in two edge views of the area similar to a parallel force system in space. The total area is their resultant, which passes through the centroid $C$ of the composite area. From the given dimensions we find that

$$A_1 = (12)(6) = 72 \qquad A_2 = (\tfrac{1}{2})(12)(6) = 36 \qquad A_3 = (\tfrac{1}{2})(3.1416)(6)^2 = 56.6$$

Hence

$$A = 72 + 36 + 56.6 = 164.6 \text{ sq in.}$$

FIG. 5-3 Location of centroid of composite area.

Then, in the front view, to find $\bar{x}$,

$[M_o]$ $\qquad\qquad$ $164.6\bar{x} = (72)(6) + (36)(8) + (56.6)(14.54)$

or

$\qquad$ $164.6\bar{x} = 432 + 288 + 823 = 1,543$ $\qquad$ and $\qquad$ $\bar{x} = 9.37$ in. $\qquad$ Ans.

Likewise, in the side view, to find $\bar{y}$,

$[M_o]$ $\qquad\qquad$ $164.6\bar{y} = (72)(3) + (56.6)(6) + (36)(8)$

or

$\qquad$ $164.6\bar{y} = 216 + 340 + 288 = 844$ $\qquad$ and $\qquad$ $\bar{y} = 5.13$ in. $\qquad$ Ans.

**5-2.** In Fig. 5-4 is shown the cross-sectional area of a composite steel beam made up of an American Standard I-beam, 12 by $5\frac{1}{4}$ in. by 50 lb per ft, with a $\frac{1}{2}$- by 8-in. cover plate riveted to its top flange. Locate the centroid of the composite area.

*Solution:* This area is symmetrical about the $Y$ axis. Hence we need determine only the distance $\bar{y}$. The cross-sectional area $A_1$ of the plate is $(\frac{1}{2})(8)$, or 4 sq in. From Table E-8 we find the cross-sectional area $A_2$ of the I-beam to be 14.6 sq in. The total area $A$ is then $14.6 + 4$, or 18.6 sq in. Then, by moments about $O$, we have

$[M_o]$ $\qquad$ $18.6\bar{y} = (4)(12.25) + (14.6)(6) = 136.6$ $\qquad$ and $\qquad$ $\bar{y} = 7.35$ in. $\qquad$ Ans.

FIG. 5-4 Location of centroid of cross-sectional area of beam.

## PROBLEMS

**5-3.** Locate the centroid with respect to the $X$ and $Y$ axes of the area shown in Fig. 5-5. (HINT: The moment of an area with a part removed equals the moment of that area *minus* the moment of the part removed. Hence represent the part removed by an arrow applied at its centroid and oppositely directed.)

*Ans.* $\bar{x} = 5.91$ in.; $\bar{y} = 2.85$ in.

FIG. 5-5 Prob. 5-3.                    FIG. 5-6 Prob. 5-4.

**5-4.** Determine the location of the centroid of the area shown in Fig. 5-6 (see hint in Prob. 5-3).

**5-5.** The irregular body shown in Fig. 5-7 is placed upon two knife-edge supports, one resting on a scale that indicates a load of 288 lb. If the weight $W$ of the body is 360 lb and if $L$ is 4 ft, compute the distance $\bar{x}$ from the left support to the vertical plane through the body containing its center of gravity.

Fig. 5-7  Prob. 5-5.

Fig. 5-8  Prob. 5-6.

**5-6.** The structural-steel angle shown in Fig. 5-8 measures 8 by 6 in. and is 1 in. thick. Compute the distances $\bar{x}$ and $\bar{y}$ from the heel of the angle to its centroid.

**5-7.** The timber beam shown in cross section in Fig. 5-9 is made of two planks 2 by 8 in. and one plank 2 by 10 in., all securely nailed together. Compute the distance $\bar{y}$ from the bottom of the beam to its centroid. $YY$ is an axis of symmetry.

Ans. $\bar{y} = 5.92$ in.

Fig. 5-9  Probs. 5-7 and 5-24.

Fig. 5-10  Probs. 5-8 and 5-25.

Fig. 5-11  Probs. 5-9 and 5-26.

**5-8.** The cast-iron beam shown in the cross section in Fig. 5-10 is symmetrical about the $YY$ axis. The two flanges and the web are $\frac{3}{4}$ in. thick. Compute the distance $\bar{y}$ from the bottom of the beam to its centroid.

**5-9.** In Fig. 5-11 is shown the cross section of a steel beam built up of two American Standard channels 10 in. by 20 lb per ft, to whose top flanges are riveted a $\frac{1}{2}$- by 12-in. plate. The area of each channel is 5.86 sq in. $YY$ is an axis of symmetry. Compute the distance $\bar{y}$ from the bottom of the beam to its centroid.

## 5-4 Moments of Inertia of Areas

To determine certain *shearing stresses* in beams, a quantity must be determined involving the *first moment of an area* Q, which, with respect to a reference axis, is the product of the area and the perpendicular distance from the axis to the centroid of the area. That is, $Q = A\bar{y}$.

Now, to determine certain *bending stresses* in beams, and also in the design of columns, a quantity must be determined which involves the *second moment of an area*, commonly called the **moment of inertia** and denoted by the symbol $I$. With respect to a reference axis in the plane of the area, **the second moment of an area, or its moment of inertia, is the sum of the products of the elemental areas, each multiplied by the square of the distance from the reference axis to its centroid.** The moment of inertia is therefore the product of the entire area multiplied by the mean value of the squares of the distances from the reference axis to the centroids of the elemental areas. When the number of areas is finite, $I = \Sigma ay^2$, by which equation the *approximate moment of inertia* may be obtained, as outlined in Art. 5-5. When each small area is infinitesimal, or $dA$, and the number of areas is infinite, $I = \int y^2 \, dA$, by which equation the *exact moment of inertia* may be obtained, as shown in Appendix B. The exact moment of inertia of a composite area is obtained as illustrated in Art. 5-7.

In practical engineering language, the moment of inertia of the cross-sectional area of a beam or a column of a given material with respect to its neutral axis may be said to represent the relative capacity of the section to resist bending or buckling in a direction perpendicular to the neutral axis.

When bending stresses are involved, the moment of inertia required is with respect to the neutral axis of the cross-sectional area and is called the **rectangular moment of inertia.** In shafts that transmit power and are therefore subjected to twisting or torsional stresses, the moment of inertia required is with respect to an axis perpendicular to the plane of the cross-sectional area; it is then called the **polar moment of inertia** of the area. See Appendix B.

**Units of Moment of Inertia.** Moment of inertia is the product of an area and the square of a distance. If the area is in square inches and the distance in inches, the product will be in *inches to the fourth power*. That is, $(\text{in.}^2)(\text{in.}^2) = \text{in.}^4$ This unit has no physical conception.

## 5-5 Approximate Determination of Moment of Inertia

The moment of inertia of an area may be obtained approximately by dividing the total area $A$ into a finite number of smaller areas $a$, then multiplying each small area $a$ by the square of the distance $(y^2)$ from the reference axis to its centroid. The moment of inertia then is the sum of these products. Consequently

$$\text{Approximate moment of inertia } I = \Sigma ay^2 \qquad (5\text{-}3)$$

The approximate method is used chiefly with irregular areas. The following example illustrates the method. A rectangular area was chosen here in order that its approximate and exact moments of inertia might be compared and thus show the probable error of the approximate method.

## ILLUSTRATIVE PROBLEM

**5-10.** Determine the approximate moment of inertia of the area shown in Fig. 5-12 with respect to axis $XX$ passing through its centroid.

FIG. 5-12 Moment of inertia of an area. Approximate method.

*Solution:* Let the area arbitrarily be divided into 10 strips of equal width running parallel with the $XX$ axis, each area then containing $(2)(10)$, or 20, sq in. We find that, of the distances to the respective centroids, $y_1 = 9$ in., $y_2 = 7$ in., $y_3 = 5$ in., $y_4 = 3$ in., and $y_5 = 1$ in. Clearly, with respect to axis $XX$, the moment of inertia of the lower half equals that of the upper half; therefore we determine only the latter and then multiply it by 2 to obtain $I$ of the entire area. That is,

$$[I = \Sigma ay^2] \qquad I_X = (2)(a_1 y_1{}^2 + a_2 y_2{}^2 + a_3 y_3{}^2 + a_4 y_4{}^2 + a_5 y_5{}^2)$$

But each of the areas $a_1$, $a_2$, etc., equals 20 sq in. Hence

$$I_X = (2)(20)(81 + 49 + 25 + 9 + 1) \qquad \text{or} \qquad I_X = 6{,}600 \text{ in.}^4 \qquad Ans.$$

The exact moment of inertia of a rectangular area is $\frac{1}{12}bh^3$, which for the area in this example is 6,667 in.$^4$, thus indicating an error in the approximate result of only 1 percent. See Appendix B.

## 5-6 Moments of Inertia of Simple Areas

The moments of inertia of many simple geometric areas are readily determined by calculus, as is illustrated in Appendix B. These areas and their moments of inertia are generally expressed in terms of simple dimensions which completely define them.

The moments of inertia and the radii of gyration of a number of simple areas are given in Table 5-1. These and many others are usually found in handbooks.

The moments of inertia of other simple areas *from which some parts have been removed,* such as those shown in Fig. 5-13, are easily obtained from the following principle:

**With respect to any axis, the moment of inertia of an area with a part removed equals the moment of inertia of the area minus the moment of inertia of the part removed.**

**Table 5-1 MOMENTS OF INERTIA AND RADII OF GYRATION OF SIMPLE AREAS**   Use $I_{xc}$

| Area | Moment of inertia | Radius of gyration |
|---|---|---|
| Rectangle | $I_{Xc} = \dfrac{bh^3}{12}$ <br><br> $I_X = \dfrac{bh^3}{3}$ | $k_{Xc} = \dfrac{h}{\sqrt{12}} = 0.289h$ <br><br> $k_X = \dfrac{h}{\sqrt{3}} = 0.577h$ |
| Triangle | $I_{Xc} = \dfrac{bh^3}{36}$ <br><br> $I_X = \dfrac{bh^3}{12}$ | $k_{Xc} = \dfrac{h}{\sqrt{18}} = 0.236h$ <br><br> $k_X = \dfrac{h}{\sqrt{6}} = 0.408h$ |
| Circle | $I_{Xc} = \dfrac{\pi R^4}{4}$ <br> $= 0.7854R^4$ <br> $= \dfrac{\pi D^4}{64}$ <br> $= 0.0491D^4$ | $k_{Xc} = \dfrac{R}{2} = 0.5R$ |
| Semicircle | $I_{Xc} = \left(\dfrac{\pi}{8} - \dfrac{8}{9\pi}\right)R^4$ <br> $= 0.110R^4$ <br> $I_X = I_{Yc} = \dfrac{\pi R^4}{8}$ <br> $= 0.3927R^4$ | $k_{Xc} = \dfrac{\sqrt{9\pi^2 - 64}}{6\pi}R$ <br> $= 0.264R$ <br> $k_X = k_{Yc} = \dfrac{R}{2} = 0.5R$ |

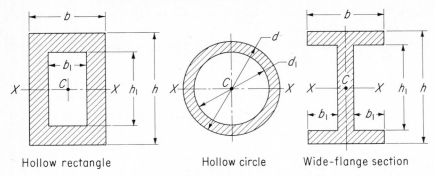

Hollow rectangle          Hollow circle          Wide-flange section

FIG. 5-13 Special equations for obtaining moments of inertia.

When the axes for the area and for the part removed coincide and both are centroidal, as in Fig. 5-13, special simple equations giving the moment of inertia may be written as shown above. Otherwise the general method outlined in the following article must be used.

### 5-7 Moments of Inertia of Composite Areas

In modern structural practice, and especially in aircraft construction, relatively few members used to resist bending are of simple geometric cross-sectional area such as rectangular or circular. Many sections, however, are so shaped that they are readily divided into two or more simple **component areas**. The entire area is then called a **composite area**. The moment of inertia of each component area with respect to a centroidal axis of the composite area is then determined by the following principle, illustrated in Fig. 5-14 and more fully explained in Appendix B:

**The moment of inertia of an area with respect to any axis not through its centroid is equal to the moment of inertia of that area with respect to its own parallel centroidal axis plus the product of the area and the square of the distance between the two axes.**

When this principle is stated in symbols, we have

$$I_X = I_C + Ad^2 \tag{5-4}$$

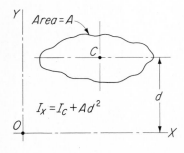

FIG. 5-14 Moment of inertia of an area with respect to a noncentroidal axis.

where $I_x$ = moment of inertia of an area with respect to any noncentroidal axis, in.$^4$

$I_C$ = moment of inertia of the area with respect to its parallel centroidal axis, in.$^4$

$A$ = area, sq in.

$d$ = perpendicular distance between the parallel axes, in.

Equation (5-4) is called the **transfer formula,** since it "transfers" the moment of inertia of an area from its centroidal axis to any other parallel axis. *The moment of inertia of a composite area* then is the sum of the moments of inertia of the component areas, all with respect to a common axis. Close study of this equation reveals that the moment of inertia of an area is always least with respect to a centroidal axis. Problems 5-11 and 5-12 will illustrate the method.

## 5-8 Radius of Gyration

In the design of columns a term called the **radius of gyration** and denoted by the symbol $k$ is very useful.

**The radius of gyration of an area is that distance from its moment-of-inertia axis at which the entire area could be considered as being concentrated without changing its moment of inertia.**

That is, if all the elemental areas $dA$ were placed side by side in a strip parallel to the moment-of-inertia axis, and if the thickness of this strip were infinitesimal, the moment of inertia $I_C$ of each elemental area $dA$ would become zero, and that term would therefore drop out of Eq. (5-4). The moment of inertia $I$ of each small area then would be $y^2\,dA$ and of the entire area $I$ would be the sum of these, or $\int y^2\,dA$. But, since $y$ now has the same value for all of the small areas, $y^2 = k^2$, and $\int dA$ (the sum of all the small areas) $= A$. (See Fig. 5-15.) Therefore

$$I_X = Ak^2 \qquad \text{or} \qquad k = \sqrt{\frac{I}{A}} \tag{5-5}$$

In many instances, the moments of inertia of the cross-sectional area of a column with respect to its two rectangular $X$ and $Y$ axes are not equal. Therefore, one of its two major radii of gyration will be smaller than the other. *A compression member tends to buckle in the direction of its least radius of gyration.*

FIG. 5-15 Radius of gyration.

## ILLUSTRATIVE PROBLEMS

**5-11.** Locate the centroid of the composite T-beam section shown in Fig. 5-16 and determine its moment of inertia $I_X$ with respect to the axis $XX$ and its radius of gyration $k_X$.

FIG. 5-16 Centroid and moment of inertia of a composite area.

*Solution:* The composite area is divided into two rectangular areas $A$ and $B$, and the centroid of each area is located by inspection, as in Fig. 5-16. Next, we draw the edge view of the entire area and apply the separate areas, represented by arrows, at their respective centroids. Then by moments about $O$

$$76\bar{y} = (36)(11.5) + (40)(5) = 614 \qquad \text{and} \qquad \bar{y} = 8.08 \text{ in.} \qquad Ans.$$

Having thus located the centroid of the composite area, we then determine distances $d_A$ and $d_B$ from axis $XX$ to the centroids of areas $A$ and $B$; that is, $d_A = 11.5 - 8.08 = 3.42$ in., and $d_B = 8.08 - 5 = 3.08$ in. Now, the moment of inertia of each of areas $A$ and $B$, with respect to axis $XX$, is given by Eq. (5-4). That is, $I_X = I_C + Ad^2$. The total moment of inertia of the composite area is then $I_X = \Sigma(I_C + Ad^2)$. Since $A_A = 36$ sq in., $A_B = 40$ sq in., $d_A = 3.42$ in., and $d_B = 3.08$ in., we have

$$I_X = \Sigma[I_C + Ad^2] \qquad \text{in which} \qquad I_C = \frac{bh^3}{12}$$

Area A:   $I_X = \dfrac{(12)(3)^3}{12} + (36)(3.42)^2 = \quad 27 + 421 = \quad 448 \text{ in.}^4$

Area B:   $I_X = \dfrac{(4)(10)^3}{12} + (40)(3.08)^2 = 333 + 380 = \quad \underline{713 \text{ in.}^4}$

Moment of inertia of the composite area = 1,161 in.⁴    *Ans.*

From Eq. (5-5), the radius of gyration with respect to axis $XX$ is

$$\left[k_X = \sqrt{\frac{I_X}{A}}\right] \qquad k_X = \sqrt{\frac{1,161}{76}} = 3.91 \text{ in.} \qquad Ans.$$

That is, if the entire area of 76 sq in. of this T-section were placed in an exceedingly thin strip parallel to axis $XX$ and at a distance $k$ of 3.91 in. from it, the moment of inertia would be $Ak^2$, or $(76)(3.91)^2 = 1{,}161$ in.[4].

**5-12.** In Fig. 5-17 is shown the cross-sectional area of a steel beam, built up of one web plate 1 by 14.5 in., four angles 5 by 5 by $\frac{3}{4}$ in., and two cover plates 1 by 12 in. Compute the moment of inertia of this composite section with respect to axis $XX$ passing through its centroid.

FIG. 5-17 Moment of inertia of a composite area.

*Solution:* This section is symmetrical about the $X$ axis, which, then, is located 8.5 in. above the bottom. From Table E-10 we find that the moment of inertia $I_C$ of each of the four angles about its own centroidal $X$ axis is 15.7 in.[4], that the area $A$ of each angle is 6.94 sq in., and that the distance $y$ from the back of the angle to its centroidal axis is 1.52 in. Considering separately the two flanges $F$, the four angles $A$, and the web $W$, we have

$$I_X = \Sigma[I_C + Ad^2]$$

Two flanges:   $I_X = 2\left[\dfrac{(12)(1)^3}{12} + (12)(8)^2\right] = 2(1 + 768) \qquad = 1{,}538$ in.[4]

Four angles:   $I_X = 4[15.7 + (6.94)(5.98)^2] = 4(15.7 + 248.2) = 1{,}055$ in.[4]

One web plate:   $I_X = \dfrac{(1)(14.5)^3}{12} = \dfrac{(1)(3{,}048)}{12} \qquad\qquad = \underline{\ \ 254}$ in.[4]

Moment of inertia of the composite area $= 2{,}847$ in.[4]

## PROBLEMS

**5-13.** In Fig. 5-18 is shown the cross-sectional area of a beam which is symmetrical about its centroidal $X$ axis. Using the transfer formula, compute its moment of inertia $I_X$ and check it by the special equation given in Art. 5-6.   *Ans. $I_X = 2{,}656$ in.[4]*

**5-14.** Locate the centroidal $XX$ axis and compute the moment of inertia $I_X$ of the angle shown in cross section in Fig. 5-19. (Check answers against values given in Table E-11.)

FIG. 5-18 Probs. 5-13 and 7-57.    FIG. 5-19 Prob. 5-14.    FIG. 5-20 Prob. 5-15.

**5-15.** Determine the location of the centroidal $XX$ axis of the area shown in Fig. 5-20 and compute the moment of inertia $I_x$ of the area.

**5-16.** The timber box beam in Fig. 5-21 is made of four 2- by 10-in. planks firmly nailed together to act as a single unit. Compute the moment of inertia $I_x$ of this section with respect to the centroidal $XX$ axis. Check it by the special equation given in Art. 5-6.        *Ans.* $I_x = 1,786$ in.[4]

**5-17.** The structural member whose cross-sectional area is shown in Fig. 5-21 is to be used as a column. This column will buckle in the direction of its *least* radius of gyration $k$. Determine the least $k$ for this section. (HINT: Obtain $I_x$ from Prob. 5-16. Then compute $I_Y$, $k_X$, and $k_Y$.)

FIG. 5-21 Probs. 5-16 and 5-17.    FIG. 5-22 Prob. 5-18.    FIG. 5-23 Prob. 5-19.

**5-18.** In Fig. 5-22 is shown the cross-sectional area of a steel-pipe column whose outside and inside diameters are 10 and 8 in., respectively. Using the special equation given in Art. 5-6, compute the moment of inertia and the radius of gyration $k$ of this section with respect to a centroidal axis.

**5-19.** Two sides of the solid wood strut whose cross-sectional area is shown in Fig. 5-23 are of semicylindrical shape. Compute the *least* moment of inertia and radius of gyration of this strut.        *Ans.* $I_Y = 44.6$ in.[4]; $k_Y = 1.1$ in.

**5-20.** The I-beam in Fig. 5-24 is 20 in. deep and its cross-sectional area is 20.42 sq in. The moment of inertia with respect to its centroidal $X$ axis is 1,214 in.[4]. A 12- by $\frac{1}{2}$-in. cover plate is riveted to its top flange. Locate the centroidal $X$ axis of the composite area and compute the moment of inertia $I_X$.

FIG. 5-24 Prob. 5-20.    FIG. 5-25 Prob. 5-21.    FIG. 5-26 Prob. 5-22.

**5-21.** The box girder whose cross-sectional area is shown in Fig. 5-25 is built up of four angles 4 by 4 by $\frac{1}{2}$ in., two web plates 23.5 by $\frac{1}{2}$ in., and two cover plates 25 by $\frac{1}{2}$ in., assembled to provide a section 25 in. deep. The centroidal moment of inertia of each angle is 5.60 in.[4], its cross-sectional area is 3.75 sq in., and the distance $\bar{y}$ from the back of the angle to its own centroid is 1.18 in. Compute the moment of inertia $I_X$ of the entire section.

**5-22.** A 2-in.-diameter hole has been bored in the upper half of the 3- by 16-in. timber joist shown in Fig. 5-26. Locate the centroidal $X$ axis. By what percentage has the moment of inertia of the solid joist been weakened by this hole?

*Ans.* 16.9 percent ($\bar{y} = 7.29$ in.; $I_X = 851$ in.[4])

**5-23.** By what percentage will the solid joist in Prob. 5-22 be weakened if the hole is bored through its middle? Compare with answer to Prob. 5-22.

**5-24.** Calculate the moment of inertia of the beam section shown in Fig. 5-9 with respect to both axes $XX$ and $YY$. Calculate also $k_X$ and $k_Y$ ($\bar{y} = 5.92$ in.).

**5-25.** The section shown in Fig. 5-10 has flange and web thicknesses of $\frac{3}{4}$ in., and $\bar{y}$ is 3.38 in. Compute the moment of inertia $I_X$ and the radius of gyration $k_X$.

**5-26.** Calculate the moment of inertia $I_X$ and the radius of gyration $k_X$ of the composite beam section shown in Fig. 5-11 ($\bar{y} = 6.78$ in.).

*Ans.* $I_X = 266.2$ in.[4]; $k_X = 3.88$ in.

## SUMMARY
### (By article number)

**5-1.** The gravitational pull of the earth on a body is its **weight**, also called the **force of gravity**. The point in the body through which this gravity force acts is called its **center of gravity**. The **centroid of an area** occupies the position of the center of gravity of a homogeneous thin plate whose thickness approaches zero.

**5-3.** A **composite area** is one made up of a number of simpler areas. The *centroid* *of a composite area* is located by use of the following principle: about any axis, *the moment of an area equals the algebraic sum of the moments of its component areas.*

**5-4.** The quantity **moment of inertia,** also called **second moment of area,** is used in the analysis and design of beams and columns. It represents the capacity of a beam or a column to resist bending or buckling. The general expression for the moment of inertia of an area is $I = \int y^2 \, dA$.

**5-5.** To obtain its moment of inertia, an irregular area is divided into narrow strips *running parallel to the reference axis.* The **approximate moment of inertia** is then found by the equation

$$I = \Sigma a y^2 \tag{5-3}$$

in which the area $a$ of each strip is multiplied by the square of the distance from its centroid to the reference axis.

**5-6.** The **exact moment of inertia of a simple geometric area** such as a rectangle, a triangle, and a circle is obtained by integration as described in Appendix B.

**5-7.** The moment of inertia of an area with respect to a noncentroidal axis is given by the following equation, called the **transfer formula:**

$$I_X = I_C + Ad^2 \tag{5-4}$$

The **moment of inertia of a composite area** is the sum of the moments of inertia of the component areas, all with respect to a common axis.

**5-8.** The **radius of gyration** $k$ of an area is given by the equation

$$k = \sqrt{\frac{I}{A}} \tag{5-5}$$

This quantity is useful in the design of all compression members, since such members tend to buckle in the direction of their *least radius of gyration.*

## REVIEW PROBLEMS

**5-27.** The timber T-beam shown in cross section in Fig. 5-27 is made of two planks 2 by 6 in. securely spiked together. Compute the distance $\bar{y}$ to its centroid $C$.

Fig. 5-27  Prob. 5-27.

Fig. 5-28  Prob. 5-28.

**5-28.** The surface shown in Fig. 5-28 is symmetrical about a vertical center line through the circular hole whose area is 10 sq in. Compute the distance $\bar{y}$ to its centroid $C$.                                                   *Ans.* $\bar{y} = 6.61$ in.

**5-29.** The cross section of a steel beam built up of a wide-flange section with a channel section riveted to its top flange is shown in Fig. 5-29. The actual depth of the wide-flange section is 18 in., and its area is 14.71 sq in. The area of the channel is 6.03 sq in., the thickness of its web is 0.28 in., and the distance from the back to its center of gravity is 0.70 in. Compute $\bar{y}$ to the centroid $C$ of the composite section.

FIG. 5-29  Prob. 5-29.                     FIG. 5-30  Probs. 5-30 and 5-35.

**5-30.** In Fig. 5-30 is shown the cross section of a steel beam, built up of one web plate $\frac{3}{4}$ by 18 in., one cover plate 1 by 12 in., and four equal-leg angles 5 by 5 by $\frac{3}{4}$ in., assembled as shown. The distance from the heel of each angle to its centroid is 1.52 in. Compute $\bar{y}$ of the composite section.

**5-31.** In Fig. 5-4, the I-beam is 12 in. deep, its cross-sectional area is 14.57 sq in., and its moment of inertia with respect to its own centroidal $X$ axis is 301.6 in.[4]. Compute the moment of inertia $I_x$ of the beam with cover plate with respect to the centroidal $X$ axis of the composite section.          *Ans.* $I_x = 424$ in.[4]

**5-32.** The beam section shown in Fig. 5-31 is built up of a 24-in. wide-flange section with a 15-in. channel riveted to its top flange. The area of the wide-flange section is 23.54 sq in., and its centroidal moment of inertia is 2,230 in.[4]. The area of the channel is 9.9 sq in., the thickness $t$ of its web is 0.4 in., the distance $y$ from its back to its horizontal centroidal axis is 0.79 in., and its moment of inertia about that axis is 8.2 in.[4]. Locate the centroidal $XX$ axis of the composite area, and compute the moment of inertia $I_x$.

**5-33.** In Fig. 5-32 is shown the cross-sectional area of a latticed column in which two 15-in. channels are held together by lattice bars riveted to their flanges. The area of each channel is 14.64 sq in., the distance $x$ from the back to its centroidal $Y$ axis is 0.8 in., and its moment of inertia about that axis is 11.2 in.[4]. The moment of inertia with respect to the $X$ axis of each channel is 401.4 in.[4]. Compute the

FIG. 5-31 Prob. 5-32.      FIG. 5-32 Prob. 5-33.      FIG. 5-33 Probs. 5-34 and 5-37.

distance $d$ required to make $I_X$ and $I_Y$ equal. (The lattice bars, indicated by dashed lines, are neglected, since they serve merely to hold the two channels in position.)

*Ans. $d = 10.2$ in.*

**5-34.** The 4- by 12-in. solid section in Fig. 5-33 has four semicircular grooves cut in its sides as shown, each having a radius of 1 in. Compute the moment of inertia of this section about the $X$ axis.

**5-35.** Compute the moment of inertia $I_X$ with respect to the centroidal $X$ axis of the section shown in Fig. 5-30 and described in Prob. 5-30. The area of each angle is 6.94 sq in., the distance from the back of the angle to its centroidal axis is 1.52 in., and its moment of inertia with respect to that axis is 15.7 in.⁴. The distance $\bar{y}$ is 11.0 in.

**5-36.** Compute $I_X$ and $I_Y$ for the area shown in Fig. 5-3.

*Ans. $I_X = 5,785$ in.⁴; $I_Y = 18,140$ in.⁴*

**5-37.** Compute $I_Y$ for the section of Fig. 5-33.

## REVIEW QUESTIONS

**5-1.** Explain the meaning of centroid of an area.

**5-2.** By what principle may the centroid of a composite area be located? Explain.

**5-3.** In what phase of engineering practice is the location of the centroid of an area required?

**5-4.** What practical use is made of the quantity moment of inertia?

**5-5.** Define the moment of inertia of an area in terms of its elemental areas and with respect to an axis in the plane of the area.

**5-6.** In practical engineering language, what does the moment of inertia of the cross-sectional area of a beam or a column represent?

**5-7.** In obtaining the rectangular moment of inertia of an area, what is the position of the reference axis with respect to the area?

**5-8.** In obtaining the polar moment of inertia of an area, what is the position of the reference axis with respect to the area?

**5-9.** What is the radius of gyration of an area, and what practical use is made of it?

**5-10.** By what method may the approximate moment of inertia of an area be determined?

**5-11.** How is the exact moment of inertia of a simple geometric area obtained?

**5-12.** State the principle sometimes used to obtain the moment of inertia of an area with some part removed.

**5-13.** Give the transfer formula and a complete word statement of the meaning of its separate terms.

**5-14.** Explain briefly how to obtain the moment of inertia of a composite area.

# CHAPTER 6

# Shear and Moment in Beams

### 6-1 Types of Beams
In this chapter we shall consider only horizontal beams supporting vertical loads.

A **beam** is a structural member which is subjected to loads acting transversely to its longitudinal axis, thus causing it to bend. Various kinds of beams are in common use. The simplest of these is one supported at both outer ends, such as that shown in Fig. 6-1a, and is called a **simple beam.**

If one or both ends of a beam overhang its supports, as shown in Fig. 6-1b and c, it is called an **overhanging beam.** A beam which is completely supported at one end by being framed into a solid wall or pier, as shown in Fig. 6-1d, is called a **cantilever beam.** The foregoing beams are said to be **statically determinate** because their support reactions may all be determined by the laws of static equilibrium, $\Sigma F = 0$ and $\Sigma M = 0$.

A cantilever beam which is also supported at its unrestrained end, as in Fig. 6-2a, is called a **supported cantilever.** A single beam extending continuously over more than two supports, as in Fig. 6-2b, is called a **continuous beam.** These two types of beams are of a group said to be **statically indeterminate** because the reacting forces and moments at the supports cannot be determined wholly by the laws of static equilibrium but depend in part upon the elastic properties of the material of which they are made. Such cases are treated in Chap. 10.

### 6-2 Loads on Beams
A **concentrated load** is one which is supported on an area of material relatively so small that it may, for convenience of calculation, be assumed to be a point.

126

FIG. 6-1 Statically determinate beams. (*a*) Simple beam. (*b*) Overhanging beam. (*c*) Overhanging beam. (*d*) Cantilever beam.

For example, the force exerted by a chair leg on the supporting floor and the load exerted by a beam on a supporting column are both considered as concentrated loads. Another example is shown in Fig. 6-3. Clearly, each of the concentrated loads is actually the resultant of a load distributed over a comparatively small but finite area. For convenience of calculation, an actual load system, such as that shown in Fig. 6-3*a*, is simplified to that shown in Fig. 6-3*b*. A concentrated load, usually expressed in pounds or kips, is conveniently represented by an arrow which indicates the direction and sense of the force.

**Distributed loads,** as the term implies, are loads distributed over larger areas. The usual variety of warehouse contents piled on a floor and pieces of furniture displayed on the floor of a store are examples of distributed floor loads.

If a floor so loaded is supported by joists whose ends in turn are supported on a beam, the load certainly is distributed along the length of each joist, while the beam is subjected to a series of closely spaced concentrated joist loads. Loads distributed over some area are generally expressed in pounds per square inch or pounds per square foot, while a load distributed along the

FIG. 6-2 Statically indeterminate beams. (*a*) Supported cantilever beam. (*b*) Continuous beam.

FIG. 6-3 Concentrated loads. (*a*) Actual load system. (*b*) Simplified load system.

length of a beam is usually expressed in pounds per inch or pounds per foot (of length of beam).

Distributed loads may be *uniform*, as on the beam in Fig. 6-1*b*, or *nonuniform*, as on the beams in Figs. 6-1*c* (trapezoidal), 6-2*a* (triangular), and 6-2*b* (varying). While distributed loads often are actually uniform, as in the case of the dead weight of a reinforced concrete floor slab of uniform thickness, the trapezoidal and triangular loads are, as a rule, merely the closest approximations of nonuniform loads which may conveniently be dealt with mathematically.

## 6-3 Beam-Support Reactions
Methods of calculating beam reactions in the most common cases have presumably been studied in previous work. The cantilever beam, however, is a special case, and the nature of the reactions at its support deserves mention.

Consider the cantilever beam *ABC* in Fig. 6-4*a*. Assume that the supported

FIG. 6-4 Reactions at support of cantilever beam. (*a*) Cantilever beam. (*b*) Probable support reactions. (*c*) Assumed support reactions.

end is imbedded solidly in a massive abutment. The probable distribution of the reaction forces on the imbedded portion is as indicated in Fig. 6-4b. For equilibrium, the algebraic sum of acting and reacting forces must equal zero. That is, $P = P + F - F = P$. The two forces $F$ constitute a couple whose moment is $Fx = M = PL$. The usual simplified manner of indicating the reacting force $R$ and the reacting moment $M$ is shown in Fig. 6-4c.

## PROBLEMS

**6-1.** Calculate the reactions $R_A$ and $R_B$ at supports $A$ and $B$ of the beam shown in Fig. 6-5 if $P_1 = P_2 = P_3 = 12$ kips.          *Ans.* $R_A = 28$ kips; $R_B = 8$ kips

**6-2.** Compute the support reactions $R_A$ and $R_B$ in Fig. 6-6.

Fig. 6-5  Probs. 6-1 and 6-42.         Fig. 6-6  Probs. 6-2 and 6-47.

## 6-4 Shear and Moment in Beams

The subjects of stresses in beams and design of beams are discussed in Chaps. 7 and 8. As will be shown in those chapters, the unit shearing stress at any section in a beam is proportional to the transverse shear $V$ at that section and the unit bending stress is proportional to the bending moment.

Beams must be designed to withstand the maximum stresses induced by loads, which, then, requires determination of the maximum shear $V$ and maximum moment $M$. Obviously, a thorough understanding of methods of obtaining the shear and moment at any section in a beam is essential to an understanding of the design of beams. Two methods are available: the free-body-diagram method described below and the semigraphical beam-diagram method described in Art. 6-6.

A basic concept of force systems is that when *all* forces acting on a body are accounted for, the force system is complete and is in equilibrium. Hence a loaded beam is in equilibrium under the action of the external loads and the consequent external support reactions. Likewise, *any part of a beam, when isolated from the remainder as a free body, must be and is in equilibrium under the action of the external forces acting on that part and the internal forces at the cut section.*

Consider a small experimental beam made up of three pieces of wood glued together at section $A$ and $B$, as indicated in Fig. 6-7a. Let this beam be

FIG. 6-7 Transverse shear $V$ and bending moment $M$ in beams. ($a$) Loaded experimental beam. ($b$) Simplified free-body diagram. ($c$) Shear and moment at section $B$. ($d$) Bending-moment couple at section $B$.

supported on two short posts under the two end sections and let it be subjected to a load $W$ resting on a short post placed on the center section. If the glued joints are weaker than the solid material itself on a parallel section, the center part will shear out from between the ends along the transverse planes $AA$ and $BB$ when the failing load is reached. Any load less than that required to cause actual shear failure will produce a *tendency* to shear and will set up transverse resisting shearing stresses in the material of the beam, as indicated by the small arrows. The total *tendency to shear* transversely may readily be visualized and measured.

A simplified free-body diagram of the beam is shown in Fig. 6-7$b$. Let this beam now be cut at section $BB$, for example, and let the part of it on the left of $B$ be isolated as a free body as in Fig. 6-7$c$. Only forces $W$ and $R$ act externally on this isolated part. Clearly now, since $R$ is necessarily less than $W$ and the equations of equilibrium, $\Sigma V = 0$ and $\Sigma M = 0$, must be satisfied, the total *internal* vertical force at the cut section is $V = W - R$. The force $V$ represents the sum of all the internal resisting unit shearing stresses and is called the **internal resisting shear.** At any section it is equal to the **external shear,** which is the *algebraic sum of the external forces on either side of the section.*

The existence of the internal resisting moment $M$ at section $B$ may be visualized as follows: Under the influence of the load $W$ the beam will sag between its end supports. Consequently the upper fibers will be shortened as a result of horizontal compressive stresses, the sum total of which is denoted by their resultant $C$, as shown in Fig. 6-7$d$. Similarly, the bottom fibers of the beam will be stretched by horizontal tensile stresses, the resultant sum of which is denoted by $T$. These resultants $C$ and $T$ are the only horizontal forces acting on the isolated part of the beam and are therefore equal, since $\Sigma H = 0$. Obviously, $C$ and $T$ form a couple, whose moment $M$ is called the **internal resisting moment,** since it is the moment of the internal stresses resisting bending of the beam. At any section this internal resisting moment is always equal to the **external bending moment,** which is the *algebraic sum of the moments of the external forces on either side of the section.*

To emphasize basic methods rather than details and to simplify solutions, the weight of the beam in the following example and in many of the problems has been disregarded. Later, in Chap. 8, Design of Beams, this weight will be accounted for.

**Signs of Forces, Shears, and Moments.** In most previous work no sign convention was used nor was any found to be essential. However, for the purpose of constructing the diagrams developed later in Art. 6-6 and also later in connection with the design of beams, the following sign convention shall hereafter be used:

**Forces.** Upward-acting forces and forces acting to the right are positive. Conversely, downward-acting forces and forces acting to the left are negative.

**Shears.** The sign of the shear at any section $c$ is the same as the sign of the algebraic sum of the external forces to the *left* of that section, as illustrated in Fig. 6-8$a$ and $b$.

**Moments.** A positive moment is one which produces tension in the bottom fibers of a beam, as illustrated in Fig. 6-8$c$. Conversely, a negative moment is one which produces tension in the top fibers, as illustrated in Fig. 6-8$d$. It is important here to realize that this convention for designating positive and negative moments in beams is purely arbitrary and is in no way connected with any mathematical convention for the determination of signs.

**In all problems to determine the internal resisting moment $M$ at a cut section of a beam, $M$ shall be assumed to be positive (counterclockwise when**

FIG. 6-8 Signs of forces, shears, and moments. ($a$) + shear. ($b$) − shear. ($c$) + moment. ($d$) − moment.

FIG. 6-9 Calculation of internal shear $V$ and resisting moment $M$ in beam.

**left end of beam is isolated as a free body, producing tension in bottom fibers). The calculated value of $M$ will then always bear the proper sign.**

A helpful aid in determining the signs of the moments at various sections of a beam is to visualize how it will bend. This is not difficult to do. In Fig. 6-9, for example, it seems likely that the beam will bend as indicated by the dash line. According to Fig. 6-8c, a "humped" section, as between $A$ and $o$ (Fig. 6-9), indicates tension on the top fibers and, hence, negative moment. A "sagged" section, as between $o$ and $C$, indicates tension on the bottom fibers and hence positive moment. At $o$, the point of contraflexure, there is no bending in the beam and the moment there is zero.

### ILLUSTRATIVE PROBLEM

**6-3.** Calculate the total internal shear $V$ and the internal resisting moment $M$ at each of two sections respectively 6 and 16 ft to the right of the left end of the beam shown in Fig. 6-9.

*Solution:* To calculate $V_6$ and $M_6$, we cut the beam 6 ft from the left end and draw a free-body diagram of the *left* portion (always) of the beam, as shown in Fig. 6-10a. This free-body diagram must show *exactly* what remains of Fig. 6-9 if all of

FIG. 6-10 Calculation of internal shear $V$ and resisting moment $M$ in beam. (*a*) Calculation of $V_6$ and $M_6$. (*b*) Calculation of $V_{16}$ and $M_{16}$.

it except the left 6 ft is covered with a sheet of paper. In addition, the $V$ and $M$ to be calculated must be shown at the cut section.

First, the resultant $W$ of the distributed load is calculated and is shown on the free-body diagram, acting at the centroid of the load area. We shall now assume $V_6$ to be a positive (upward-acting) force. Then

$$[\Sigma V = 0] \qquad V_6 + 70 = 16 + 24 \qquad \text{and} \qquad V_6 = -30 \text{ kips} \qquad Ans.$$

The negative sign indicates that $V_6$ *is* a downward-acting force.

The moment center for calculating $M$ should *always* be a point $c$ on the cut section. We shall assume $M$ to be positive, that is, producing tension in the bottom fibers of the beam (counterclockwise when *left* portion of beam is isolated). Then, equating the counterclockwise moments to the clockwise moments, we have

$$[\Sigma M_c = 0] \qquad\qquad M_6 + (24)(3) + (16)(6) = (70)(2)$$

or

$$M_6 = 140 - 72 - 96 \qquad \text{and} \qquad M_6 = -28 \text{ kip-ft} \qquad Ans.$$

The negative sign indicates $M_6$ to be a negative moment rather than positive, as originally assumed. Consequently the curved moment arrow must act clockwise so as to produce tension in the top fibers.

As soon as they are determined, the values of $V$ and $M$ should be recorded on the free-body diagram, and a check should be made to assure that $\Sigma V$ and $\Sigma M$ both equal zero. Note here that the calculated $V$ and $M$ are actually the *internal resisting shear* and *internal resisting moment*, respectively.

To determine $V_{16}$ and $M_{16}$ assume $V_{16}$ to be a positive (upward-acting) force. Then

$$[\Sigma V = 0] \qquad V_{16} + 70 = 16 + 64 \qquad \text{and} \qquad V_{16} = 10 \text{ kips} \qquad Ans.$$

The positive sign indicates that $V_{16}$ is an upward-acting force, as assumed. We shall assume that $M_{16}$ is positive, that is, acting counterclockwise so as to produce tension in the bottom fibers.

$$[\Sigma M_c = 0] \qquad\qquad M_{16} + (64)(8) + (16)(16) = (70)(12)$$

or

$$M_{16} + 512 + 256 = 840 \qquad \text{and} \qquad M_{16} = 72 \text{ kip-ft} \qquad Ans.$$

The positive sign of $M_{16}$ indicates that it was correctly assumed.

## PROBLEMS

NOTE: In each of the following problems, cut the beam at the section or sections indicated, draw a *complete* free-body diagram of the portion of the beam on the *left* or the cut section, and calculate the total internal shear $V$ and resisting moment $M$ at the section.

**6-4.** Calculate the shear $V$ and moment $M$ at sections respectively 5 and 9 ft from the left end of the beam shown in Fig. 6-11.

134   SHEAR AND MOMENT IN BEAMS

Fig. 6-11  Probs. 6-4, 6-12, and 6-23.     Fig. 6-12  Probs. 6-5, 6-13, and 6-24.

**6-5.** Determine $V$ and $M$ at sections 4 and 10 ft from the left end of the beam shown in Fig. 6-12.

Ans. $V_4 = 20$ kips ↓; $M_4 = 120$ k-ft↑; $V_{10} = 10$ kips ↑; $M_{10} = 150$ kip-ft↑

**6-6.** Calculate the reactions $R_A$ and $R_B$ at supports $A$ and $B$ of the beam in Fig. 6-13. Then compute $V$ and $M$ at sections 6 and 14 ft from the left end of the beam. $P = 6$ kips.

Fig. 6-13  Probs. 6-6, 6-14, and 6-28.     Fig. 6-14  Probs. 6-7 and 6-15.

**6-7.** In Fig. 6-14, calculate the support reactions $R_A$ and $R_B$. Then compute $V$ and $M$ at sections 4 and 14 ft from the left end of the beam.

**6-8.** Calculate reactions $R_A$ and $R_B$ of the beam in Fig. 6-15. Then calculate $V$ and $M$ at a section 6 ft from its left end.

Fig. 6-15  Probs. 6-8, 6-16, and 6-30.     Fig. 6-16  Probs. 6-9, 6-17, and 6-33.

**6-9.** Compute the reactions at supports $A$ and $B$ of the beam shown in Fig. 6-16. Then calculate $V$ and $M$ at a section 6 ft from its left end.

_Ans._ $R_A = 1,680$ lb $\uparrow$, $R_B = 2,160$ lb $\uparrow$; $V_6 = 120$ lb $\downarrow$, $M_6 = 5,760$ lb-ft $\uparrow$

wrong?

## 6-5 Critical Sections in Beams. Sections of Maximum Moment

A critical section is one at which is found a critical or maximum stress. Two types of critical stresses are generally found in beams: (1) tensile and compressive _bending stresses_ and (2) _shearing stresses_. In most ordinary beams and in comparatively long beams, the bending stresses are usually critical and thus determine the design. In relatively short beams, shearing stresses often are critical and thus become the governing factor in design.

As will be shown in Chap. 7, the bending stress at any section in a beam is proportional to the bending moment $M$, and the shearing stress is proportional to the shear $V$. Obviously, then, since beams must be designed to resist the _maximum_ probable stresses in bending and shear induced by loads, we must locate the sections of maximum moment $M$ and maximum shear $V$ in order to evaluate those quantities.

**Section of Maximum Shear.** Since the shear $V$ at any transverse section of a beam is the algebraic sum of the transverse forces to the _left_ of the section (left, in order to obtain the proper sign), the shear can, in most cases, be evaluated at a glance. To obtain the maximum, it should be evaluated at the beginning and ending of each distributed load and _barely_ to the left and right of each concentrated force.

**Sections of Maximum Moment.** The problem of finding the section at which the bending moment is greatest is equally simple. It can readily be shown mathematically[1] that _a maximum bending moment occurs at every section where the shear is zero or changes sign._ A change of sign of shear may or may not be found _at a concentrated force._ When a change of sign occurs _between the concentrated forces,_ a section of zero shear lies between them which then must be located. Zero shear usually occurs at some section through a distributed load. We must realize here that, mathematically speaking, several sections of "maximum moment," positive or negative, may occur in a beam. Only one maximum moment occurs in simple beams. Two may be found in beams overhanging at one end, and three may be found in beams overhanging at both ends. Of these, the moment having the greatest numerical value generally determines the design of the beam.

### ILLUSTRATIVE PROBLEMS

**6-10.** Determine the maximum shear $V$ and moment $M$ in the beam shown in Fig. 6-17.

---

[1] See Appendix C.

FIG. 6·17 Determination of maximum shear in beam.

*Solution:* To determine the maximum shear, we must evaluate the shears at the beginning and end of the distributed load and at the left and right of each concentrated force. Upward-acting forces are positive. Thus we obtain

At $A$ right, $V = 0$
At $B$ left, $V = -(3)(4) = -12$ kips
At $B$ right, $V = 36 - (3)(4) = +24$ kips $\leftarrow$ max $V$      *Ans.*
At $C$ left, $V = 36 - (3)(8) = +12$ kips
At $C$ right, $V = 36 - (3)(8) - 6 = +6$ kips
At $D$ left, $V = 36 - 3(16) - 6 = -18$ kips

To obtain the maximum moment in the beam, we must first locate all sections at which the shear is zero or changes sign. In the calculations above, we found a change of sign at $B$, from $-12$ kips at $B$ left to $+24$ kips at $B$ right. We also note that between $C$ and $D$ the shear changes from $+6$ to $-18$ kips, indicating a section of zero shear somewhere between $C$ and $D$. Letting $x$ be the unknown distance from $C$, we obtain

$$V = 36 - 6 - (3)(8) - 3x = 0 \qquad \text{and} \qquad x = 2 \text{ ft}$$

Consequently maximum moments occur at sections 4 and 10, and we must calculate both $M_4$ and $M_{10}$ to determine which has the greatest numerical value. The beam is cut at sections 4 and 10, and free-body diagrams are drawn as in Fig. 6-18.

(a)                                           (b)

FIG. 6-18 Determination of maximum moment in beam. (*a*) Calculation of $M_4$. Calculation of $M_{10}$.

The shears are calculated and shown merely in order to complete the free-body diagrams. At section 4

$[\Sigma V = 0]$     $V_4 = 12$ kips

As before, we shall assume $M$ to be positive (producing tension on bottom fibers, acting counterclockwise when *left* portion of beam is isolated). Then

$[\Sigma M_c = 0]$     $M_4 + (12)(2) = 0$     and     $M_4 = -24$ kip-ft$\downarrow$

The negative moment is now indicated by a clockwise arrow. At section 10, $V$ is known to be zero. Again assuming $M_{10}$ to be positive, we obtain

$[\Sigma M_c = 0]$     $M_{10} + (6)(2) + (30)(5) = (36)(6)$     and     $M_{10} = 54$ kip-ft$\uparrow$     *Ans.*

**6-11.** Show that the maximum moment $M$ in a simple beam supporting a uniform load occurs at the center and is $M = wL^2/8$, where $w$ is the load per unit of length of beam and $L$ is the span.

*Solution:* At the section of maximum moment the shear $V$ is zero. Hence $W = R$. To locate this section, we solve for $x$. That is, from Fig. 6-19,

$[W = R]$     $wx = w\dfrac{L}{2}$     and     $x = \dfrac{L}{2}$     *Ans.*

The maximum moment then is

$$M + \frac{wL}{2} \cdot \frac{L}{4} = \frac{wL}{2} \cdot \frac{L}{2}$$

from which

$$M = \frac{wL^2}{4} - \frac{wL^2}{8} \quad \text{and} \quad M = \frac{wL^2}{8} \quad Ans.$$

FIG. 6-19 Maximum moment in simple beam with uniform load. (*a*) Loaded beam. (*b*) Section of maximum moment.

## PROBLEMS

NOTE: In each of the following problems, calculate the maximum external shear $V$ and maximum internal moment $M$ in the beam in the illustration referred to, using the free-body diagram method illustrated in Fig. 6-18.

**6-12.** See Fig. 6-11.

**6-13.** See Fig. 6-12.  *Ans.* $V_A = 40$ kips; $M_8 = 160$ kip-ft⌐

**6-14.** See Fig. 6-13.  $p = 6$ kips.

**6-15.** See Fig. 6-14.  $V_{12} = 24$ kips  $M_{6.93} = 55.4$ k-ft

**6-16.** See Fig. 6-15.  *Ans.* $V_{10} = -20$ kips; $M_{5.77} = 38.5$ kip-ft⌐

**6-17.** See Fig. 6-16.

**6-18.** Calculate the reactions at supports $A$, $B$, and $D$ of the beam shown in Fig. 6-30. Then calculate the maximum external shear $V$ and bending moment $M$.

**6-19.** Derive an equation giving the maximum moment $M$ in a simple beam of length $L$ supporting a concentrated load $P$ at its center.

**6-20.** Show that in a simple beam of length $L$ supporting a concentrated load $P$ placed at distances $a$ from the left support and $b$ from the right support, the maximum moment $M = Pab/L$.

**6-21.** Prove that in a simple beam of length $L$ supporting a uniformly distributed load $w$ per unit of length, the moment at any distance $x$ from the left support is $M_x = (wx/2)(L - x)$ (see Prob. 6-11).

## 6-6 Shear and Moment Diagrams

A semigraphical method of determining shears and moments in beams will now be developed. A distinct, and perhaps the greatest, advantage of these diagrams is that not only do they show and evaluate the maximum shear and moment in a beam, but also present at a glance a clear and full graphic picture of the *variations* of shear and moment along the entire beam, thus greatly facilitating a clear understanding of the subject. In developing the beam-diagram method, we shall make use of the free-body-diagram method illustrated in the preceding article.

A simple example is shown in Fig. 6-20. This beam carries a single concentrated load of 18 kips at $B$. By inspection, the shear $V$ at any section between $A$ and $B$ is seen to be $+6$ kips and between $B$ and $C$ is $-12$ kips, which, in each case, is the algebraic sum of the forces to the left of each section. The *shear diagram* is obtained by plotting these values respectively as positive and negative *ordinates* above and below a horizontal reference axis, finally connecting the far ends of all ordinates with a line which in this case is straight.

Similarly, if the beam is cut at 2, 4, 6, and 8 ft from $A$, the bending moments are respectively 12, 24, 36, and 48 kip-ft, and all are positive. Note here that the moment $M_8$ at $B$ is $+48$ regardless of whether the beam is cut at $B$ left or $B$ right. If cut at 10 ft from $A$, $M_{10} + (18)(2) = (6)(10)$, or $M_{10} = 60 - 36 = 24$ kip-ft. At $C$ left, $M_{12} + (18)(4) = (6)(12)$, or $M_{12} = 72 - 72 = 0$. The *moment diagram* is now obtained by plotting these values as ordinates, positive and negative, after which their far ends are connected, resulting in straight, sloping lines. Clearly, now, these diagrams give the designer in a single glance not only the maximum shear and moment values, but also a complete and graphic picture of the *variations* in shear and moment along the entire beam.

FIG. 6-20 Shear and moment diagrams.

A second example is that of a uniformly loaded simple beam, as illustrated in Fig. 6-21. By inspection, if the beam is cut at $A$ right, the external shear is $+36$ kips. A little study will show that this shear is reduced by 3 kips for every foot to the right of $A$ at which a section might be cut. Then at 4, 8, 12, 16, 20, and 24 ft from $A$, the shears will be $+24$, $+12$, 0, $-12$, $-24$, and $-36$ kips, respectively. When these values are plotted as ordinates and their far ends are connected, the shear diagram is obtained.

For the purpose of obtaining the moments at various sections in this example, the free-body diagrams of the parts to the left should be visualized. At 4 ft from $A$, for example, the resultant of the distributed load is 12 kips concentrated 2 ft from the cut section. Hence $M_4 + (12)(2) = (36)(4)$ or $M_4 = 144 - 24 = 120$ kip-ft. Similarly, at 8, 12, 16, 20, and 24 ft from $A$, the moments are respectively $+192$, $+216$, $+192$, $+120$, and 0. When these moment values then have been plotted as ordinates, their far ends are connected by a *smooth curve* which in this instance is a second-degree parabola.

The process of constructing shear and moment diagrams in this manner is, of course, slow and cumbersome and will not be used hereafter. But, from the two examples thus constructed and shown in Figs. 6-20 and 6-21, certain rules may now be formulated by means of which similar beam diagrams may

FIG. 6-21 Shear and moment diagrams.

be constructed rapidly and accurately. These rules, hereafter referred to as the laws of beam diagrams, are as follows:

**First Law of Beam Diagrams:**

$$\begin{bmatrix} \text{Slope of curve at any point} \\ \text{in any diagram} \end{bmatrix} = \begin{bmatrix} \text{length of ordinate at correspond-} \\ \text{ing point in next higher diagram} \end{bmatrix}$$

**Second Law of Beam Diagrams:**

$$\begin{bmatrix} \text{Difference in length of any} \\ \text{two ordinates in any diagram} \end{bmatrix} = \begin{bmatrix} \text{area between corresponding ordi-} \\ \text{nates in next higher diagram} \end{bmatrix}$$

The first law is used to determine the *shape* of a curve, the second to determine values of ordinates representing either shear or moment. The load, shear, and moment diagrams are usually constructed in that order from top to bottom of the sheet. The word "higher" in the two laws then means *higher* on the sheet. See Appendix C for mathematical derivations of the laws of beam diagrams.

**Definitions.** Four terms appearing in these two laws will now be defined:

**Curve.** Any line that may be precisely defined by a mathematical equation is called a **curve.** (Thus, a straight line may, mathematically speaking, be referred to as a curve.)

## FIRST LAW OF BEAM DIAGRAMS

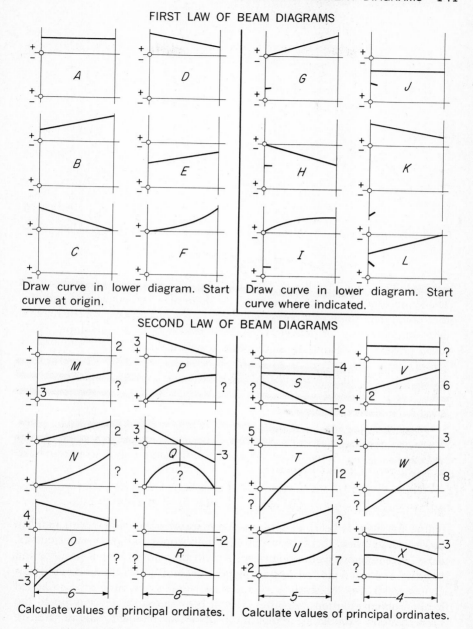

Draw curve in lower diagram. Start curve at origin.

Draw curve in lower diagram. Start curve where indicated.

## SECOND LAW OF BEAM DIAGRAMS

Calculate values of principal ordinates.

Calculate values of principal ordinates.

**Practice Problems.** In order to gain proficiency in the application of the first and second laws of beam diagrams, students should copy and complete the four practice sheets shown above. Draw curves proportionately correct, but not to scale.

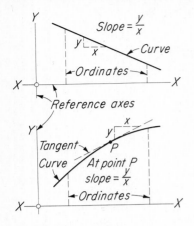

FIG. 6-22 Slope of curves.

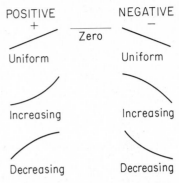

FIG. 6-23 Seven possible slopes of curves.

**Slope.** The **slope** of a line is the ratio of $y$ to $x$, as indicated in Fig. 6-22. The slope of a curve at any point is the slope of a tangent to the curve at that point. The slope of a curve may be positive or negative, and it may be uniform, increasing, or decreasing. A horizontal line has a zero slope. These seven possible slopes of curves are shown in Fig. 6-23.

**Ordinate.** With reference to a system of rectangular axes ($X$ and $Y$), an ordinate is a line drawn parallel to the $Y$ axis and extending from the $X$ axis to the curve. **Principal ordinates** are the *initial, final,* and *maximum ordinates* of each segment of a curve.

**Areas under Curves.** By **area under a curve** is meant that area lying between the curve and the horizontal $X$ axis and bounded on the left and right by vertical ordinates. Areas under the curves most commonly encountered are shown in Fig. 6-24.

## 6-7 Applications of Shear and Moment Diagrams

Before the laws of beam diagrams can successfully be applied with rapidity and accuracy, they, together with the various slopes and areas under curves, must be reasonably well memorized. A load diagram in which positive and negative loads are plotted respectively below and above the horizontal axis may be added to the shear and moment diagrams and should be so added in all problem solutions hereafter. It will bear the same relationship to the shear diagram that the shear diagram bears to the moment diagram; that is, the load diagram conforms in every respect to the laws of beam diagrams. The following illustrative problem presents difficulties somewhat in excess of those normally encountered in practice.

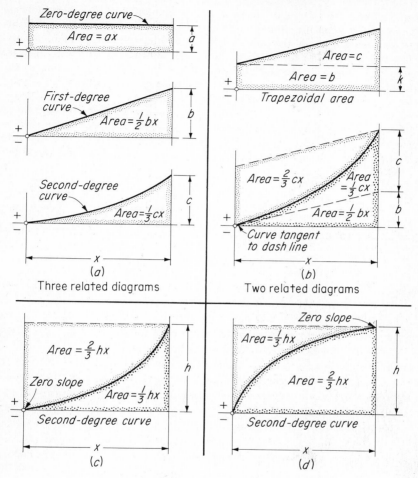

FIG. 6-24 Areas under curves.

## ILLUSTRATIVE PROBLEM

**6-22.** Construct the load, shear, and moment diagrams for the loaded beam shown in Fig. 6-25a. Evaluate all principal ordinates and determine the maximum shear $V$ and moment $M$.

*Solution:* First, a diagram of the loaded beam should always be drawn at the top. Second, the necessary framework of vertical and horizontal axes should be drawn, with $+$ and $-$ indicating the positive and negative sides of the horizontal axes. Third, the load diagram is then constructed. The downward-acting (negative) loads are plotted below the horizontal axis and the upward-acting (positive) reactions above.

Then, the shear diagram for each segment (*AB*, *BC*, and *CD*) of the beam is constructed and the principal ordinates are evaluated, using the laws of beam diagrams, first in segment $AB$, then in $BC$ and $CD$, followed by the three segments of the moment diagram in the same order. The two laws are used simultaneously, the first law to ascertain the shape of the curves and the second law to evaluate the principal ordinates. The explanatory notes on Fig. 6-25b, pertaining to the relationship between slope of a curve and the ordinate at the corresponding point in the next higher diagram, should be studied closely.

The thought processes while constructing the diagrams might be as follows: Segment $AB$ of shear curve, by first law, increasing negative load ordinates starting with zero ordinate, hence increasing negative *slope* of shear curve starting with zero slope. Shear at $A$ is zero; by second law, *difference* in shear at $A$ right and $B$ left

FIG. 6·25 Construction of load, shear, and moment diagrams.

equals load area $AB$, or $(\frac{1}{2})(-3)(6) = -9$. Also, *difference* between shear $B$ left and $B$ right equals concentrated force at $B$, or $-9 + 25 = +16$. (EXPLANATION: Every concentrated force actually is the resultant of a load distributed over a small area; see Art. 6-2.) Next, segment $BC$ of shear curve, first law: constant negative load ordinates and constant negative slope of shear curve; second law, difference between shear $B$ right and $C$ left equals load area $BC$, or $(-2)(7) = -14$; hence shear $C$ left $= 16 - 14 = 2$.

The remaining segments in the shear and moment diagrams are completed in the same manner. Note here that the difference in moments at $A$ and $B$ left equals the shear area $AB$, which is $(\frac{1}{3})(-9)(6) = -18$. Also, the difference in moments at $B$ right and $C$ left equals the shear area $BC$, which is $(\frac{1}{2})(16 + 2)(7) = +63$. Hence the moment at $C$ left is $-18 + 63 = +45$, which also is the moment at $C$ right, since the shear area between $C$ left and $C$ right is zero.

When all diagrams are completed, we see at a glance that $V_{max} = 16$ kips at $B$ right and $M_{max} = 45$ kip-ft at $C$. We note also that, in this example, the two maximum moments both occur at sections where the shear changes sign.

The tremendous value of these diagrams now becomes clearly evident when we notice that a single glance at them reveals the variations of shear and moment along the entire length of the beam.

A few hints helpful for constructing beam diagrams may be noted here:

1. The value of the initial shear ordinate is most easily determined by the usual algebraic summation of external forces.
2. An abrupt change in values of adjacent shear ordinates occurs at every concentrated force, load, or reaction.
3. The moment at the end of a beam is always zero unless the support there is such that it restrains rotation of the beam, as, for example, at the supported end of a cantilever beam.
4. There can never be an abrupt change in value of adjacent moment ordinates. However, an abrupt change in slope of the moment curve occurs at every concentrated force.

## PROBLEMS

NOTE: In each of the following problems, draw a diagram of the beam in the illustration referred to. Then construct the load, shear, and moment diagrams, calculate the values of all principal ordinates, and determine the maximum shear $V$ and moment $M$.

**6-23.** See Fig. 6-11.    *Ans.* $V_8 = -6$ kips; $M_8 = 24$ kip-ft
**6-24.** See Fig. 6-12.
**6-25.** See Fig. 6-26.    *Ans.* $V_{15} = -12$ kips; $M_{10} = 80$ kip-ft
**6-26.** See Fig. 6-27.
**6-27.** See Fig. 6-28.    *Ans.* $V_9 = 24$ kips; $M_9 = 48$ kip-ft
**6-28.** See Fig. 6-13.   $P = 9$ kips.

FIG. 6-26   Prob. 6-25.

FIG. 6-27   Prob. 6-26.

FIG. 6-28   Prob. 6-27.

FIG. 6-29   Prob. 6-29.

FIG. 6-30   Probs. 6-18 and 6-31.

FIG. 6-31   Prob. 6-32.

**6-29.** See Fig. 6-29.
**6-30.** See Fig. 6-15.
**6-31.** See Fig. 6-30.
**6-32.** See Fig. 6-31.
**6-33.** See Fig. 6-16.

*Ans.* $V_{12} = -24$ kips; $M_{6.92} = 55.3$ kip-ft
$(R_A = 26$ kips ↑; $R_B = 70$ kips ↑; $R_D = 16$ kips ↑)

## 6-8 Moving Loads

When a system of concentrated wheel loads, such as those of a truck or locomotive, moves across a simple beam, the problem arises of finding the particular position of the loads at which the bending moment in the beam will be an absolute maximum. Likewise, the position of the loads which will produce the absolute maximum shear must be determined before the beam can be designed. The following rules apply:

1. **The absolute maximum shear** $V$ **occurs at one of the two end supports and numerically equals the maximum support reaction.**
2. **The absolute maximum shear** $V$ can be evaluated only by trial, alternately placing each load over a support and calculating the reactions. The critical position usually occurs when one of the largest loads is over a support while as many as possible of the remaining loads are still on the beam. Remember that the load can move in either direction.
3. **The absolute maximum moment** in a simple span due to a system of concentrated moving loads always occurs under a *large* (critical) load located *near* the resultant of all the loads on the beam. In unusual cases, maximum moment may occur with one or more relatively small end loads off the beam.
4. The critical load for maximum moment must be determined by trial. **The absolute maximum moment will be found when the distance between the critical load and the resultant of all the loads on the beam is bisected by the center line of the beam.** (The truth of this statement may be proved by the use of calculus. Also, any number of trials using finite values will verify it.)

### ILLUSTRATIVE PROBLEM

**6-34.** In Fig. 6-32 is shown a simple beam subjected to a system of three moving loads, $B = 5$ kips, $C = 10$ kips, and $D = 15$ kips. Position the loads so they will produce the absolute maximum shear and moment. Then calculate the magnitude of each.

FIG. 6-32 A system of moving, concentrated wheel loads.

*Solution:* The resultant $R$ of the loads equals $5 + 10 + 15 = 30$ kips, and may be located by moments about $B$ in Fig. 6-33a. Since the moment of the resultant equals the sum of the moments of the separate forces, we obtain

$$[\Sigma M_B = 0] \qquad 30\bar{x} = (10)(6) + (15)(16) = 300 \qquad \text{and} \qquad \bar{x} = 10 \text{ ft}$$

FIG. 6-33 A system of moving, concentrated wheel loads. (a) Load position for maximum moment at C. (b) Calculation of maximum moment at C.

According to rule 2, the absolute maximum shear $V$ apparently occurs when the loads are moved to the right until load $D$ is over the right support. The distance from $A$ to the resultant $R$ is then 24 ft. By moments about $A$ we obtain

$$[\Sigma M_A = 0] \qquad 30R_E = (30)(24) \qquad \text{and} \qquad R_E = 24 \text{ kips} = V_{max} \qquad Ans.$$

Study of Fig. 6-32 will quickly show the following results. When load $C$ is over support $E$, $V = 10 + (\frac{4}{5})(5) = 14$ kips; when $B$ is over $A$, $V = 5 + (\frac{4}{5})(10) + (\frac{14}{30})(15) = 20$ kips; when $C$ is over $A$, $V = 10 + (\frac{2}{3})(15) = 20$ kips, none of which exceeds 24 kips.

According to rule 3 the maximum moment should be found either under load $C$ or under load $D$. Load $C$ is 10 kips and acts 4 ft to the left of the resultant $R$. Load $D$ is 15 kips and acts 6 ft to the right of $R$. Two variables determine which is the critical load: (1) *nearness* to the resultant $R$ and (2) *magnitude* of the load. In this case the critical load is not apparent. Therefore, we shall compute the moments under both $C$ and $D$.

Assuming first that $C$ is the critical load, we position the three loads $B$, $C$, and $D$ so that the center of the beam will bisect the distance between load $C$ and the resultant $R$, as in Fig. 6-33a. By moments about $E$ we evaluate $R_A$:

$$[\Sigma M_E = 0] \qquad 30R_A = (30)(13) = 390 \qquad \text{and} \qquad R_A = 13 \text{ kips}$$

The beam is now cut at $C$, the left portion is isolated as in Fig. 6-33b, and the moment at $C$ is calculated as follows:

$$[\Sigma M_C = 0] \qquad M_C + (5)(6) = (13)(13) \qquad \text{or} \qquad M_C = 169 - 30 = 139 \text{ kip-ft}$$

To obtain the moment at section $D$ we now position loads $B$, $C$, and $D$ so that the center of the beam bisects the distance between load $D$ and the resultant $R$, as shown in Fig. 6-34a. The reaction $R_E$ is now evaluated by a moment summation about $A$. That is,

$$[\Sigma M_A = 0] \qquad 30R_E = (30)(12) = 360 \qquad \text{and} \qquad R_E = 12 \text{ kips}$$

We cut the beam at $D$, isolate the part to the right, as shown in Fig. 6-34b, and calculate the moment at $D$ as follows:

$$[\Sigma M_D = 0] \qquad M_D = (12)(12) = 144 \text{ kip-ft} = M_{max} \qquad Ans.$$

FIG. 6-34 A system of moving, concentrated wheel loads. (*a*) Load position for maximum moment at *D*. (*b*) Calculation of maximum moment at *D*.

which is the absolute maximum moment for which the beam should be designed. It is clearly unnecessary to calculate the moment under load *B*, since *B* is low in value and is farthest removed from the resultant *R*. ($M_B$ = 100 kip-ft.)

## PROBLEMS

**6-35.** In Prob. 6-34 let the loads at *B*, *C*, and *D* be 20, 20, and 25 kips, respectively, then solve.

**6-36.** Two wheel axles are 18 ft apart. One wheel load is 10 kips and the other is 20 kips. Calculate the absolute maximum shear and moment developed in a 36-ft simple span.      *Ans.* $V_{max}$ = 25 kips; $M_{max}$ = 187.5 kip-ft

**6-37.** The wheel loads for the truck of Fig. 6-35 are $P_1 = P_2 = 4$ kips, $P_3 = 6$ kips, and $P_4 = 2$ kips. Calculate the absolute maximum moment in the supporting beam directly under the wheels if the span length is 40 ft.

FIG. 6-35 Probs. 6-37 and 6-50.

FIG. 6-36 Probs. 6-38 and 6-51.

**6-38.** The ore car shown in Fig. 6-36 and its contents weigh 32 tons. Assume that this load is distributed equally to all eight wheels. Calculate the absolute maximum moment in each of the two supporting beams if their span is 32 ft.

*Ans.* $M_{max}$ = 98 kip-ft

114  k-ft

## SUMMARY

*(By article number)*

**6-1.** Several **types of beams** are in common use: the simple beam, the overhanging beam, and the cantilever beam, all of which are **statically determinate.** Single beams extending continuously over more than two supports are **statically indeterminate** and are treated in Chap. 10.

**6-2. Loads on beams** may be concentrated or distributed. Distributed loads may be uniform or nonuniform. Nonuniformly distributed loads of considerable variation are generally approximated into triangular or trapezoidal forms of loads.

**6-3. Beam-support reactions** are readily calculated by the principle of moments. At the single support of a cantilever beam are found both a reacting force and a reacting moment.

**6-4.** At any section in a beam, the **transverse shear** $V$ is the algebraic sum of the transverse forces acting on either side of the section. Similarly at any section in a beam, the **bending moment** $M$ is the algebraic sum of the moments of the transverse forces acting on either side of the section.

**6-5.** The **critical section in a beam** is one at which either the transverse shear $V$ or the bending moment $M$ is maximum. A maximum moment is found at all sections at which the shear $V$ is minimum; that is, where the shear is zero or changes sign.

**6-6.** Values of shears and moments at all sections in a beam may be quickly determined from **beam diagrams** drawn in accordance with the following laws:

**First Law of Beam Diagrams:**

$$\begin{bmatrix} \text{Slope of curve at any point in any} \\ \text{diagram} \end{bmatrix} = \begin{bmatrix} \text{length of ordinate at corresponding} \\ \text{point in next higher diagram} \end{bmatrix}$$

**Second Law of Beam Diagrams:**

$$\begin{bmatrix} \text{Difference in length of any two} \\ \text{ordinates in any diagram} \end{bmatrix} = \begin{bmatrix} \text{area between corresponding ordinates} \\ \text{in next higher diagram} \end{bmatrix}$$

The diagrams are to be drawn in this order: load, shear, and moment, of which the load diagram is the highest.

**6-8. Movable loads** should be positioned for absolute maximum shear and moment. While some trials must be made, in general, maximum shear will be obtained when a large load is over a support and as many loads as possible on the beam. Maximum moment will be obtained under a large interior load with the loads positioned so that the critical load is as far on one side of the center line of the beam as the resultant of the loads on the beam is on the other side.

## REVIEW PROBLEMS

**6-39.** Calculate the reactions at supports $B$ and $E$ of the beam shown in Fig. 6-37.

*Ans.* $R_B = 16$ kips ↑; $R_E = 5$ kips ↑

**6-40.** Using the free-body-diagram method, calculate $V_5$ and $M_5$ at section 5 and $V_{14}$ and $M_{14}$ at section 14 of the beam shown in Fig. 6-37 (see answer to Prob. 6-39).

**6-41.** Locate by inspection and calculate the maximum shear $V$ in the beam in

FIG. 6-37  Probs. 6-39 and 6-40.

FIG. 6-38  Probs. 6-41 and 6-45.

Fig. 6-38. Then, using the free-body-diagram method, calculate the maximum moment $M$. *Ans.* Max $V_{16} = -18.25$ kips; max $M_8 = 62$ kip-ft⌐

**6-42.** Locate by inspection and calculate the maximum shear $V$ in the beam shown in Fig. 6-5. Then, using the free-body-diagram method, calculate the maximum moment $M$. $P_1 = 12$ kips, $P_2 = P_3 = 18$ kips.

*Ans.* Max $V_5 = 22$ kips; max $M_5 = 70$ kip-ft⌐

**6-43.** Calculate the reactions at the supports of the beam in Fig. 6-39. By inspection, locate and calculate the maximum external shear $V$. Then, by the free-body-diagram method, calculate the maximum moment $M$. $P = 9$ kips.

*Ans.* $R_A = 3$ kips ↑; $R_D = 18$ kips ↑; max $V_8 = -10$ kips; max $M_8 = -16$ kip-ft⌐

FIG. 6-39  Probs. 6-43 and 6-48.

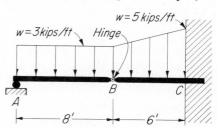

FIG. 6-40  Probs. 6-44 and 6-49.

**6-44.** The two sections of the beam shown in Fig. 6-40 are joined with a frictionless hinge at $B$. Calculate all reactions at supports $A$ and $C$ and determine the maximum shear $V$. Then, using the free-body-diagram method, calculate maximum $M$.

**6-45.** Draw a complete set of load, shear, and moment diagrams of the beam shown in Fig. 6-38, and determine maximum $V$ and $M$ (see Prob. 6-41 for answers).

**6-46.** For the beam shown in Fig. 6-17, draw a complete set of load, shear, and moment diagrams and verify the maximum shear and moment found in Prob. 6-10.

**6-47.** Draw complete load, shear, and moment diagrams for the beam in Fig. 6-6, and determine maximum $V$ and $M$ ($R_A = 21$ kips; $R_B = 45$ kips).

*Ans.* Max $V_{12} = -27$ kips; max $M_{5.25} = 55.13$ kip-ft

**6-48.** For the beam in Fig. 6-39 draw complete load, shear, and moment diagrams and determine maximum $V$ and $M$. $P = 15$ kips.

**6-49.** Draw complete load, shear, and moment diagrams for the beam in Fig. 6-40, and determine maximum $V$ and $M$ ($R_A = 12$ kips ↑; $R_C = 36$ kips ↑).

*Ans.* Max $V_{14} = -36$ kips; max $M_{14} = 138$ kip-ft⌐

**6-50.** Refer to Prob. 6-37. Let $P_1 = P_2 = 8$ kips, $P_3 = 5$ kips, $P_4 = 3$ kips, and the span length $L = 60$ ft and solve.

**6-51.** The wheel loads shown in Fig. 6-36 are 10 kips each. Calculate the absolute maximum shear and moment if the span is 40 ft.        *Ans.* $M_{max} = 214$ kip-ft

**6-52.** Two equal vertical loads of 10 kips must remain 12 ft apart. They can be placed in any position along a simple beam of 30-ft span. Calculate the absolute maximum shear and moment.

## REVIEW QUESTIONS

**6-1.** What is meant by (*a*) a simple beam, (*b*) an overhanging beam, and (*c*) a cantilever beam?

**6-2.** Explain the meaning of the terms "statically determinate" and "statically indeterminate." Give examples of beams in each category.

**6-3.** Define the terms "concentrated load" and "distributed load."

**6-4.** Discuss the reactions at the support of a cantilever beam.

**6-5.** Define and discuss the terms "external shear" and "internal shear."

**6-6.** Define and discuss the terms "external bending moment" and "internal resisting moment."

**6-7.** When is the external shear at a section positive? When is the internal resisting moment at a section positive?

**6-8.** What is meant by a critical section in a beam?

**6-9.** At what section in a beam is the shear critical?

**6-10.** At what section in a beam is the moment critical?

**6-11.** At what sections in a beam is the moment a maximum?

**6-12.** Describe the general functions of shear and moment diagrams, also called beam diagrams.

**6-13.** State the first law of beam diagrams. What is its function?

**6-14.** Give the second law of beam diagrams. What is its function?

**6-15.** Where is absolute maximum shear found for beams supporting wheel loads?

**6-16.** How are loads positioned for absolute maximum shear?

**6-17.** Where is absolute maximum moment found for beams supporting wheel loads?

**6-18.** How should loads be positioned on beams for absolute maximum moment?

# CHAPTER 7

# Stresses in Beams

## 7-1 Tensile and Compressive Stresses in Beams

When a beam is bent, the material within it is subjected to both tensile and compressive stresses and, in most cases, also to shearing stresses. In this chapter, we shall show the distribution of these stresses over a cross-sectional plane of a beam. We shall also develop two general equations, of which the first expresses the relationship at any given section of a prismatic beam between (1) the unit tensile and compressive stresses $s_t$ and $s_c$, (2) the bending moment $M$, and (3) the moment of inertia $I$ of the beam. The second equation expresses the relationship at any section of a beam between (1) the unit shearing stress $s_s$, (2) the total transverse shear $V$, and (3) the cross-sectional area $A$ of the beam. These two basic equations are used in various forms in the design of all beams.

As a general rule, beams are bent by transverse loads applied between their supported ends, as illustrated in Fig. 7-1. As a result of such loading, internal tensile, compressive, and shearing stresses are produced. Bending of a beam may also be accomplished by the application at its ends of two equal and opposing moments, as illustrated in Fig. 7-2. When it is so bent, no transverse shearing forces act at any section and the member is said to be in **pure bending.** Hence the material is subjected only to tensile and compressive *bending stresses.*

Let us now consider a straight rectangular experimental beam as shown in Fig. 7-3a. We shall assume that this beam is made up of a large number of "fibers," all lying parallel to its longitudinal axis. The stresses in these fibers

153

FIG. 7-1 Beam bent by transverse forces.

FIG. 7-2 Beam bent by end moments. Pure bending.

are referred to as **fiber stresses.** We shall also assume that the material of the beam is homogeneous and is of equal strength and modulus of elasticity in tension and compression. If this experimental beam is made of firm but flexible rubber and is subjected to two equal and oppositely directed bending moments acting in a plane which is also a plane of symmetry of the beam, a condition of pure bending is produced and a number of observations may be made. From these, and from the above-stated assumptions, certain conclusions which explain the distribution of internal normal tensile and compressive bending stresses in such a beam may then be made. Some of them are as follows:

| *Observation or assumption* | *Conclusions* |
|---|---|
| 1. The top fibers have been shortened an amount equal to the lengthening of the bottom fibers. | 1. According to Hooke's law, stress is proportional to deformation. Therefore, the compressive stress in the top fibers must equal the tensile stress in the bottom fibers. |

FIG. 7-3 Experimental beam subjected to pure bending. (a) Straight beam. (b) Bent beam.

| *Observation or assumption* | *Conclusions* |
|---|---|
| 2. A plane section (such as $AA$, $BB$, $CC$, and $DD$) before bending remains a plane section after bending. Each plane revolves (from 1 above) about its intersection with central plane $xx$. | 2. The deformations, and consequently the stresses, must then increase uniformly from zero at the central plane to a maximum at the outer fibers. |
| 3. From 1 and 2 it follows that there is no longitudinal deformation of the fibers along the central plane $xx$. Hence line $xx$ equals arc $xx$ in length. | 3. Since no deformation has taken place, there can be no tensile or compressive stress due to bending along this plane, called the "neutral plane." |

These conclusions are illustrated graphically in Fig. 7-4, in which is shown enlarged the portion of the beam between sections $BB$ and $CC$. The intersection of the plane of no stress, the *neutral* plane, with the plane of the cut section is called the **neutral axis of bending** or simply **neutral axis.** The small arrows indicate compressive stresses varying from zero at the neutral axis to a maximum $s_c$ at the top and tensile stresses varying from zero at the neutral axis to a maximum $s_t$ at the bottom.

The forces $C$ and $T$ are the equal and opposite resultants of the compressive and tensile stresses, respectively, and each is applied at the centroid of its stress volume. Consequently $C$ and $T$ form a couple whose resisting moment RM represents the internal resistance offered by the fibers of the beam to the external bending moment $M$. Since, in pure bending, the stresses on the cross-sectional plane $BB$ are equal and opposite to those on plane $CC$, this portion of the beam is in equilibrium as, of course, it must be. That is, $\Sigma V = 0$, $\Sigma H = 0$, and $\Sigma M = 0$. Because this beam is subjected to pure bending, no shearing stresses exist.

Fig. 7-4 Distribution of tensile and compressive stresses in bent rectangular beam.

## 7-2 Relationship among Bending Moment, Fiber Stress, and Moment of Inertia: The Flexure Formula

A relationship among the moment $M$, the fiber stress $f$, and the moment of inertia $I$ (Chap. 5) of the cross-sectional area of a beam may now be established.

Consider again the rectangular beam shown in Fig. 7-4 and now shown in two-dimensional form in Fig. 7-5. If the cross-sectional area of this beam is arbitrarily divided into a number of equal smaller areas, such as $a_1, a_2, a_3, \ldots, a_n$, as shown in Fig. 7-5, and if $p_1, p_2, p_3, \ldots, p_n$ are the resultants of the fiber stresses on those areas, the sum of the moments about the neutral axis of these resultant forces is the total internal resisting moment RM, which at any section equals the bending moment $M$. That is,

$$M = p_1v_1 + p_2v_2 + p_3v_3 + \cdots + p_nv_n \tag{a}$$

But $p$ is the resultant of the stresses acting on area $a$. If $f$ is the unit fiber stress on area $a$, $p = fa$ and

$$M = f_1a_1v_1 + f_2a_2v_2 + f_3a_3v_3 + \cdots + f_na_nv_n \tag{b}$$

These unit stresses $f$, acting at varying distances $v$ from the neutral axis, may now be expressed in terms of the maximum fiber stress $s$ in the extreme fiber acting at a maximum distance $c$ from the neutral axis. That is, from Fig. 7-5, by similar triangles,

$$\frac{f}{v} = \frac{s}{c} \tag{c}$$

or

$$f = \frac{s}{c}v \tag{d}$$

Elevation of beam                    Cross section

FIG. 7-5 Derivation of the flexure formula.

Consequently

$$f_1 = \frac{s}{c}v_1 \qquad f_2 = \frac{s}{c}v_2 \qquad \cdots \qquad f_n = \frac{s}{c}v_n \qquad (e)$$

Substituting these values in Eq. (b) gives

$$M = \frac{s}{c}v_1(a_1v_1) + \frac{s}{c}v_2(a_2v_2) + \cdots + \frac{s}{c}v_n(a_nv_n) \qquad (f)$$

or

$$M = \frac{s}{c}(a_1v_1{}^2 + a_2v_2{}^2 + \cdots + a_nv_n{}^2) \qquad (g)$$

Note here that the terms $a_1v_1$, $a_2v_2$, etc., represent the *first moment* of these areas about the neutral axis and that the terms $a_1v_1{}^2$, $a_2v_2{}^2$, etc., represent the *second moment,* also called the **moment of inertia,** of the areas about that axis. The sum of all the $av^2$ terms in Eq. (g) then represents the moment of inertia of the entire cross-sectional area of the beam. If the symbol $I$ designates this moment of inertia (see also Chap. 5),

$$\textbf{Moment of inertia } I = \Sigma av^2 \qquad (h)$$

Should we desire a practical definition of this term, we might say that *the moment of inertia of the cross-sectional area of a prismatic beam is a measure of the stiffness of the beam, or of its capacity to resist bending.*
When $I$ is substituted for the sum of the $av^2$ terms in Eq. (g), we obtain

**The flexure formula:**    $M = s\dfrac{I}{c}$    or    $s = \dfrac{Mc}{I}$    (7-1)

where $M$ = bending moment at section, usually in lb-in.
$s$ = fiber stress at distance $c$ from neutral axis, psi
$I$ = moment of inertia, in.$^4$
$c$ = distance from neutral axial to fiber stress $s$, in.

Equation (7-1) is called the **flexure formula.** According to Fig. 7-5 and Eq. (c), $s$ may, of course, be the stress at any variable distance $c$ from the neutral axis. *When $c$ is the distance to the extreme fiber, $s$ is the maximum stress.* The flexure formula, Eq. (7-1), was here developed by using a specific rectangular section and by visualizing certain necessary physical relationships existing at any cross section of a bent member. The exact mathematical derivation of the flexure formula, using the calculus, is shown in Appendix D.

According to Eq. (h), the moment of inertia $I$ of an area may be obtained by dividing it into a finite number of strips all lying parallel to the neutral axis, as illustrated in Fig. 7-5 (see also Prob. 5-9). The moment of inertia of each strip then is $av^2$ and of the entire area comprising all of the strips is $\Sigma av^2$. This method is approximate only and is not generally used except in connection with irregularly shaped areas. The error involved in its use is decreased as the number $n$ of areas is increased, and it becomes exact when $n$ equals infinity, in which case each area is infinitesimally small, or $dA$. The exact mathematical derivations of the expressions giving the moments of inertia of a rectangular area, a triangular area, and a circular area, using the calculus, are shown in Appendix B. In Appendix B is also shown the exact derivation of the transfer formula used for obtaining the moment of inertia of composite areas.

The magnitude of the moment of inertia of an area depends upon two things: the magnitude of the area and its distribution. Of these, the distribution has the greatest effect. For example, let us consider that 18 sq in. of area is available to form the cross section of a beam and that this area is supplied in the form of three ½- by 12-in. steel plates.

If these plates are combined in the three ways illustrated in Fig. 7-6, in each case so that they act as a single unit, striking differences appear in their respective moments of inertia. Clearly, the highest moment of inertia is obtained by placing as much of the material as possible as far away from the neutral axis as is consistent with maintaining a stable cross section. This fact is also disclosed by a study of Eq. ($h$). Consequently, we find most steel and other metal beam sections to be of the wide-flange type shown in Fig. 7-6$c$. Because of its relative weakness in shear, timber cannot generally be used in this manner. Hence timber beams are usually of the rectangular type shown in Fig. 7-6$b$.

Unless otherwise specifically stated or shown, the presumption is that all beams are horizontal and that all loads are vertical and are applied to the narrow face of the beam. Under these conditions, the neutral axis will always be horizontal.

**Location of Neutral Axis.** As previously stated, the neutral axis of bending is the line, or axis, formed by the intersection of the neutral plane with the plane of any cross section of a structural member. Its location is readily determined because *the neutral axis of bending passes through the centroid of the cross-sectional area* and lies perpendicular to the plane of bending. The mathematical proof of this fact is shown in Appendix D.

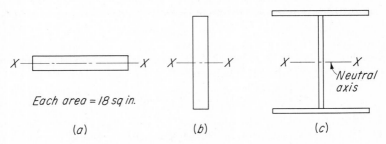

*Each area = 18 sq in.*

(a)          (b)          (c)

FIG. 7-6 Effect of distribution of area upon moment of inertia.

The centroids of various geometric and symmetrical areas are easily located, of course, without calculation. The methods outlined in Chap. 5 are used to locate centroids of nonsymmetrical areas. Problem 7-1 shows the calculations involved in the location of the centroid, or neutral axis, of a composite area.

### ILLUSTRATIVE PROBLEM

**7-1.** In Fig. 7-7 is shown the cross section of a beam built up of two 2- by 6-in. wood planks, securely spiked together to act as a single unit. It is used as a simple span 10 ft long supporting a load of 200 lb per ft including its own weight. Calculate the maximum bending stress $s$ in this beam.

FIG. 7-7 Location of neutral axis. Calculation of moment of inertia and of bending stress in beam.

*Solution:* Since this section is unsymmetrical with respect to its horizontal neutral axis, we must first locate that axis. Then the moment of inertia $I$ with respect to the neutral axis must be calculated, the dimension $c$ must be determined, and finally the bending moment $M$ must be computed, after which the maximum bending stress $s$ can be calculated.

To determine $\bar{y}$ (see Chap. 5), using the principle that about any point the moment of an area equals the algebraic sum of the moments of its component areas, we take moments about the bottom $O$ of the section. That is,

$$[M_O] \qquad 24\bar{y} = (12)(7) + (12)(3) = 84 + 36 = 120 \qquad \text{and} \qquad \bar{y} = 5 \text{ in.}$$

From Chap. 5, the moment of inertia of a composite area with respect to its neutral (centroidal) axis is $I_X = \Sigma(I_C + Ad^2)$. For a rectangular area, $I_C = bh^3/12$. Then

Area $A$:
$$I_X = \frac{(6)(2)^3}{12} + (12)(2)^2 = 4 + 48 = 52 \text{ in.}^4$$

Area $B$:
$$I_X = \frac{(2)(6)^3}{12} + (12)(2)^2 = 36 + 48 = 84 \text{ in.}^4$$

Moment of inertia of composite area $= I_X = 136$ in.[4]

The flexure formula, $s = Mc/I$, clearly shows that $s$ is maximum when $c$ is greatest. Also, Fig. 7-5 shows that the stress is maximum in the extreme fibers lying farthest away from the neutral axis. Consequently, for maximum stress, $c = 5$ in.

Finally, since $w = 200$ lb per ft and $L = 10$ ft, the maximum bending moment is

$$\left[M = \frac{wL^2}{8}\right] \qquad M = \frac{(200)(10)^2}{8} = 2,500 \text{ lb-ft} = 30,000 \text{ lb-in.}$$

To be used in the flexure formula, $M$ must be expressed in pound-inches, $c$ in inches, and $I$ in inches[4]. Then the maximum stress in the bottom fibers is

$$\left[s = \frac{Mc}{I}\right] \qquad s = \frac{(30,000)(5)}{136} = 1,111 \text{ psi} \qquad Ans.$$

The maximum stress in the top fibers, using a value of $c = 3$ in., is 667 psi.

## PROBLEMS

NOTE: In solving the following problems, full use should be made of the tables in the Appendix giving properties of sections.

**7-2.** A timber beam of cross section as shown in Fig. 7-8 is full 6 in. wide by 12 in. deep. It is subjected to a bending moment $M$ of 72,000 lb-in. Calculate the maximum bending stress $s$ in this beam.  *Ans. $s = 500$ psi*

**7-3.** The beam shown in section in Fig. 7-8 is a 6- by 12-in. timber, S4S. It is subjected to a maximum bending moment $M$ of 15 kip-ft. Calculate the maximum bending stress $s$. What is the percentage decrease in stress if the beam is made the full nominal size?

**7-4.** In a problem similar to 7-1, let us assume that the beam is made of a 3- by 12-in. flange and a 4- by 10-in. stem. If the beam is 20 ft long and supports a uniform load of 300 lb per ft including its own weight, what are the maximum tensile and compressive stresses ($\bar{y} = 8.08$ in. and $I_x = 1,161$ in.[4])?

**7-5.** Assume that the box beam mentioned in Prob. 5-16 and shown in Fig. 5-21 is used as a simple span with a length between supports of 20 ft. If the uniformly distributed load is 400 lb per ft including the weight of the beam, what is the value of the maximum fiber stress $s$ ($I_x = 1,786$ in.[4])?

**7-6.** A structural-steel wide-flange section, W 16 × 40, whose cross section is shown in Fig. 7-9, is used as a simple beam 30 ft long. It supports a uniform load of 800

FIG. 7-8 Probs. 7-2 and 7-3.     FIG. 7-9 Prob. 7-6.     FIG. 7-10 Prob. 7-8.

lb per ft, which includes the weight of the beam. Calculate the maximum bending stress $s$ in this beam.

**7-7.** A cast-iron beam of the cross section illustrated in Fig. 5-20 and also mentioned in Prob. 5-15 is used as a simple-span beam 8 ft long. It supports a uniform load of 4,500 lb per ft including its own weight. Calculate the maximum stresses, $s_t$ in tension and $s_c$ in compression ($\bar{y} = 4$ in.; $I_x = 576$ in.$^4$).

*Ans.* $s_t = 3,000$ psi; $s_c = 4,500$ psi

**7-8.** Two structural-steel angles 5 by $3\frac{1}{2}$ by $\frac{1}{2}$ in. placed with their long legs back to back, as illustrated in Fig. 7-10, are to be used as a lintel beam in a brick wall to span an opening 12 ft wide between supports. The load is triangular and decreases from a maximum of 400 lb per ft at the center of the beam to zero at both ends, giving a total load $W$ of 2,400 lb, which includes the weight of the beam. Calculate the maximum bending stress $s$ in this beam. *Ans.* $s_c = 9,600$ psi

**7-9.** A standard steel pipe of 2-in. nominal diameter, used as a clothesline post, extends 72 in. vertically out of a solid concrete base in which it is firmly imbedded. What maximum stress is caused in this pipe by a horizontal pull $P$ at its top of 100 lb?

**7-10.** A footbridge with a clear, simple span of 16 ft uses logs with an average diameter of 12 in. Each log supports a uniform load of 60 lb per ft, which includes its own weight and weight of deck planking, and is also capable of supporting a concentrated load $P$ of 2,400 lb at its center. Calculate the maximum bending stress $s$ in one log under this load. *Ans.* $s = 815$ psi

**7-11.** A 2-in.-diameter steel shaft is 6 ft long horizontally between centers of its two supporting bearings. At 2 ft from the left end is a downward-acting belt pull of 140 lb, and at 4 ft from the left is an upward-acting belt pull of 220 lb, both vertical. Calculate the maximum bending stress in the shaft due to these loads only.

**7-12.** A wood plank 2 in. thick by 12 in. wide and 16 ft long is supported at its outer ends. What stress is produced in this plank by a man weighing 200 lb and standing at its middle when the plank is (a) placed on edge and (b) laid flat? Assume full dimensions. *Ans.* (a) $s = 200$ psi; (b) $s = 1,200$ psi

**7-13.** A 1-in.-diameter unfinished bolt is used to fasten together three steel straps, as illustrated in Fig. 3-5. Each outer strap is $\frac{3}{8}$ in. thick, and the middle strap is $\frac{5}{8}$ in. thick. The bolt holes are $1\frac{1}{8}$ in. in diameter. Assuming that the resultants of the bearing stresses against the bolt act along the center lines of the three straps, calculate the maximum bending stress in the bolt caused by a pull $P$ of 14,000 lb.

## 7-3 Section Modulus

In general structural practice, the vast majority of beams are selected from available standard stock sizes, the properties of which ($A$, $I$, $c$, etc.) are listed in various easily available handbooks. The properties of nonstandard sections and of built-up sections may be calculated by the methods outlined in Chap. 5. The quantity $I/c$ in Eq. (7-1) is called the **section modulus** and is denoted by the symbol $Z$. That is,

$$\text{Section modulus } Z = \frac{I}{c} \tag{7-2}$$

Equation (7-1) then becomes

**The flexure formula:**   $M = sZ$   or   $s = \dfrac{M}{Z}$   (7-3)

Since $I$ and $c$ of standard sections are known, their section moduli ($Z$ or $S$ in some literature) are also listed in handbooks. For nonstandard sections and for regular geometric areas, the section modulus may be obtained by calculating the moment of inertia $I$ of the area and then dividing $I$ by $c$, the distance from the neutral axis to the *extreme fiber*. In symmetrical sections, $c$ has only one value, but in unsymmetrical sections $c$ will have two values. In the analysis and design of beams, however, we are usually interested only in the *maximum stress* which occurs in the extreme fiber. In all such problems, the greatest value of $c$ must be used.

Consider, for example, a rectangular area, such as the cross-sectional area of a rectangular timber beam whose width is $b$ and whose depth is $h$. From Table 5-1 its moment of inertia $I$ with respect to its neutral (centroidal) axis is $bh^3/12$. Clearly $c = h/2$ and

**For rectangular areas:**   $Z = \dfrac{bh^3/12}{h/2} = \dfrac{bh^2}{6}$   (a)

As an aid in understanding this value of section modulus, an excellent physical concept of the term may be obtained from the following development. From Fig. 7-4, we note that

$$M = C(\tfrac{2}{3}h)$$   (b)

But $C$ is the resultant of a "stress volume." That is,

$$C = \frac{1}{2}\left(\frac{h}{2}\right)(s)(b)$$   (c)

Consequently

$$M = \frac{1}{2}\left(\frac{h}{2}\right)(s)(b)(\tfrac{2}{3}h)$$   (d)

or

$$M = s\frac{bh^2}{6} = sZ$$   (7-4)

from which again we find that the section modulus $Z$ of a rectangular section is $bh^2/6$. We note here also that the section modulus is purely a function of the area of the section. The equation $M = sZ$ is the section-modulus form of the flexure formula which is most generally used by designers and should hereafter be used in all problems whenever possible and suitable.

**PROBLEMS**

**7-14.** Calculate the section modulus $Z$ of a rough, rectangular timber beam whose cross-sectional area is 8 in. wide by 12 in. deep with respect to its horizontal neutral axis.
Ans. $Z = 192$ in.$^3$

**7-15.** If the timber beam in Prob. 7-14 is surfaced on all four sides (S4S), by what percentage has its section modulus, and hence its strength, been reduced from that of the rough beam?

**7-16.** A round timber log with a diameter of 10 in. is used as a beam supporting a footbridge which spans a small stream. Calculate its section modulus $Z$ with respect to its neutral axis.
Ans. $Z = 98.18$ in.$^3$

**7-17.** Consider the beam section discussed in Prob. 7-1, for which $I_x = 136$ in.$^4$. Calculate the value of section modulus $Z$ that must be used to obtain the maximum stress (*a*) in tension and (*b*) in compression. (*c*) Which of these two stresses will be the greater?

**7-18.** A beam having the section shown in Fig. 5-16 is to be used on a simple span. Calculate the section modulus $Z$ that would be used in determining the maximum (*a*) tensile stress and (*b*) compressive stress. (*c*) Which of these two stresses will be the greater?

**7-19.** Assume that the beam shown in Fig. 7-7 is built up of two 3- by 8-in. rough planks. Then calculate its section modulus $Z$ with respect to the horizontal neutral axis that must be used to obtain the maximum stress in the beam.

## 7-4 Shearing Stresses in Beams

The subject of transverse shearing forces in beams was discussed in Art. 6-4. A discussion of the unit shearing stresses in beams is now in order. In the following development, we shall see that both transverse and longitudinal shearing stresses exist and that at any point in a beam the unit transverse and longitudinal shearing stresses are equal. Since beams are normally horizontal and the cross sections upon which bending stresses are investigated are therefore vertical, these shearing stresses in beams are generally referred to as horizontal (longitudinal) and vertical (transverse).

The existence of horizontal shearing stresses in a bent beam can readily be visualized by bending a flexible book or a deck of cards. The sliding of one surface over another, which is plainly visible, is a shearing action, which, if prevented, will set up horizontal shearing stresses on those surfaces.

Consider now a beam made of three identical planks of rectangular cross section, one placed directly above the other, as shown in Fig. 7-11*a*. Assume the surfaces of contact to be smooth and frictionless. When subjected to a load *P*, as in Fig. 7-11*b*, all three pieces will bend similarly and independently. Obviously, in this position, the bottom surface of each upper plank is longer than the top surface of the plank below it. Should the movement of these surfaces over one another now be prevented, as, for example, by gluing the

FIG. 7-11 Illustration of horizontal shear in beams. (*a*) Unbent beam of three identical separate planks. (*b*) Deformation when planks bend separately. (*c*) Deformation when planks are glued together.

three planks together in the unbent position so that they would act as a single beam, as indicated in Fig. 7-11*c*, longitudinal shearing stresses on the glued planes, and on any other parallel plane within each plank, would result upon reapplication of force *P*.

## 7-5 Relationship between Vertical and Horizontal Shearing Stresses in Beams

From the discussions presented in previous articles it is now seen that in any stressed beam both vertical and horizontal shearing stresses are present as well as the normal tensile and compressive stresses. We shall now show that at any point in a stressed beam the vertical and horizontal unit shearing stresses are equal.

Let us consider the loaded beam shown in Fig. 7-12*a*. When this beam is cut on section *cc*, the shear force *V* at the cut section (Fig. 7-12*b*) represents the sum total of all unit vertical shearing stresses on the cut section, as indicated in Fig. 7-12*c*.

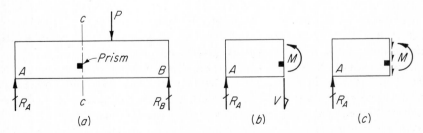

FIG. 7-12 Vertical, or transverse, shearing stresses in beams.

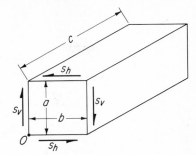

FIG. 7-13 Shearing stresses on elemental prism.

If now we cut out and isolate a small rectangular prism of this beam, as in Fig. 7-13, the following facts become evident. If the dimensions $a$ and $b$ are infinitesimally small, the normal stresses (not shown) on the right and left sides of the prism are equal and opposite and therefore by themselves form a system of forces in equilibrium which may be entirely removed from the prism without altering any remaining forces.

Let now $s_v$ and $s_h$, respectively, be the vertical and horizontal unit shearing stresses on the right and top sides of the prism, the areas of which are $(a)(c)$ and $(b)(c)$. Then the total shear force on the right is $(s_v)(a)(c)$ and on the top side is $(s_h)(b)(c)$. Since the forces acting on the prism must be and are in equilibrium, it follows that if point $O$ is chosen as a moment center, the moments of the shearing forces on the left and bottom sides of the prism will be zero, and

$$[\Sigma M_O] \qquad\qquad s_v ac \cdot b = s_h bc \cdot a \qquad\qquad (a)$$

from which

$$s_v = s_h \qquad\qquad (b)$$

*That is, at any point in a stressed beam, the unit vertical and horizontal shearing stresses are equal.* Hence the symbol $s_s$ generally denotes *all* shearing stresses.

### 7-6 The General Shear Formula

The development of the flexure formula in Art. 7-2 provided a means of determining the normal bending stresses, tensile and compressive, at any point in a stressed beam. We shall now develop a formula which will give the shearing stress at any point in a stressed beam.

Let us consider a portion of a stressed beam of constant cross section, as shown in Fig. 7-14. From this portion we shall cut and remove the small piece from the notch in the top surface extending across the beam and part way down to the neutral axis and lying between cross-sectional planes 1 and 2, as shown in Fig. 7-15. We shall assume that the moment $M_2$ at section 2 is greater than is $M_1$ at section 1. Hence $C_2$, the resultant of the normal stresses

FIG. 7-14  Portion of a stressed beam.

on surface 2, is greater than $C_1$, the resultant of the normal stresses on section 1. But since the algebraic sum of the horizontal forces must be zero, an additional horizontal force must be present. Such a force can be exerted only on the bottom surface of the piece and is, of course, the total shear resistance S offered by the fibers on the bottom surface. This shear force equals $C_2 - C_1$.

Consider now the small area $\Delta A$ on the stressed surface 2. According to the flexure formula, Eq. (7-1), the unit normal stress on this small area is $s = My/I$, in which $y$ has replaced $c$ as the distance from the area $\Delta A$ to the neutral axis. Then the total stress on area $\Delta A = (My\,\Delta A)/I$ and the total normal force on surface 2 is

$$C_2 = \Sigma \frac{M_2 y\,\Delta A}{I} = \frac{M_2}{I}\Sigma\Delta A\,y \qquad (a)$$

since both $M$ and $I$ are constants at any given section. But $\Sigma\Delta A\,y$ is the sum of all $\Delta A$ areas on surface 2, each multiplied by its distance $y$ to the neutral axis (see Fig. 7-15b). This finally becomes $A\bar{y}$, in which $A$ is the stressed area and $\bar{y}$ is the distance from the neutral axis to the centroid of that area. This quantity $A\bar{y}$ is usually denoted by the symbol $Q$ and may be defined as the first or **statical moment** of the stressed area $A$.

It is interesting to note here that the quantity $\Sigma\Delta A\,y$, the *first moment* of the stressed area about the neutral axis, arrived at in the development of

(a) Side view of piece    (b) Cross-sectional plane 2

FIG. 7-15  Horizontal shear force S equals difference between resultants of opposing normal stresses.

the *shear formula*, is analogous to the quantity $\Sigma \Delta A\, y^2$, Eq. (*h*), Art. 7-2, the *second moment* of the stressed area about the neutral axis, arrived at in the development of the *flexure formula*.

When $Q$ is substituted for $\Sigma \Delta A\, y$ in Eq. (*a*), we obtain for both surfaces 1 and 2

$$C_1 = \frac{M_1 Q}{I} \qquad (b)$$

and

$$C_2 = \frac{M_2 Q}{I} \qquad (c)$$

From Fig. 7-15 the difference between $C_2$ and $C_1$ is therefore

$$\frac{M_2 Q}{I} - \frac{M_1 Q}{I} = S \quad \text{or} \quad \frac{(M_2 - M_1)Q}{I} = S \qquad (d)$$

When the beam under consideration is of constant cross section, $Q_1 = Q_2 = Q$ and $I_1 = I_2 = I$.

Before obtaining the final shear formula, we shall need to establish one more relationship: the piece shown in Fig. 7-16 is the full top-to-bottom slice of the beam lying between the vertical cross-sectional planes 1 and 2.

On each of the two faces of the slice is shown the shear force $V$ and the moment $M$. Using point $E$ as a moment center, we obtain

$$[\Sigma M_E = 0] \qquad M_2 - M_1 - V\, \Delta x = 0 \qquad (e)$$

or

$$M_2 - M_1 = V\, \Delta x \qquad (f)$$

When $V\, \Delta x$ is substituted for $M_2 - M_1$ in Eq. (*d*), we obtain

$$\frac{V\, \Delta x\, Q}{I} = S$$

FIG. 7-16 Vertical transverse slice of stressed beam.

from which

$$\frac{S}{\Delta s} = \frac{QV}{I} \tag{7-5}$$

where $S/\Delta x$ is the shear flow (usually in pounds per inch of length) on the longitudinal shear surface.

Refer to Fig. 7-15a. The shear force on the bottom surface of the piece is $S$. The unit shearing stress $s_s$ on this surface is the force $S$ divided by the area $(b \, \Delta x)$, or $s_s = S/b \, \Delta x$. Dividing both sides of Eq. (7-5) by $b$ gives

$$\frac{S}{b \, \Delta x} = \frac{QV}{Ib}$$

from which we obtain

**The general shear formula:**   $$s_s = \frac{QV}{Ib} \tag{7-6}$$

where $s_s$ = unit shearing stress on horizontal (or vertical) plane at any given cross section of a beam

$V$ = total transverse shear at section

$Q$ = statical moment, about neutral axis, of that part of cross-sectional area lying on either side of horizontal plane at which shearing stress is desired

$I$ = moment of inertia of entire cross-sectional area of beam with respect to its neutral axis

$b$ = width of beam at plane where $s_s$ is calculated

Equation (7-6) is called the **general shear formula** and will give the shearing stress, horizontal or vertical, at any point in a stressed beam.

## ILLUSTRATIVE PROBLEMS

**7-20.** The beam shown in Fig. 7-17 is built up of three 2- by 8-in. planks securely spiked together to act as a single unit. Calculate the unit shearing stresses at plane

FIG. 7-17 (*a*) Determination of shearing stresses in a beam. (*b*) Calculation of $Q$ to determine $s_s$ at plane *aa*. (*c*) Calculation of $Q$ to determine $s_s$ at neutral axis.

$aa$ and at the neutral axis caused by a total shear $V$ of 2,000 lb. The moment of inertia $I_x$ of the entire section is 895 in.[4].

*Solution:* Since $V$ and $I$ are given and $b$ may be obtained by inspection, $Q$ remains the only quantity to be calculated. Now, $Q$ is the first moment about the neutral axis of the area above (or below) the plane on which the shearing stress is desired. The area above plane $aa$ has been isolated and is shown in Fig. 7-17b, and the area above the neutral axis is shown in Fig. 7-17c. At plane $aa$, to illustrate a point, we shall use both $b = 2$ in. and $b = 8$ in. Then at plane $aa$, $Q = (8)(2)(5) = 80$ in.[3]. Hence, for $b = 2$ in.,

$$\left[ s_s = \frac{QV}{Ib} \right] \qquad s_s = \frac{(80)(2,000)}{(895)(2)} = 89.4 \text{ psi} \qquad Ans.$$

When $b = 8$ in.,

$$\left[ s_s = \frac{QV}{Ib} \right] \qquad s_s = \frac{(80)(2,000)}{(895)(8)} = 22.4 \text{ psi}$$

These calculations show that at an abrupt change in width of a section there is a correspondingly abrupt change in the unit shearing stress. At the neutral axis, $Q = (8)(2)(5) + (2)(4)(2) = 96$ in.[3]. Hence, for $b = 2$ in.,

$$\left[ s_s = \frac{QV}{Ib} \right] \qquad s_s = \frac{(96)(2,000)}{(895)(2)} = 107 \text{ psi} \qquad Ans.$$

**7-21.** The beam shown in Fig. 6-25 has the section shown in Fig. 7-18a. The moment of inertia of the entire cross section is 760 in.[4]. Each rivet is capable of transmitting a shear force of 2.44 kips. Find the required rivet spacing at the support $B$.

Fig. 7-18 Determination of longitudinal rivet spacing in cover plate. (*a*) Beam section. (*b*) Longitudinal forces on cover plate. (*c*) Shear diagram (see Fig. 6-25).

*Solution:* Because the moment varies along the beam, the tensile stress in the top cover plate varies in a similar manner. The tensile forces $T_1$ and $T_2$ at half the rivet spacing each side of the rivet will differ by the value of the shear force $S$ (see Fig. 7-18b). Since each rivet can transmit a 2.44-kip shear force, a pair of rivets can transmit a force $S = 4.88$ kips.

From the shear diagram of Fig. 6-25 (reproduced in Fig. 7-18c), the maximum shear at $B$ is $V_B = 16$ kips. Substituting the known values in Eq. (7-5) gives $\Delta x$, the longitudinal spacing of the cover-plate rivets. That is,

$$\left[\frac{S}{\Delta x} = \frac{QV}{I}\right] \qquad \frac{4.88}{\Delta x} = \frac{(8)(0.5)(7.25)(16)}{760} \qquad \text{or} \qquad \Delta x = 8 \text{ in.} \qquad \textit{Ans.}$$

## PROBLEMS

**7-22.** Assume in Fig. 7-17 that the web of the I-beam is 2 by 14 in. and that each of the two flanges is 2 by 10 in. The vertical shear force on the section is 6 kips. Calculate $s_s$ at plane $aa$ and at the neutral axis.

**7-23.** A rough timber beam 8 in. wide and 12 in. deep, as illustrated in Fig. 7-19, is subjected to a load which causes a maximum transverse shear $V$ of 6,400 lb. The maximum unit shearing stress occurs at the neutral axis. Calculate this stress.

$\textit{Ans. } s_s = 100 \text{ psi}$

FIG. 7-19 Prob. 7-23.  FIG. 7-20 Prob. 7-24.  FIG. 7-21 Prob. 7-25.

**7-24.** The T-beam shown in Fig. 7-20 is made of two 2- by 6-in. planks securely fastened together. Determine the unit shearing stress $(a)$ at the top of the flange, $(b)$ at the bottom of the flange using both $b = 6$ in. and $b = 2$ in., and $(c)$ at the neutral axis, caused by a transverse shear $V$ of 1,000 lb $(I_X = 136 \text{ in.}^4)$.

**7-25.** The timber box beam shown in Fig. 7-21 is made of four planks securely spiked together. Calculate the unit shearing stress at the neutral axis (maximum) caused by a total transverse shear $V$ of 4,000 lb $(I_X = 1,786 \text{ in.}^4$, from Prob. 5-16).

$\textit{Ans. } s_s = 95.1 \text{ psi}$

**7-26.** A structural-steel wide-flange section 16 in. deep by 50 lb per ft (W $16 \times 50$) is used as a simple beam 30 ft long. It supports a uniformly distributed load of 650 lb per ft, which includes the weight of the beam. Calculate the maximum unit shearing stress $s_s$ in this beam which occurs at the neutral axis.

**7-27.** A structural T-section 9 in. deep by 25 lb per ft (WT $9 \times 25$) is used as a simple lintel beam, flange down, in a brick wall. The span is 12 ft between centers of supports, and the beam carries a load of 480 lb per ft including its own weight. Calculate the unit shearing stress $s_s$ at the neutral axis (maximum) in this beam. Dimensions are those of the lower half of a W $18 \times 50$. *Ans.* Max $s_s = 1,258$ psi

**7-28.** The two 2- by 6-in. planks of Prob. 7-24 could be spiked together. If the allowable lateral force on each spike is 120 lb, what spacing would be required? Does this seem to be a practical spacing?

**7-29.** Assume that the T-beam in Prob. 7-24 is used on a 10-ft simple span and supports a total uniform load of 50 lb per ft. Determine the maximum permissible spike spacing $(a)$ at the end of the beam and $(b)$ at a point 2.5 ft from the end of the beam, if the allowable lateral load on each spike is 150 lb.

Ans. (a) $\Delta x = 3.4$ in.; (b) $\Delta x = 6.8$ in.

**7-30.** A W 8 × 17 section is used as a simple beam with a 10-ft span, supporting a total load of 3,000 lb per ft. Cover plates ½ by 10 in. (similar to those shown in Fig. 7-18a) are to be welded to the top and bottom flanges of the wide-flange section with intermittent $\frac{7}{32}$-in. fillet welds having an effective length of 1½ in. Calculate the required longitudinal center-to-center spacing of the welds (a) at the supports and (b) at the quarter-points of the span.

## 7-7 Variation of Shearing Stresses in Beams

Beams must be so designed that they can safely withstand the maximum stresses to which they may be subjected, tensile, compressive, and shearing. The variation of tensile and compressive bending stresses over the cross-sectional area of a beam was discussed and illustrated in Art. 7-2. We shall now discuss and illustrate the variation of shearing stresses at any section of a beam and shall show that, *with few exceptions, the maximum shearing stress generally occurs at the neutral axis.*

Consider, for example, a rectangular beam of cross section as shown in Fig. 7-22a. If we now calculate the unit shearing stresses on several equally spaced horizontal planes in such a beam caused by a total transverse shear $V$ and then plot their intensities horizontally from a vertical axis, as illustrated by the graph also shown in Fig. 7-22a, we note that in a rectangular beam the shearing stresses increase from zero at the top to a maximum at the neutral axis and then decrease again to zero at the bottom.

Proof that the shearing stress is maximum at the neutral axis of a rectangular beam may be shown as follows. The stress on any random horizontal plane, such as that indicated by the dash line in Fig. 7-19, is $s_s = QV/Ib$, in which $Q$ is the moment of the (shaded) area above the plane with respect to the neutral axis. That is, $Q = A\bar{y}$. Since $A = b(\frac{1}{2}h - v)$ and $\bar{y} = v + \frac{1}{2}(\frac{1}{2}h - v)$, or $\frac{1}{2}(\frac{1}{2}h + v)$, it follows that

$$Q = b\left(\frac{h}{2} - v\right)\frac{1}{2}\left(\frac{h}{2} + v\right) = \frac{b}{2}\left(\frac{h^2}{4} - v^2\right) \tag{g}$$

FIG. 7-22 Variation of shearing stresses in beams. (a) Rectangular beam. (b) I-beam. (c) T-beam.

This equation clearly shows that $Q$, and consequently the unit shearing stress $s_s$, is maximum when $v$ is zero, which it is when the stress desired is on a plane which contains the neutral axis.

The shearing-stress variation over the cross section of an I-beam or a wide-flange section is illustrated in Fig. 7-22$b$. The dashed curve indicates what the stress variation would be if the beam area had remained rectangular with a constant width $b$. This variation, then, would be similar to that shown in Fig. 7-22$a$. The sudden increase in stress at the bottom of the top flange comes from the change of the width $b$ from $b$ to $t$ in $s_s = QV/Ib$. A similar change occurs at the flange of the T-beam in Fig. 7-22$c$, but here the curve below the neutral axis follows the usual pattern for a rectangular section.

If we multiply the intensity of stress at each of a number of planes in a beam by the width of the beam at that plane and then plot the resulting values horizontally from a vertical reference axis, we obtain a diagram such as shown at the right of Fig. 7-23, in which the central area is a measure of the portion of the total transverse shear resisted by the web of the beam, and the smaller end areas measure the portion resisted by the flanges, which is seen to be relatively small. If the beam shown in Fig. 7-6$c$ is used as an example, we find that the web will resist about 92 percent of the total shear, while the two flanges resist only about 8 percent. As an interesting comparison we find that the web of this beam will resist only about 13 percent of a bending moment to which it may be subjected, while the two flanges will resist about 87 percent.

In some sections, which, however, are rarely used as beams, the maximum shearing stress does not occur at the neutral axis. Consider the H-section illustrated in Fig. 7-24$a$. The centroidal neutral axis of bending lies at $h/2$ from the bottom, while the planes of maximum shearing stress lie at the top and bottom of the horizontal web of the section. In the beam section of rather unusual shape shown in Fig. 7-24$b$, the neutral axis passes through the centroid, as always, but the maximum shearing stress occurs on the plane between the narrow upper portion and the wider central part.

FIG. 7-23 Distribution of shear flow in beam.

FIG. 7-24 Maximum shearing stress not always at neutral axis. ($a$) H-beam. ($b$) Cross.

## PROBLEMS

**7-31.** In Fig. 7-19 is shown the cross section of a rectangular beam 6 in. wide and 12 in. deep. Using the general shear formula, $s_s = QV/Ib$, calculate the unit shearing stresses at the top of the beam and at horizontal planes 2, 4, and 6 in. from the top caused by a total transverse shear $V$ of 4,320 lb. Draw a cross-sectional view of the beam. On the right, plot the stresses calculated above to obtain a stress-variation diagram as in Fig. 7-22a.

**7-32.** Assume that the beam in Fig. 7-22b is a wide-flange section 8 in. deep by 17 lb per ft (W 8 × 17). This beam is subjected to a total transverse shear $V$ of 6,000 lb. Calculate the unit shearing stresses at the neutral axis, at 2 in. above the neutral axis, at the bottom of the top flange, using both $b = 0.230$ in. and $b = 5.25$ in., and at the top of the upper flange. Draw a cross-sectional view of the beam. On the right, plot the stresses obtained above to obtain a stress-variation diagram as in Fig. 7-22b ($I_x = 56.4$ in.⁴).                                    *Ans.* Max $s_s = 3,600$ psi

**7-33.** A beam has the section shown in Fig. 7-24a. All thicknesses are 2 in., $h = 10$ in., and the total width of the beam is 14 in. Calculate and plot the shearing stress at 2-in. intervals from top to bottom of the beam. The shear $V$ is 6,800 lb.

## 7-8 Special Shear Formulas

The problem of determining the *maximum shearing stresses* in beams is one that constantly confronts designers. The rectangular and circular cross sections occur frequently. For these simple geometric shapes, two greatly simplified maximum-shear formulas may readily be derived. Also, we shall develop here a very useful formula giving the approximate, but sufficiently accurate, shearing stresses in channels, I-beams, and wide-flange sections.

**Rectangular-Area Shear Formula.** As was shown in Art. 7-7, the maximum shearing stress in a beam of rectangular cross section occurs at the neutral axis. For calculating $s_s$ at the neutral axis, $v$ in Fig. 7-25 becomes zero. The shaded area then is $A = bh/2$, $\bar{y} = h/4$, and

$$Q = A\bar{y} = \frac{bh}{2} \cdot \frac{h}{4} = \frac{bh^2}{8}$$

Also, from Table 5-1, $I_x = bh^3/12$. Consequently for rectangular sections the maximum shearing stress is

$$\left[ s_s = \frac{QV}{Ib} \right] \qquad s_s = \frac{(bh^2/8)V}{(bh^3/12)b} = \frac{3V}{2bh} \qquad \text{or} \qquad s_s = \frac{3V}{2A} \qquad (7\text{-}7)$$

FIG. 7-25 Rectangular-area shear formula.

in which $A$ is the entire cross-sectional area. Equation (7-7) indicates that the maximum stress in a beam of rectangular cross section is 50 percent greater than the average stress, which would be $s_s = V/A$.

**Circular-Area Shear Formula.** The shearing stress in a circular section, see Fig. 7-26, is also maximum at the neutral axis, as can be shown. Hence, in similar manner, with respect to its centroidal neutral axis, the area above the neutral axis is $A = \pi R^2/2$, $\bar{y} = 4R/3\pi$ (from Fig. 5-2), and therefore $A\bar{y} = Q = 2R^3/3$. Also, from Table 5-1, $I_x = \pi R^4/4$ and at the neutral axis $b = 2R$. Consequently for *circular sections the maximum shearing stress is*

$$\left[ s_s = \frac{QV}{Ib} \right] \qquad s_s = \frac{(2R^3/3)V}{(\pi R^4/4)2R} = \frac{4V}{3\pi R^2} \qquad \text{or} \qquad s_s = \frac{4V}{3A} \qquad (7\text{-}8)$$

in which $A$ is the entire circular area. That is, the maximum shearing stress in a circular beam is one-third greater than the average.

**Web-Shear Formula.** The problem of determining the precise shearing stress in some of the commonly used structural beams such as channels, I-beams, and wide-flange sections by use of the general shear formula is a laborious one at best. However, in most cases in practice, the actual shearing stresses in such beams seldom exceed 50 percent of the allowable shearing stress. We know also that the web of such a beam usually resists about 85 to 90 percent of the total transverse shear, as is indicated in Fig. 7-22b and c. Consequently the following web-shear formula is extremely useful, since, in most cases, it will give an *approximate* shearing stress which is seldom more than 10 to 15 percent *below* that obtained by the precise general shear formula. This web-shear formula, based on the assumption that the web of the beam (see Fig. 7-27), resists all of the transverse shear $V$, is

**The web-shear formula:** $\qquad\qquad s_s = \dfrac{V}{th} \qquad\qquad\qquad (7\text{-}9)$

where $V$ = transverse shear, lb
$\qquad t$ = thickness of web of beam, in.
$\qquad h$ = total depth of beam, in.

Fig. 7-26 Circular-area shear formula.

Fig. 7-27 Average web-shearing stress. The web-shear formula.

Clearly, Eq. (7-9) gives only an *average* web-shearing stress. This average stress is generally not more than 10 percent below the actual maximum stress in channels, I-beams, and wide-flange sections and is, therefore, satisfactory for investigative purposes. In T-beams, the average web-shearing stress may be as high as 33 percent below the maximum.

### PROBLEMS

**7-34.** Let the beam in Fig. 7-25 be 6 in. wide and 10 in. deep, full size, and assume it to be subjected to a total transverse shear $V$ of 4,800 lb. Using the rectangular-area shear formula, calculate the maximum shearing stress $s_s$ (see also Prob. 7-23).

*Ans.* $s_s = 120$ psi

**7-35.** In Prob. 7-34 calculate $s_s$ if the beam is 6 by 10 in. S4S. (NOTE: The maximum shearing stress in this S4S beam is more than 14 percent greater than in the rough beam of the same nominal dimensions.)

**7-36.** A contractor must provide a rectangular timber beam for a simple span of 8 ft to carry a uniform load. He has available three rough beams, 8 by 12 in., 10 by 10 in., and 10 by 12 in., all known to be sufficiently strong in bending. Allowable shearing stress is 100 psi. Which beam should he use for greatest economy (*a*) if the load is 1,600 lb per ft including weight of beam and (*b*) if the load is 1,800 lb per ft?

**7-37.** Assume that Fig. 7-26 represents the cross section of a steel shaft whose diameter is 2 in. and that this shaft is subjected to a transverse shear $V$ of 6,000 lb. Then calculate the maximum longitudinal shearing stress $s_s$.

*Ans.* $s_s = 2,550$ psi

**7-38.** In Prob. 7-37 let the diameter of the shaft be $2\frac{1}{2}$ in. and let $V = 3,750$ lb. Then calculate the maximum $s_s$.

**7-39.** The floor joists in a rustic building are 8-in.-diameter logs 18 ft long carrying a uniform load of 160 lb per ft. Calculate the maximum unit shearing stress $s_s$.

*Ans.* $s_s = 38.2$ psi

**7-40.** A steel channel 10 by $2\frac{5}{8}$ in. by 15.3 lb per ft used as a beam is subjected to a maximum transverse shear $V$ of 4,800 lb. (*a*) Using the web-shear formula, calculate the average web-shearing stress $s_s$ in this channel. (*b*) The precise maximum shearing stress in this beam is 2,360 psi. By what percentage is the average web-shearing stress lower than the precise maximum?

**7-41.** A standard steel I-beam 12 in. deep by 31.8 lb per ft (S 12 × 31.8) resists a transverse shear $V$ of 12,000 lb. (*a*) Calculate the average web-shearing stress. (*b*) The precise maximum shearing stress in this beam is 3,300 psi. Calculate the percentage by which the web-shearing stress is lower than the maximum.

## 7-9 Load Capacity of Given Beams

Beam problems are generally of three types: to determine (1) the unknown stresses in a given-size beam due to a given load, (2) the unknown load capacity of a given-size beam, using given allowable stresses, or (3) the required size of beam to support a given load, using given allowable stresses. The first type is covered in Arts. 7-2 and 7-6, the third type is covered in Chap. 8, and the second type is discussed here.

## ILLUSTRATIVE PROBLEM

**7-42.** Find the value of the loads $P$ that can safely be supported by a rough 6-by 12-in. Douglas fir beam, as shown in Fig. 7-28a. The maximum allowable stress in bending is 1,450 psi and in shear is 120 psi.

FIG. 7-28 Prob. 7-42. (a) Load and beam. (b) Shear and moment diagrams.

*Solution:* The loads $P$ will be determined by the ability of the beam to carry the loads in bending, $s = M/Z$, or in shear, $s_s = 3V/2A$. Both computations should be made, and the least load so computed will be the controlling one. Shear will control in the case of comparatively short beams and bending in the case of long beams.

The shear and bending-moment diagrams are shown in Fig. 7-28b. The maximum bending moment is $PL/3 = P(18)/3 = 6P$ lb-ft. The section modulus $Z$ of the beam is $bh^2/6 = (6)(12)(12)/6 = 144$ in$^3$. From the flexure formula, Eq. (7-3),

$$[M = sZ] \qquad 6P(12) = (1,450)(144) \qquad \text{or} \qquad P = 2,900 \text{ lb}$$

The special shear formula for rectangular beams, Eq. (7-7), can be used. Since $V_{\max} = P$,

$$\left[ s_s = \frac{3V}{2A} \right] \qquad 120 = \frac{3}{2} \cdot \frac{P}{72} \qquad \text{or} \qquad P = 5,760 \text{ lb}$$

The maximum permissible value of $P$ is the least of these two computed values, or

$$P = 2,900 \text{ lb} \qquad Ans.$$

If the weight of the beam is to be considered, the effect may be approximated by subtracting its weight from the allowable total load. Assuming the beam weighs 20 lb per ft,

$$P = 2,900 - (20)(9) = 2,720$$

## PROBLEMS

**7-43.** Solve Prob. 7-42 if the maximum allowable stress in bending is 1,800 psi and in shear is 140 psi.                     *Ans. P* = 3,600 lb

**7-44.** Solve Prob. 7-42 if the beam is a rough 6- by 18-in. timber.

**7-45.** A Douglas fir timber beam 6 by 12 in. S4S and 12 ft long between supports must support a concentrated load $P$ at a point 4 ft from its left end. If allowable stresses are 1,700 psi in bending and 145 psi in shear, calculate the maximum permissible value of $P$.

**7-46.** Calculate the maximum load per foot $w$ which may safely be carried by a southern pine No. 1 stress rated beam 12 by 18 in. S4S 20 ft long. The beam weighs 55.4 lb per ft. (See Table 8-1.)                *Ans.* $w = 1,272$ lb per ft

**7-47.** A warehouse floor is supported by 3- by 14-in. joists of Douglas fir common S4S 16 ft long between centers of end supports and spaced 16 in. on centers. Calculate the allowable uniformly distributed load $w$ in pounds per square foot of floor area if weight of joists is disregarded. (See Table 8-1 and Appendix E-2.)

**7-48.** If the joists in Prob. 7-47 are increased to 4 by 16 in. (S4S) in size and are spaced to 2 ft on centers, what maximum uniformly distributed load $w$ can the floor safely support if weight of joist is disregarded?

## SUMMARY
### (By article number)

**7-1.** In a bent beam, tensile stresses are produced on the convex side, where the fibers have been stretched, and compressive stresses are produced on the concave side, where the fibers have been compressed. The plane at which the bending stresses are zero, and also change from tensile to compressive, is called the **neutral plane.**

These tensile and compressive stresses increase uniformly from zero at the neutral plane to a maximum at the extreme fiber. The intersection of the neutral plane and any cross-sectional plane of a beam is called the **neutral axis.**

**7-2.** The relationship among bending moment $M$, fiber stress $s$, and moment of inertia $I$ at any cross section of a beam is expressed by the **flexure formula,** used in all beam design:

**Flexure formula:**         $M = s\dfrac{I}{c}$     or     $s = \dfrac{Mc}{I}$         (7-1)

in which $c$ is the distance from the neutral axis to the extreme fiber.

The neutral axis of the cross-sectional area of a beam always passes through the centroid of the area, and it is therefore readily located.

**7-3.** The quantity $I/c$ is called the **section modulus** and is denoted by the symbol $Z$. The flexure formula then becomes

**Flexure formula:**         $M = sZ$     or     $s = \dfrac{M}{Z}$         (7-3)

**7-4.** Shearing stresses, horizontal and vertical, are induced in all beams that are subjected to transverse loads.

**7-5.** At any point in any beam subjected to external shearing forces, the unit vertical and horizontal shearing stresses are equal.

**7-6.** The shear flow $S$ that accumulates in the distance $\Delta x$ can be found from

$$\frac{S}{\Delta x} = \frac{QV}{I}$$         (7-5)

The **general shear formula** gives the unit shearing stress at any point in any beam. It is

**General shear formula:**
$$s_s = \frac{QV}{Ib} \qquad (7\text{-}6)$$

**7-7.** Shearing stresses in beams vary from zero at the top of the beam to a maximum, which generally but not always occurs at the neutral axis, and then to zero again at the bottom.

**7-8.** **Two special shear formulas** which will give the *maximum unit shearing stresses at the neutral plane* of a rectangular beam and of a circular beam are

**Rectangular-area shear formula:**
$$s_s = \frac{3V}{2A} \qquad (7\text{-}7)$$

**Circular-area shear formula:**
$$s_s = \frac{4V}{3A} \qquad (7\text{-}8)$$

In beams with sections such as standard types of channels, I-beams, and wide-flange sections, approximately 85 to 90 percent of the total transverse shear is resisted by the web.

**Web-shear formula:**
$$s_s = \frac{V}{th} \qquad (7\text{-}9)$$

Assuming that the web resists all the shear, the web-shear formula will give an average web-shearing stress which seldom is more than 10 to 15 percent low.

### REVIEW PROBLEMS

**7-49.** A round, tapering, straight wood pole, used as a radio aerial mast, extends 27 ft vertically upward from a concrete base into which it is solidly imbedded. Its diameter at the base is 6 in. and at the top is 3 in. If the horizontal pull $P$ at the top of the mast is 100 lb, (*a*) what maximum bending stress is produced in the pole and (*b*) what is the maximum shearing stress? (*c*) Why is the shearing stress so low?

**7-50.** A 1-in. round steel reinforcing bar is 20 ft long and weighs 2.67 lb per ft. (*a*) What maximum bending stress is produced in this bar by its own weight if a man holds it horizontally, grasping the bar with one hand at its mid-point, and (*b*) what is the maximum shearing stress? (*c*) What are these same stresses if the bar is held horizontally by two men, one at each end? (*d*) Why is the shearing stress so low?        *Ans.* (*a*) $s = 16{,}300$ psi; (*b*) $s_s = 44.9$ psi; (*c*) same

**7-51.** Calculate the maximum unit bending stress in a simple timber beam 6 by 12 in. S4S caused by its own weight of 17.5 lb per ft when its length between end supports is (*a*) 10 ft, (*b*) 20 ft, and (*c*) 40 ft. If the allowable bending stress in this timber is 1,200 psi, what percentage of the capacity of the beam is nullified by its own weight for each given length?

*Ans.* (*a*) $s = 21.7$ psi $= 1.81$ percent; (*b*) $s = 86.7$ psi $= 7.22$ percent; (*c*) $s = 347$ psi $= 28.9$ percent

**7-52.** Calculate the maximum unit horizontal shearing stress in the timber box beam shown in Fig. 7-21 at the junction between the top plank and the side planks caused by a transverse shear $V$ of 4,000 lb ($I_x = 1,786$ in.⁴).

**7-53.** Assume that the timber T-beam mentioned in Prob. 5-11 is used as a simple beam 20 ft long and supports a uniform load of 300 lb per ft including its own weight. Calculate (a) the maximum bending stress in this beam and (b) the maximum shearing stress.

**7-54.** A structural-steel wide-flange section 16 by 7 in. by 40 lb per ft (W 16 × 40) is subjected to a maximum transverse shear $V$ of 15,000 lb. Its flange is 7 in. wide by 0.503 in. thick. Its web is 0.307 in. thick. Calculate (a) the average web-shearing stress; (b) the precise maximum shearing stress; (c) the percentage by which the average stress is lower than the maximum.

**7-55.** Let the beam of Prob. 7-21 be a W 12 × 36 and the cover plates be ¾ by 8 in. Solve for the maximum rivet spacing at support $D$.

**7-56.** A 3½-in. standard steel pipe extends 10 ft horizontally out of a solid concrete pier. It supports a vertical load of 400 lb at its extreme outer end. Calculate (a) the maximum bending stress in this beam and (b) the maximum shearing stress. Disregard weight of pipe.          *Ans.* (a) $s = 20,050$ psi; (b) $s_s = 298$ psi

**7-57.** The beam whose cross section is shown in Fig. 5-18 and is further described in Prob. 5-13 is built up of three 2- by 12-in. wood planks securely spiked together. If the maximum allowable shearing stress in this wood is 120 psi, calculate the greatest value of transverse shear $V$ to which this beam may safely be subjected.

**7-58.** A beam of cross section similar to that shown in Fig. 7-24b is subjected to a total transverse shear $V$ of 1,600 lb. Assume that the vertical web is 2 in. wide and 10 in. deep, the horizontal web is 8 in. wide and 4 in. deep, and the section is symmetrical. Calculate the unit shearing stress (a) at the horizontal neutral axis and (b) at the horizontal plane where the two webs join. (NOTE: In this case the maximum shearing stress does not occur at the neutral axis.)
          *Ans.* (a) $s_s = 37.2$ psi; (b) $s_s = 84.6$ psi

**7-59.** The height $h$ in Fig. 7-24b is 12 in., and the width at the neutral axis is 10 in. All thicknesses are 2 in. If the section is subjected to a vertical shear force $V$ of 4 kips, calculate the maximum shear stress $s_s$ and compare with the shear stress at the neutral axis.     *Ans.* Max $s_s = 238.7$ psi; 438 percent of $s_s$ at neutral axis

**7-60.** Assume that the bolt in Fig. 3-5 is 1 in. in diameter, that the center plate is ½ in. thick, each side plate is ¼ in. thick, and the resultant of the bearing stresses against each plate acts along its center line. Calculate the maximum bending and shearing stresses in this bolt due to a pull $P$ of 12,000 lb.
          *Ans.* $s = 2,870$ psi; $s_s = 10,180$ psi

**7-61.** Assume that the beam whose cross section is shown in Fig. 5-18 is built up of three 2- by 12-in. rough wood planks securely spiked together and that it is supported at its outer ends. The maximum allowable stresses in this wood are 1,200 psi in bending and 120 psi in shear. The length of the beam is 20 ft. What maximum concentrated load $P$ may it safely support at its center (a) based on bending and (b) based on shear ($I_x = 2,656$ in.⁴)?

**7-62.** The wood joists supporting the floor in a small building are of Douglas fir No. 1 common 2 by 12 in. S4S 18 ft long between end supports and spaced 16 in. on centers. If the floor itself, without joists, weighs 12 psf, what additional load $w$ in pounds per square foot may safely be imposed on it?     *Ans.* $w = 48$ psf

## REVIEW QUESTIONS

**7-1.** Name the types of internal stresses to which a transversely loaded beam is subjected.

**7-2.** What is meant by the term "pure bending"? What stresses are induced in a beam subjected only to pure bending?

**7-3.** Describe in general the deformations and stresses in a beam subjected to pure bending.

**7-4.** Define the terms "neutral plane" and "neutral axis."

**7-5.** The flexure formula defines the relationship between which three quantities?

**7-6.** By what other name is the moment of inertia of an area also known?

**7-7.** Give a practical definition of the term "moment of inertia of an area."

**7-8.** Through what point of an area does the neutral axis of bending always pass?

**7-9.** How is the section modulus of an area with respect to its neutral axis of bending most easily obtained?

**7-10.** When the neutral axis of an unsymmetrical beam section lies nearer the top than the bottom, in which of the two section moduli is a designer generally interested?

**7-11.** Describe a demonstration by which the existence of horizontal shearing stresses in beams may easily be visualized.

**7-12.** What relationship exists between the unit vertical and horizontal shearing stresses at any point in a beam?

**7-13.** In the general shear formula $s_s = QV/Ib$, what are the quantities $Q$, $V$, $I$, and $b$?

**7-14.** Describe the variation of shearing stresses from top to bottom at any cross section of a rectangular beam.

**7-15.** What causes the sudden change in intensity of shearing stress at the junction of the flange and the web of an I-beam?

**7-16.** Name several beam sections in which the shearing stress is maximum at the neutral axis and some section in which it is not.

**7-17.** Name the three special shear formulas and describe the application of each.

**7-18.** By what percentage is the maximum shearing stress greater than the average (a) in a rectangular beam and (b) in a circular beam?

**7-19.** In channels, I-beams, and wide-flange sections, by what approximate percentage is the average web-shearing stress lower than the precise maximum?

# CHAPTER 8

# Design of Beams

## 8-1 Considerations in Beam Design

Beams must be so designed that they can safely withstand the maximum stresses to which they are subjected by the imposed loads, both in bending and in shear. The design formulas are the flexure formula $s = Mc/I$ or its section-modulus form $s = M/Z$ and the shear formula $s_s = QV/Ib$ or the appropriate one of its special simplified forms.

Since, in the design of a beam, the bending moment $M$ and allowable stress $s$ are known, its size and shape are the unknown quantities to be determined. Because size and shape are functions of $I/c$, or $Z$, the design forms of the flexure formula become $I/c = M/s$ and $Z = M/s$. The latter of these is generally used.

To accomplish in a single mathematical operation the design of a beam which would meet the strength requirements both in bending and in shear is a laborious process. Instead, the designer usually estimates which of the two stresses will be the critical one, and then proceeds with the design to satisfy that requirement. The chosen beam is thereafter investigated to determine the maximum value of the other stress. If the latter does not exceed the allowable, the design is complete. If it does, a stronger beam must be selected.

In general, the bending stresses are usually critical in "relatively long" beams, while shearing stresses are critical in "relatively short" beams. For example, of four beams all 16 ft in length, a 6- by 8-in. timber beam and a 7-in. steel I-beam are relatively long, while a 10- by 20-in. timber beam and a 20-in. I-beam are relatively short. The usual practice, however, is to design beams for bending and then to investigate them for maximum shearing stresses. In some cases, the deflection of a beam may be the governing factor. This problem is discussed in Chap. 9.

The **weight of a beam** must be considered in its design. That is, the beam finally selected must be capable of supporting its own weight in addition to the imposed load. In some cases, however, the weight of a beam may be disregarded with insignificant error. In the special case of simple-span timber beams carrying uniformly distributed loads, an estimated weight of beam of 5 percent of the imposed load may be added for design purposes, as illustrated in Prob. 8-1. If the actual weight of the beam selected differs from the estimated weight by more than 50 percent, a revision of the design may be necessary. For all other beams, timber and steel, it is advisable first to select a trial beam based on the moment $M$ and required section modulus $Z$ due to load only and then to calculate the additional moment and additional required section modulus due to the weight of the trial beam. A final selection is then based on the total required section modulus, as will be illustrated in Probs. 8-2, 8-3, 8-22, and 8-23. While not true in all cases, the critical section due to both imposed load and weight of beam may be considered to coincide with the critical section due to load only. Hence the additional moment $M$ due to weight of beam should be that at the section of critical load moment.

The **most economical beam** is the one which has the least cross-sectional area (or the least weight per foot of length) and still satisfies all strength requirements and other specified limitations. In general, it will be found that a deep beam is more economical than a shallower beam of equal strength. Unless otherwise specified, it is taken for granted in general practice that the beam selected shall be the most economical (lightest) one. This rule shall apply also to all problems in this text.

**Lateral Support of Beams.** All beams under load have a tendency to buckle sidewise (laterally) between supports, especially beams that are relatively narrow and deep. In the vast majority of cases, however, such buckling is prevented by furnishing lateral support for the beams. In buildings of the usual type, for example, all floor beams are laterally supported, since the floors they support are fastened to them. Laterally unsupported beams such as crane rails and others are occasionally found in industrial buildings and yards. Such beams require special treatment in their design as illustrated in Art. 8-4. Unless otherwise stated, all beams are presumed to be laterally supported.

## 8-2 Design of Timber Beams

Because timber is not a manufactured material, like steel and other metals, wide variances in its strength under different conditions of stress are to be expected. Timber has its greatest usable strength in tension and in bending.

In compression parallel to the grain, as in columns, its estimated strength is about 75 percent of that in bending, while in bearing (compression perpendicular to the grain) and in horizontal shear its strength is approximately 20 and 5 percent, respectively, of that in bending. Consequently, shearing stresses in timber beams must be closely watched.

The **span of a timber beam** is considered to be from face to face of girder if the beam is supported on steel hangers fastened to the sides of the supporting girders, or from center to center of girders if the beams rest on top of the supporting girders. Floor girders generally rest on steel column caps especially made for that purpose, and girder spans are then considered to extend from face to face of supporting columns. In general, the span of a simple beam is considered to be the distance between the centers of its two supports. Correct determination of span is important, since the bending moment and hence the required section modulus vary as the square of the span ($M = wL^2/8$). For example, for a span of 20 ft the bending moment is $400w/8$, while for a 21-ft span it is $441w/8$, a difference of 10 percent.

The actual dimensions of rough timbers are presumed to equal their nominal dimensions. Actual dimensions and other properties of dressed timbers (S4S) are given in Table E-2. Three design methods are illustrated below. Methods A and B are basic in concept, while method C makes use of available tabulated information. While experienced designers use such tables extensively, beginners should not use them until they have first mastered the technique of beam design using only fundamental concepts and formulas.

The **depth of a timber beam** is most often about twice its width. In relation to span, a workable rule of thumb is that the depth should be about 1 in. per ft of span (20 in. for a 20-ft span) for uniformly loaded beams and about 1.25 in. for beams and girders subjected to concentrated loads. Narrow, deep timber beams are more economical than are wide, shallow beams, but they are often weak in shear. In general structural practice, however, the width-depth ratio ($b/h$) of structural timber beams is usually not less than $\frac{1}{3}$ nor more than $\frac{3}{4}$. Unless otherwise specified, these limits should be observed in the following timber-beam-design problems. Obviously, the limits apply only to structural beams and not to joists or other "special" beams.

It is important to realize here that, in any beam-design problem, the maximum shear $V$ and moment $M$ must always be determined in the usual and most suitable manner. Thereafter, the required section modulus ($Z = M/s$) and cross-sectional area ($A = 1.5V/s_s$) are calculated. From this point on, several methods are available for completing the design, as illustrated in Probs. 8-1 to 8-3.

**Design Method A** solves the equation $Z = bh^2/6$ (or $6Z = bh^2$) by the substitution of different values of $b$ (of *available* timber sizes) and uses for convenience a tabulation of the results. This method is especially useful to the beginner in that it shows that several beam sizes are available which satisfy the requirements of strength but that only a few of these are suitable and that other logical restrictions generally reduce the choice to a single beam.

**Design Method B,** illustrated in Prob. 8-2, makes use of a predetermined width-depth ratio $(b/h)$, usually $\frac{1}{3}$, $\frac{1}{2}$, $\frac{2}{3}$, or $\frac{3}{4}$, whichever best satisfies the requirements as they are known to the designer. This predetermination serves two functions: (1) it represents the designer's estimate of the most desirable ratio for the particular beam being designed and (2) it permits solution of any otherwise insoluble equation containing two unknowns, $b$ and $h$. Any ratio so predetermined and used may, however, in the beam finally selected, vary by one size in either dimension, $b$ or $h$. It must be realized that this predetermination of a ratio of $b$ and $h$ is entirely up to the designer and cannot

**Table 8-1 ALLOWABLE UNIT STRESSES FOR STRESS-GRADE LUMBER, PSI[a]**

| Species and commercial grade | Extreme fiber in bending[b] | Tension | Longi-tudinal shear | Compression[c] | | Modulus of elasticity |
|---|---|---|---|---|---|---|
| | | | | Perpendicular to grain | Parallel to grain | |
| Cypress, southern: | | | | | | |
| select | 1,700 | . . . | 145 | 360 | 1,425 | 1,132,000 |
| common | 1,300 | . . . | 120 | 360 | 1,125 | |
| Douglas fir, coast region: | | | | | | |
| dense select structural | 1,900 | 1,100 | 85 | 455 | 1,300 | |
| select structural | 1,600 | 950 | 85 | 385 | 1,100 | 1,750,000 |
| dense No. 1 | 1,550 | 775 | 85 | 455 | 1,100 | |
| No. 1 | 1,300 | 675 | 85 | 385 | 925 | |
| Hemlock, mountain: | | | | | | |
| select structural | 1,350 | 900 | 90 | 370 | 875 | 1,540,000 |
| No. 1 | 1,100 | 750 | 90 | 370 | 750 | |
| Oak, white and red: | | | | | | |
| select | 1,900 | . . . | 145 | 600 | 1,375 | 1,650,000 |
| common | 1,450 | . . . | 120 | 600 | 1,050 | |
| Pine, southern: | | | | | | |
| No. 1 stress rated | 1,300 | 850 | 110 | 270 | 925 | |
| No. 1 dense SR | 1,500 | 1,000 | 110 | 315 | 1,050 | 1,760,000 |
| No. 2 stress rated | 1,100 | 725 | 95 | 270 | 675 | |
| No. 2 SR | 1,300 | 850 | 95 | 315 | 775 | |
| Spruce, eastern: | | | | | | |
| select structural | 1,150 | 775 | 60 | 255 | 850 | 1,320,000 |
| No. 1 | 950 | 600 | 60 | 255 | 725 | |

[a] Some differences exist among building codes in the matter of allowable unit stresses for timber. Those given in this table are for lumber used under conditions continuously dry, such as in most covered structures, and for "normal" load conditions. Allowable stresses for other conditions are given in "National Design Specification," published by the National Forest Product Association, Washington, D.C.

[b] Bending stresses in beams and stringers.

[c] Compression parallel to grain in "short" columns. See column formulas, Chap. 12.

be fixed by any text or code. When used in problems in this text, it represents, of course, only the judgment of the author.

**Design Method C** makes full use of tables of available beam sizes and their properties, such as area, weight, moment of inertia, and section modulus. Where the required values of $Z$ and $A$ have been calculated, a suitable beam is quickly selected from such a table.[1] See Appendix Table E-2.

### ILLUSTRATIVE PROBLEMS

**8-1. Method A.** Design the most economical rectangular timber beam 20 ft long simple span of Douglas fir dense select structural grade S4S to support a uniformly distributed load of 845 lb per ft. Weight of beam must be considered. For practical reasons, the ratio of $b/h$ must not be less than $\frac{1}{3}$ nor more than $\frac{3}{4}$.

Fig. 8-1 Design of timber beam.

*Solution:* While this method of design is not generally used in practice, its purpose here is to illustrate that several beams may be available which satisfy the strength requirements but that only one, of course, will be the most economical. We shall first design this beam for bending based only on the given load plus the estimated weight of the beam and shall later consider its safety in shear.

First we solve for the support reactions $R$ and the maximum bending moment $M$ and transverse shear $V$. Clearly, for the conditions of this problem, $R = wL/2$, $M = wL^2/8$, and $V = R$. The estimated weight of the beam is $(845)(0.05) = 42.2$, say 45 lb per ft. Then, since $w = 845 + 45 = 890$ lb per ft and $L = 20$ ft,

$$[R = \tfrac{1}{2}wL] \qquad\qquad R = (\tfrac{1}{2})(890)(20) = 8{,}900 \text{ lb}$$

$$\left[M = \frac{wL^2}{8}\right] \qquad M = \frac{(890)(20)(20)}{8} = 44{,}500 \text{ lb-ft} \qquad \text{or} \qquad 534{,}000 \text{ lb-in.}$$

From Table 8-1, we find the allowable stress for Douglas fir dense select structural to be 1,900 psi in bending and 85 psi in horizontal shear. Since $s = M/Z$, $Z = M/s$, and the required section modulus then is

$$\left[Z = \frac{M}{s}\right] \qquad\qquad Z = \frac{534{,}000}{1{,}900} = 281 \text{ in.}^3$$

For a rectangular section, $Z = bh^2/6$. Consequently

$$Z = \frac{bh^2}{6} = 281 \qquad \text{or} \qquad bh^2 = (6)(281) = 1{,}686$$

---

[1] Among the many available handbooks, designers will find the following especially useful: "Wood Structural Design Data," published by the National Forest Products Association, and the "Douglas Fir Use Book," published by the West Coast Lumbermen's Association.

### Table 8-2  DESIGN OF TIMBER BEAM

| $b$ | $h^2$ | $h$ | Use S4S | Area |
|-----|-------|-----|---------|------|
| 5.5 | 307 | 17.5 | 6 × 18 | 96.25 |
| 7.5 | 225 | 15.0 | 8 × 16 | 116.25 |
| 9.5 | 177 | 13.3 | 10 × 14 | 128.25 |

Having found that $bh^2 = 1,686$, we may then substitute various values for $b$ and solve for $h^2$ and $h$. Table 8-2 gives a clear view of the results. In the fourth column are listed the nearest available sizes, obtained from Table E-2. The fifth column gives the cross-sectional area of each beam, also from Table E-2.

Since all four of these beams satisfy the strength requirements and the width-depth ratio limitations, we shall select the most economical one whose stress in horizontal shear is within the allowable limit of 85 psi.

Using the special shear equation $s = 3V/2A$, we now may solve for $s$ in each beam. But it will be simpler here to solve for the required area $A$, which then is compared with the already known areas of the three tabulated beams. That is, $A = 3V/2s$. Since $V = 8,900$ lb and $s = 85$ psi,

$$\left[A = \frac{3V}{2s}\right] \qquad \text{Required } A = \frac{(3)(8,900)}{(2)(85)} = 157.2 \text{ sq in.}$$

Obviously, the first three beams would be weak in shear, since they have insufficient area. The 12- by 14-in. beam is the logical choice. (About one percent overstressed in shear.) From Table E-2 its weight is 43.1 lb per ft, 1.9 lb less than its estimated weight of 45 lb per ft. Hence no revision of the design is necessary.

Use 12- by 14-in. S4S Douglas fir beam.      Ans.

**8-2. Method B.** Design the most economical rough timber beam of Douglas fir No. 1 grade 24 ft long to support the load shown in Fig. 8-2a plus its own weight. A ratio of width $b$ to depth $h$ of approximately $\frac{1}{2}$ is considered suitable for this beam.

*Solution:* Because the load is irregular while the weight of the beam is uniformly distributed, we shall tentatively design this beam for the given load only, leaving its weight for later consideration. In this case, a set of load, shear, and moment diagrams will be helpful, although the maximum moment may, of course, be determined by any other suitable method. The maximum shear $V$ is found to be 10 kips, and the maximum moment $M$ is 50 kip-ft at a point 14 ft from the left end.

From Table 8-1, the allowable stresses for Douglas fir No. 1 are 1,300 psi in bending and 85 psi in shear. In the flexure formula, $M$ must always be given in pound-inches. Hence, since $M = 50$ kip-ft, the required section modulus is

$$\left[Z = \frac{M}{s}\right] \qquad \text{Required } Z = \frac{(50)(1,000)(12)}{1,300} = 462 \text{ in.}^3$$

(a)

FIG. 8-2  Design of timber beam. (a) Loaded beam. (b) Load, shear, and moment diagrams.

(b)

Based on a maximum shear $V$ of 10 kips, which must be converted to pounds, the required cross-sectional area is

$$\left[ A = \frac{3V}{2s} \right] \qquad \text{Required } A = \frac{(3)(10)(1,000)}{(2)(85)} = 177 \text{ sq in.}$$

Letting $b = \frac{1}{2}h$, we solve for $h$ and then for $b$. That is,

$$\frac{bh^2}{6} = \frac{h}{2} \cdot \frac{h^2}{6} = \frac{h^3}{12} = 462 \text{ in.}^3$$

or $h^3 = (12)(462) = 5,540$ and $h = 17.7$ in. Hence $b = 8.85$ in., which suggests the use of a 10- by 20-in. rough beam weighing $(10)(20)(12)(40)/1,728 = 56$ lb per ft. Its section modulus is $(10)(20)(20)/6 = 667$ in.$^3$ and its area is $(10)(20) = 200$ sq in.

The additional moment due to the weight of the beam must be evaluated *at the section of maximum moment due to load,* which occurred 14 ft from the left end. When $x = 14$ ft and $L = 24$ ft, this moment (see Prob. 6-21), in a simple beam with a uniformly distributed load, is

$$\left[ M_x = \frac{wx}{2}(L - x) \right] \qquad M_{14} = \frac{(56)(14)}{2}(24 - 14) = 3,920 \text{ lb-ft}$$

The additional required section modulus is

$$\left[ Z = \frac{M}{s} \right] \qquad\qquad Z = \frac{(3,920)(12)}{1,300} = 36.2 \text{ in.}^3$$

and the additional required area is

$$\left[ A = \frac{3V}{2s} \right] \qquad\qquad A = \frac{(3)(56)(12)}{(2)(85)} = 11.9 \text{ sq in.}$$

Hence
$$\text{Total required } Z = 463 + 36 = 499 \text{ in.}^3$$
$$\text{Total required } A = 177 + 11.9 = 188.9 \text{ in.}^2$$

Use 10- by 20-in. beam, Douglas fir rough.  *Ans.*
$(Z = 667 \text{ in.}^3; A = 200 \text{ sq in.})$

Note that shear controls, not bending.

**8-3. Method C.** In Fig. 8-3 is shown a section of a typical timber beam-and-girder floor system, such as is commonly used in mill-construction buildings for light manufacture. All columns are 12- by 12-in. timbers. Beams and girders are of Douglas fir No. 1 grade S4S. The beams rest on steel hangers fastened to the sides of the girders, and beam spans are from face to face of girders. The girders rest on steel column caps, and girder spans are from face to face of columns. Let $a = 17$ ft, $b = 21$ ft, and $c = 7$ ft.

The floor load to be supported is 125 lb per sq ft uniformly distributed, which includes the weight of the subfloor planking and a finish floor covering only. Design one beam ($B1$) and one girder ($G1$).

FIG. 8-3 Design of timber beam-and-girder floor system. Probs. 8-3 and 8-9.

*Solution:* In each case, the design load must be calculated from the given floor load. From Table 8-1, the allowable stress in bending is 1,300 psi and in shear is 85 psi.

**Design of Beam *B*1.** Each beam *B*1 clearly supports all the floor load from face to face of girders and halfway to each adjacent beam. If we assume that each of girders *G*1 is 12 in. wide, the span of the beam is 16 ft. The design load per foot of beam will then be as follows:

$$
\begin{aligned}
\text{Floor load} &= (125)(7) & &= 875 \text{ lb per ft} \\
\text{Est. weight of beam} &= (0.05)(875) &= &\underline{\phantom{0}43} \\
\text{Design load} & & &= 918 \text{ lb per ft}
\end{aligned}
$$

For a uniformly distributed load on a simple beam the maximum moment is

$$\left[ M = \frac{wL^2}{8} \right] \qquad\qquad M = \frac{(918)(16)(16)}{8} = 29{,}400 \text{ lb-ft}$$

Since $s = 1{,}300$ psi, the required section modulus is

$$\left[ Z = \frac{M}{s} \right] \qquad\qquad \text{Required } Z = \frac{(29{,}400)(12)}{1{,}300} = 272 \text{ in.}^3$$

Since $s_s = 85$ psi and $V = (918)(8) = 7{,}344$ lb, the required area is

$$\left[ A = \frac{3V}{2s_s} \right] \qquad\qquad \text{Required } A = \frac{(3)(7{,}344)}{(2)(85)} = 129.6 \text{ sq in.}$$

From Table E-2, we find that an 8- by 18-in. beam S4S weighing 36.4 lb per ft meets these requirements. Since the error in estimating its weight is less than 50 percent, no revision is necessary. Hence

Use 8- by 18-in. beam Douglas fir S4S.      *Ans.*
$(Z = 383 \text{ in.}^3;\ A = 131.3 \text{ sq in.})$

**Design of Girder *G*1.** The load on the girder is made up of three parts: (1) the concentrated loads from the four beam ends, (2) the floor load directly above the girder, which was not considered in the design of beam *B*1, and (3) its own weight. We shall leave item 3, its weight, for later consideration.

FIG. 8-4 Girder loads and reactions.

Each beam transmits half of its load to the girder at the third-points of the 21-ft span. Since two beam ends are supported at each third-point, $P$ equals the total load on one beam. The previously calculated beam load is 875 lb per ft, and the beam weight is 27 lb per ft. Consequently the concentrated load at each third-point is

$$P = (875 + 27)(16) = (902)(16) = 14,400 \text{ lb}$$

FIG. 8-5 Calculation of maximum bending moment.

The girders having been previously assumed to be 12 in. wide, the floor load still unaccounted for and carried directly by the girder is 125 lb per ft. Then, to calculate the maximum bending moment and the required section modulus and area, we have

Girder end reaction:

$$R = 14,400 + 1,250 = 15,650 \text{ lb}$$

Maximum bending moment, assumed positive:

$$M + (14,400)(3.5) + (1,250)(5) = (15,650)(10)$$

or

$$M = 156,500 - 50,500 - 6,250 = 99,750 \text{ lb-ft}$$

Required section modulus:

$$\left[ Z = \frac{M}{s} \right] \qquad Z = \frac{(99,750)(12)}{1,300} = 920 \text{ in.}^3$$

Required cross-sectional area for $V = 15,650$ lb:

$$\left[ A = \frac{3V}{2s_s} \right] \qquad A = \frac{(3)(15,650)}{(2)(85)} = 277 \text{ sq in.}$$

From Table E-2, we find that a 14- by 22-in. beam S4S most nearly satisfies these requirements, since its section modulus is 1,040 in.³ and its area is 290 sq in. We shall need to determine now the additional required section modulus and area due to its own weight of 80.6 lb per ft. That is,

$$\left[ M = \frac{wL^2}{8} \right] \qquad \text{Additional } M = \frac{(80.6)(20)(20)}{8} = 4,030 \text{ lb-ft}$$

$$\left[ Z = \frac{M}{s} \right] \qquad \text{Additional required } Z = \frac{(4,030)(12)}{1,300} = 37.2 \text{ in.}^3$$

$$\left[A = \frac{3V}{2s}\right]$$  Additional required $A = \dfrac{(3)(80.6)(10)}{(2)(85)} = 14.2$ sq in.

Total required $Z = 920 + 37 = 957$ in.$^3$

Total required $A = 277 + 14.2 = 291.2$ in.$^2$

Use 14- by 22-in. girder, Douglas fir S4S.  *Ans.*

($Z = 1{,}040$ in.$^3$; $A = 290$ sq in.; less than $\frac{1}{2}$ percent overstressed in shear.)

## PROBLEMS

NOTE: Unless otherwise specifically stated, (*a*) the allowable stresses given in Table 8-1 shall be used, (*b*) the weight of the beam shall be considered (at 40 pcf), (*c*) the width-depth ratio shall not be less than $\frac{1}{3}$ nor more than $\frac{3}{4}$ (joists and "special" beams excepted), (*d*) the most economical (lightest) beam shall be selected, and (*e*) a maximum overstress of 2 percent is permitted.

**8-4.** Design a timber beam of Douglas fir dense No. 1 S4S with a simple span of 15 ft to support a uniformly distributed load of 500 lb per ft plus its own weight. Use Design Method A as illustrated in Prob. 8-1.  *Ans.* 6 by 14 in. S4S

**8-5.** A timber beam of southern pine No. 2 stress rated rough with a simple span of 20 ft must support a uniformly distributed load of 1,600 lb per ft plus its own weight. Design this beam using Method A as illustrated in Prob. 8-1.

**8-6.** A timber beam of Douglas fir No. 1 rough in a warehouse floor system spans 18 ft between centers of supports and must carry an estimated uniform load of 670 lb per ft. A width-depth ratio of $\frac{1}{2}$ is considered desirable. Design this beam using Method B, as illustrated in Prob. 8-2.  *Ans.* 8 by 14

**8-7.** Design a beam of mountain hemlock No. 1 S4S to support the load shown in Fig. 8-6. Let $b = \frac{1}{2}h$. Use Method B.

FIG. 8-6 Prob. 8-7.                    FIG. 8-7  Prob. 8-8.

**8-8.** The beam shown in Fig. 8-7 is to be designed of select white oak S4S. Use Method C as illustrated in Prob. 8-3. Note here that the additional required section modulus and area due to weight of beam may easily be obtained by simple proportion.

**8-9.** In Fig. 8-3, let $a = 21$ ft, $b = 24$ ft, and $c = 6$ ft. Note that when $c = 6$ ft, three beams frame into the sides of the interior girders $G1$, instead of two as shown. All columns are assumed to be 12 by 12 in. The floor load is 125 psf, which includes weight of subfloor planking and top finish floor. Design one beam $B1$ and one girder

G1. Let width $b$ of girder be 12 in. Use Method C, as illustrated in Prob. 8-3. Use S4S timber with $s = 1,700$ psi and $s_s = 145$ psi.

**8-10.** A section of a common type of floor system is shown in Fig. 8-8. The joist span is 15 ft between centers of end supports, and the joist spacing is 16 in. on centers. What size of 2-in. joist, southern pine No. 2 SR S4S, should be used to support a uniform floor load of 60 psf, which includes the estimated weight of the subfloor and top floor?                                    *Ans.* 2 by 12 in. S4S

FIG. 8-8 Probs. 8-10 and 8-11.

**8-11.** The joists supporting a warehouse floor are to be of S4S southern pine No. 1 SR. The joist span is 20 ft and spacing is 2 ft on centers. The maximum practical depth of joist is 16 in. The floor must support a live load of 200 psf plus 17.5 psf for the dead weight of the heavy plank subfloor and the top floor. Design the most economical joist. See Fig. 8-8.

**8-12.** A builder requires 2-in. floor joists to span 15 ft. Only 2- by 10-in. S4S red oak common joists are available. Calculations show that when spaced the normal 16 in. on centers, the maximum bending stress in them is 1,655 psi. (*a*) If a 16-in. spacing is used, by what percentage are the joists overstressed? (*b*) What spacing should be used to bring this maximum stress down to the allowable?

*Ans.* (*a*) 14 percent; (*b*) 14 in.

**8-13.** The springboard shown in Fig. 8-9 is of softwood with $s = 2,400$ psi and must be 14 in. wide. Calculate its minimum required thickness $h$ based on bending stress and load only. Then consider weight at 36 pcf and recalculate $h$. Round off to nearest higher multiple of $\frac{1}{4}$ in.

FIG. 8-9 Prob. 8-13.                              FIG. 8-10 Prob. 8-14.

**8-14.** A pedestrian walk along the side of a timber bridge is supported by timber beams cantilevered out 4 ft, as illustrated in Fig. 8-10. Design one beam of dense structural Douglas fir S4S.

**8-15.** A round cedar pole used as a radio mast extends vertically 36 ft out of a concrete footing in which it is solidly imbedded. The allowable stress in bending is 1,200 psi and in shear is 100 psi. Calculate the exact minimum required diameter $D$ of this pole at its base and at its mid-point to withstand a horizontal force of 60 lb at the top. Disregard weight of pole.

**8-16.** The temporary earth-retaining wall illustrated in Fig. 8-11 consists of planks driven vertically into the ground. The soft, muddy earth creates a triangular load against the planks with a maximum rate of load at the bottom of $w = 240$ lb per ft against a 12-in.-wide plank. Disregard weight of planks. If allowable stresses are 1,000 psi in bending and 100 psi in shear, what minimum thickness $h$ of planks can be used?                     *Ans. $h = 1.47$ in.*

FIG. 8-11 Prob. 8-16.        FIG. 8-12 Prob. 8-17.        FIG. 8-13 Prob. 8-18.

**8-17.** A semipermanent earth-retaining wall uses horizontal planks supported by vertical timber cantilever posts spaced 5 ft on centers, as illustrated in Fig. 8-12. The horizontal pressure of the earth produces a triangular load against each post which reaches a maximum $w$ of 800 lb per ft at the bottom. Design a square post of Douglas fir using $s = 1,450$ psi and $s_s = 120$ psi.

**8-18.** A timber flume for carrying water is constructed as illustrated in Fig. 8-13. The water pressure produces a triangular hydrostatic load against each vertical post, reaching a maximum $w$ of 1,500 lb per ft at the bottom. Each post is supported at the bottom as shown and at the top by a steel rod extending through the opposite post. Design the most economical rough timber post. Allowable stress in bending is 750 psi and in shear is 100 psi.                     *Ans. 6 by 8 in. rough*

**8-19.** A rustic footbridge in a park is to span 32 ft between centers of end supports. Two logs of approximately constant diameter are to be used. Each log must be capable of supporting a uniform load of 400 lb per ft plus its own weight. If the allowable stresses are 1,200 psi in bending and 100 psi in shear, what diameter logs (to nearest inch) are required?

**8-20.** A contractor has to provide shoring in a trench excavation which will withstand the pressure of a wet earth mixture. Vertical planks are held in place by wales (beams) and struts as illustrated in Fig. 8-14. The horizontal load against the planks reaches a maximum $w$ of 720 lb per ft against the bottom of a 12-in. plank. If the allowable stresses are 1,200 psi in bending and 100 psi in shear, what minimum thickness $h$ of planks can be used?

**8-21.** In order to design the horizontal wales in the trench shoring shown in Fig. 8-14, the reacting forces in struts $A$ and $B$, spaced 6 ft on centers along the trench,

FIG. 8-14  Probs. 8-20 and 8-21.

must be calculated. Then the total load, uniformly distributed, on a 6-ft section of the upper wale equals $R_A$ and on the lower wale equals $R_B$. Although the wales are continuous beams, assume that maximum $M = wL^2/8$. Using square rough common timbers, design both wales. Refer to Prob. 8-20 for load against vertical planks. Use $s = 1,450$ psi and $s_s = 120$ psi.

*Ans.* Upper, 6 by 6 in.; lower, 10 by 10 in.

## 8-3 Design of Steel Beams

A complete treatment of the design of steel beams is rather extensive, since it includes the design of built-up beams and of plate girders, which is clearly beyond the scope of this text. This article, therefore, is confined to the design or, rather, "selection" of regular rolled structural-steel beam sections and other beam sections of simple geometric form.

Some codes allow slightly higher bending stresses for *compact* sections than those permitted for members with thin compression flanges and webs. Most rolled sections qualify as compact sections. Beams that do not meet this requirement are noted in the steel handbooks.

ASTM A7 steel and ASTM A373, a steel of superior weldability, usually are permitted a bending stress of 20,000 psi for noncompact sections and 22,000 psi for compact sections. ASTM A7 steel was for many years the most suitable carbon steel available for construction of both buildings and bridges. ASTM A36, an all-purpose steel of greater strength and improved weldability, is now generally available. For this steel the allowable bending stresses are usually 22,000 psi for noncompact sections and 24,000 psi for compact sections. Higher-strength steels are available for special applications.

The **span of a steel beam** must be carefully determined. In the beam-and-girder floor system shown in Fig. 8-15, for example, the beams $B1$ and $B2$ extend between webs of girders or webs of columns, as the case may be, and their spans are considered to equal the dimension $a$ between centers of columns. The girder spans are from flange to flange of the columns. When so arranged, the beams carry all of the distributed floor load. The girders

FIG. 8-15 Design of steel beam-and-girder floor system.

support only the concentrated loads from the beam ends plus their own weight. No floor load comes directly on the girder.

The **depth of steel floor beams,** according to a common specification, should be not less than $\frac{1}{24}$ of the span ($\frac{1}{30}$ for roof beams), which is equal to $\frac{1}{2}$ in. of depth per foot of span. In all cases where beams and girders support plastered ceilings, they must be so proportioned that the deflection will not exceed $\frac{1}{360}$ of the span (1 in. in 30 ft).

## ILLUSTRATIVE PROBLEMS

**8-22.** The roof of a one-story retail-store building 30 by 80 ft is to be supported by wide-flange steel beams spanning the 30 ft and spaced 16 ft on centers longitudinally. The loads, all uniformly distributed, are (*a*) live load (snow, etc.), 25 psf; (*b*) estimated dead weight of roof structure exclusive of beam, 15 psf; (*c*) estimated dead weight of plastered ceiling, 10 psf. Select a suitable W beam. Allowable stresses: bending, 22,000 psi; shear, 14,500 psi. See Table E-1.

*Solution:* Since the beams are spaced 16 ft on centers, the design load per foot of beam is

$$\begin{aligned}
\text{Live load} &= (25)(16) &&= 400 \text{ lb per ft} \\
\text{Dead load} &= (15 + 10)(16) &&= \underline{400} \\
\text{Beam design load} & &&= 800 \text{ lb per ft}
\end{aligned}$$

Since this is a simple span, the maximum moment is

$$\left[ M = \frac{wL^2}{8} \right] \qquad M = (0.125)(800)(30)(30) = 90,000 \text{ lb-ft}$$

and the required section modulus is

$$\left[ Z = \frac{M}{s} \right] \qquad \text{Required } Z = \frac{(90,000)(12)}{22,000} = 49.2 \text{ in.}^3$$

Try a W 16 × 36 (Z = 56.3 in.³).

The moment and the additional required section modulus due to the weight of 36 lb per ft of this trial beam are

$$\left[ M = \frac{wL^2}{8} \right] \qquad M = (0.125)(36)(30)(30) = 4,050 \text{ lb-ft}$$

$$\left[ Z = \frac{M}{s} \right] \qquad \text{Additional required } Z = \frac{(4,050)(12)}{22,000} = 2.21 \text{ in.}^3$$

The total required section modulus is then

$$\text{Total required } Z = 49.2 + 2.21 = 51.41 \text{ in.}^3$$

Several other wide-flange sections will carry the load. Trial will show, however, that no lighter wide-flange section will do.

The maximum average web-shearing stress is

$$\left[ s_s = \frac{V}{th} \right] \qquad s_s = \frac{(836)(15)}{(0.299)(15.85)} = 2,650 \text{ psi} \qquad \text{OK}$$

Use W 16 × 36 steel beam (Z = 56.3 in.³).     *Ans.*

**8-23.** In Fig. 8-15 is shown a typical structural-steel beam-and-girder floor system. The floor is to be designed for a live load of 100 psf. The estimated dead weight of top and subfloor is 15 psf. The columns are wide-flange sections, W 12 × 65. Design one beam B1 and one girder G1, when a = 20 ft, b = 28 ft, and c = 7 ft.

*Solution:* Since all beams and girders are laterally supported by the subfloor, which fastens to the beams through nailing strips bolted to their top flanges, the allowable bending stress is 22,000 psi. Allowable shearing stress is 14,500 psi.

**Design of Beam B1:**

$$\text{Live load} = (100)(7) = 700 \text{ lb per ft}$$
$$\text{Dead load} = (15)(7) = \underline{105}$$
$$\text{Design load} = 805 \text{ lb per ft}$$

$$\left[ M = \frac{wL^2}{8} \right] \qquad M = (0.125)(805)(20)(20) = 40,250 \text{ lb-ft}$$

$$\left[ Z = \frac{M}{s} \right] \qquad \text{Required } Z = \frac{(40,250)(12)}{22,000} = 21.95 \text{ in.}^3$$

Try a W 10 × 25; Z = 26.4 in.³.

$$\left[ M = \frac{wL^2}{8} \right] \qquad \text{Additional } M = (0.125)(25)(20)(20) = 1,250 \text{ lb-ft}$$

$$\left[ Z = \frac{M}{s} \right] \qquad \text{Additional required } Z = \frac{(1,250)(12)}{22,000} = 0.69 \text{ in.}^3$$

$$\text{Total required } Z = 21.95 + 0.69 = 22.64 \text{ in.}^3 \qquad \text{OK}$$

The maximum average web-shearing stress must now be calculated. That is,

$$\left[ s_s = \frac{V}{th} \right] \qquad s_s = \frac{(805 + 25)(10)}{(0.252)(10.08)} = 3,270 \text{ psi} \qquad \text{OK}$$

Use W $10 \times 25$ ($Z = 26.4$ in.$^3$).  *Ans.*

**Design of Girder G1.** Since the columns are 12 in. deep and $b = 28$ ft, the girder span is 27 ft with 7-ft spacing of beams. Each concentrated load equals the load on one beam. That is,

$$P = (830)(20) = 16,600 \text{ lb}$$

and each reaction is one-half of the total load, or        *HOW ?*

$$R = (0.5)(3)(16,600) = 24,900 \text{ lb}$$

Calculating the maximum moment $M$, we obtain

$$M + (16,600)(7) = (24,900)(13.5)$$

or

$$M = 336,150 - 116,200 = 219,950 \text{ lb-ft}$$

The required section modulus then is

$$\left[ Z = \frac{M}{s} \right] \qquad \text{Required } Z = \frac{(219,950)(12)}{22,000} = 120 \text{ in.}^3$$

Try W $21 \times 62$ ($Z = 126.4$ in.$^3$).
The additional moment due to the weight of the beam is

$$\left[ N = \frac{wL^2}{8} \right] \qquad \text{Additional } M = (0.125)(62)(27)(27) = 5,650 \text{ lb-ft}$$

and the additional required section modulus is

$$\text{Additional } Z = \frac{(5,650)(12)}{22,000} = 3.08 \text{ in.}^3$$

The total required section modulus is

$$Z = 120 + 3.08 = 123.08 \text{ in.}^3 \qquad \text{OK}$$

$$\left[ s_s = \frac{V}{th} \right] \qquad \text{Maximum average web } s_s = \frac{24,900 + (62)(13.5)}{(0.40)(20.99)} = 3,063 \text{ psi} \qquad \text{OK}$$

Use W $21 \times 62$ steel beam ($Z = 126.4$ in.$^3$).  *Ans.*

FIG. 8-16 Design of steel girder. (*a*) Loaded girder. (*b*) Maximum moment.

## PROBLEMS

NOTE: Unless otherwise stated, (a) all beams are presumed to be supported against lateral buckling, (b) the allowable stresses are 22,000 psi in bending and 14,500 psi in shear, (c) the weight of the beam shall be considered, (d) the most economical beam shall be selected, and (e) a maximum overstress of 2 percent is permitted.

**8-24.** Assume in Prob. 8-22 that the building is 40 by 100 ft, the beam span is 40 ft, the spacing of beams is 20 ft on centers, and the loads per square foot are the same. Then (a) select a suitable steel wide-flange beam. (b) Calculate the average web-shearing stress.

**8-25.** (a) Select a structural-steel wide-flange beam to carry the load shown in Fig. 6-5 (see and use maximum V and M given with Prob. 6-42). (b) Calculate the average web-shearing stress.      Ans. (a) W 16 × 36; (b) $s_s$ = 5,120 psi

**8-26.** A structural-steel beam is to be selected to carry the load shown in Fig. 6-25. Assume that the load includes the weight of the beam. Select (a) the most economical American Standard I-beam and (b) the shallowest available wide-flange section. In each case, calculate the average web-shearing stress.

**8-27.** A steel beam is to be selected to carry the load shown in Fig. 6-17. Assume that load includes weight of beam. Select (a) the lightest wide-flange section; (b) the lightest American Standard I-beam; (c) the shallowest available beam. In each case, calculate the average web-shearing stress.

Ans. (a) W 12 × 27, $s_s$ = 8,360 psi; (b) S 12 × 31.8, $s_s$ = 5,720 psi;
(c) W 8 × 35, $s_s$ = 9,490 psi

**8-28.** In Fig. 8-15 is illustrated a typical structural-steel beam-and-girder floor system. A similar system for the same building supports the roof, which must be designed for a live load of 40 psf. The estimated dead weight of roof planking and four-ply felt tarred and graveled is 20 psf. (a) Design one beam B1 and one girder G1. (b) Calculate the average web-shearing stresses.

**8-29.** Assume that beam B3 in Fig. 8-15 supports only its share of the uniformly distributed live floor load of 85 psf plus an estimated dead weight of 35 psf for the floor. Let a = 26 ft, b = 40 ft, c = 10 ft, d = 6 ft, and e = 20 ft. (a) Select a suitable wide-flange beam. (b) Calculate the average web-shearing stress.

**8-30.** Beam B4 in Fig. 8-15 supports only the end-reaction loads from the two beams B3. The uniformly distributed floor load is 100 psf live load plus 15 psf dead weight of sub- and top floors; a = 20 ft, b = 28 ft, c = 7 ft, d = 6 ft, and e = 14 ft. (a) Select a suitable wide-flange beam for B4. (b) Calculate the average web-shearing stress.

**8-31.** To form a wide beam, two American Standard channels are placed back to back but about 16 in. apart. Together they support a uniformly distributed load of 1,600 lb per ft over a simple span of 22 ft. (a) What size channels are most suitable? (b) Calculate the average web-shearing stress.

Ans. (a) Two 15-in. by 33.9-lb American Standard channels; (b) $s_s$ = 1,550 psi

**8-32.** The load shown in Fig. 6-15 is to be supported by a beam made up of two American Standard channels placed back to back. Consider also weight of channels. (a) Select the most economical channels to use. (b) Calculate the average web-shearing stress.

**8-33.** Two angles placed with long legs back to back are to be used as a simple lintel (beam) spanning an opening in a brick wall 10 ft between centers of supports. The estimated uniformly distributed load is 800 lb per ft, which may be assumed to include the weight of the angles. Select the most suitable angles.

**8-34.** A lintel in a brick wall spans 8 ft between centers of supports. The load is considered to be triangular, increasing uniformly from zero at the left end to a maximum of 450 lb per ft at the center, then decreasing uniformly to zero at the right end. Select two angles to carry this load, back to back. The outstanding legs must not be less than $3\frac{1}{2}$ in. wide.        *Ans.* Two $3\frac{1}{2}$- by $3\frac{1}{2}$- by $\frac{1}{4}$-in. angles

**8-35.** A solid circular steel shaft 6 ft long between centers of supporting bearings must be capable of withstanding a thrust perpendicular to the shaft of 1,472 lb at any point between the bearings without the maximum stress exceeding 10,000 psi. Calculate the appropriate shaft diameter $D$. Neglect weight of shaft.

*Ans.* $D = 3$ in.

**8-36.** Design a solid circular steel shaft, 6 ft long between centers of supporting bearings and acting as a simple beam, to withstand a thrust of 524 lb perpendicular to the shaft at any point along its length. Maximum allowable stress is 12,000 psi. Neglect weight of shaft.

**8-37.** A section of standard steel pipe is to serve as a clothesline pole extending 72 in. vertically out of a solid concrete base. Select a suitable diameter capable of withstanding a horizontal force at the top of 300 lb without exceeding a unit stress of 20,000 psi.

**8-38.** A steel bar is used as a lever to operate a jackscrew. A horizontal force $P$ of 200 lb is applied perpendicular to the bar at a point 16 in. from the vertical center line of the jackscrew. If the elastic limit stress of 32,000 psi must not be exceeded, what diameter ($a$) of solid bar and ($b$) of standard pipe should be used?

**8-39.** A steel pole supporting several trolley wires extends 21 ft vertically out of a concrete footing into which it is solidly imbedded. The pole is made up of three 7-ft sections of standard pipe and must be capable of withstanding a horizontal force of 650 lb at its top. Determine the required nominal diameter $D$ of ($a$) the top 7-ft section, ($b$) the middle 7-ft section, and ($c$) the bottom 7-ft section. Use $s = 20,000$ psi.

## 8-4 Laterally Unsupported Beams

A laterally unsupported beam is one in which the compression side or flange is free to buckle or deflect sidewise under loads, which normally are vertical gravity loads. The vast majority of beams, such as floor and roof beams in buildings, are laterally supported by the floor or roof structures attached to and supported by them. However, some roof beams that support rather light-weight roof coverings are not considered to be laterally supported. Other beams such as crane rails, or beams supporting crane rails, are often laterally unsupported. All beams are presumed to be supported laterally, as well as vertically, at their supports.

Lateral buckling of timber beams has been thoroughly investigated by the Forest Products Laboratory, and is fully and ably discussed in their "Wood Handbook," obtainable from the U.S. Government Printing Office, Washington, D.C.

In order to visualize what causes a beam to buckle sidewise, under a vertical load, say, let us consider a simple beam, supported at the ends and subjected to a downward-acting vertical load, as illustrated in Fig. 8-17a. Clearly now, the lower portion of the beam is in tension and will therefore tend to remain straight, as do all tension members. The upper portion, however, is in compression. If this upper portion could now be isolated from the lower portion, the internal compressive stresses in it would tend to make it behave in a manner similar to that of a slender column subjected to axial end loads which would produce similar internal compressive stresses. Under such conditions, this upper portion would buckle sidewise quite readily.

These two opposing tendencies within the same beam, the tendency of the compression side to buckle or deflect sidewise and of the tension side to remain straight, result in a compromise effect: (1) vertical deflection of the beam, due to load, as illustrated in Fig. 8-17a, (2) a certain lateral deflection of the upper portion of the beam, as illustrated in Fig. 8-17b, and (3) a smaller deflection of the lower portion of the beam, caused solely by the sidewise pull of the upper portion upon the lower portion. The combined effect, at

FIG. 8-17 Laterally unsupported simple steel beams. (a) Side view of simple beam showing vertical deflection. (b) Top view of simple beam showing lateral deflection. (c) Top view of simple beam supported vertically at $A$ and $B$ and laterally at $A$, $B$, and $C$. (d) Section $AA$. (e) Compressive area used in computing flange area.

section $AA$, of the three separate effects is a vertical deflection plus a lateral deflection plus a twist, as is illustrated in section in Fig. 8-17$d$.

If this simple beam is given lateral support at some point $C$, between the vertical supports $A$ and $B$, the lateral deflections will be as indicated in Fig. 8-17$c$, and the laterally unsupported length to be considered in its design will then be the greater of $L_1$ or $L_2$.

## 8-5 Design of Laterally Unsupported Simple Steel Beams[2]

The tendency of the compression flange of a laterally unsupported steel beam to deflect or buckle sidewise produces in that flange compressive stresses in addition to those produced by normal bending in a laterally supported beam. This fact necessitates the use of allowable design bending stresses somewhat lower than the usual 22,000 psi in bending. Of the several allowable stress formulas proposed, the ones recommended by the American Institute of Steel Construction are, perhaps, most widely used. The allowable bending stress for laterally unsupported beams of ASTM A36 steel when the compressive flange is solid and approximately rectangular in shape and its area is not less than that of the tension flange is:

**Allowable stress in laterally unsupported steel beams**

$$s = \frac{12,000,000\ C_b}{Ld/A_f} \text{ (but not more than 22,000 psi)} \qquad (8\text{-}1)$$

where $L$ = laterally unsupported length, in.

$d$ = depth of beam, in.

$A_f$ = area of compression flange, sq in.

$C_b$ = a coefficient, conservatively taken as unity

The quantity $Ld/A_f$ in Eq. (8-1) is a "slenderness" factor. When $Ld/A_f$ is less than 546, the computed value of $s$ would exceed 22,000 psi, and therefore 22,000 psi would be used instead of the computed value. If $Ld/A_f$ exceeds 546, the reduced stress computed by Eq. (8-1) would be used. The coefficient $C_b$ may conservatively be taken as unity, or, if a more exact value is desired, it may be evaluated from a rather complex formula given in the code.

Since the allowable stress depends upon the dimensions of the section to be used and since these dimensions are obviously unknown, the design of a laterally unsupported beam must be accomplished by successive trials. A suitable procedure is as follows:

---

[2] In the handbook "Steel Construction" are found a series of charts from which certain laterally unsupported steel beams may be selected directly (American Institute of Steel Construction, New York, 1970, pp. 2-87 to 2-105).

## PROCEDURE

1. Calculate the maximum moment $M$ in the beam.
2. Estimate the probable allowable stress, calculate the probable required section modulus, and select a trial section having a comparatively wide flange.
3. Calculate the allowable stress in the trial section, and also the actual stress, and compare the two. If they differ by more than 10 percent, select a more suitable trial section.
4. Consider the effect on the actual stress of the weight of the beam.

NOTE: Failure of laterally unsupported beams is invariably due to buckling of the compression flange. In such beams, the average web-shearing stress is comparatively low and, as a rule, need not be calculated.

### ILLUSTRATIVE PROBLEM

**8-40.** A simply supported beam with a span of 28 ft is to carry a uniformly distributed load of 1,500 lb per ft. What lightest wide-flange section may be used if the beam is laterally unsupported?

*Solution:* We shall assume the weight of the beam to be 5 percent of the imposed load. The total load is then $(1.05)(1,500) = 1,575$ lb per ft and the maximum moment is

$$\left[ M = \frac{wL^2}{8} \right] \qquad M = \frac{(1,575)(28)(28)}{8} = 154,400 \text{ lb-ft, or } 1,852,000 \text{ lb-in.}$$

If we now assume the allowable stress to be 18,000 psi, the probable required section modulus is

$$\left[ Z = \frac{M}{s} \right] \qquad \text{Required } Z = \frac{1,852,000}{18,000} = 103 \text{ in.}^3$$

The section should be rather compact with a very wide flange to better resist buckling. From Table E-6, we might select an W 18 × 64 with a depth of 17.87 in., a radius of gyration about the web axis of 1.93 in., and with flanges 0.686 by 8.715 in. Since the length $L = (12)(28) = 336$ in., Eq. (8-1) gives

$$\left[ s = \frac{12,000,000}{Ld/A_f} \right] \qquad \text{Allowable } s = \frac{12,000,000}{(336)(17.87)/5.98} = 11,950 \text{ psi}$$

Obviously this is a low allowable stress, and we should seek a section with a better depth-flange area ratio. Possible beams with their dimensions are shown in Table 8-3.

Since we may use the larger of the stresses given by Eqs. (8-1) and (8-2), the former is applicable here.

Although the calculations indicate that a W 12 × 65 section with a computed stress of 21,050 psi is adequate from a strength standpoint, its use would violate a common

**Table 8-3 DESIGN OF LATERALLY UNSUPPORTED STEEL BEAM**

| Section | Flange width | Flange thickness | Section modulus | Radius of gyration | Allowable stress Eq. (8-1) | Computed stress $M/Z$ |
|---------|--------------|------------------|-----------------|--------------------|----------------------------|-----------------------|
| W 18 × 64 | 8.715 | 0.686 | 117.0 | 1.93 | 11,950 | 15,850 |
| W 16 × 64 | 8.50 | 0.715 | 104.2 | 1.91 | 13,570 | 17,800 |
| W 14 × 68 | 10.04 | 0.718 | 103.0 | 2.46 | 18,310 | 17,970 |
| W 12 × 65 | 12.00 | 0.606 | 88.0 | 3.02 | 21,400 | 21,050 |

specification that the depth of a beam should be at least ½ in. per ft of span (see Art. 8-3). A W 14 × 68 has an allowable stress of 18,310 psi and a computed stress ($s = M/Z$) of 17,970 psi. Therefore,

Use W 14 × 68     Ans.

## 8-6 Design of Laterally Unsupported Overhanging and Cantilever Steel Beams

Cantilever beams, or the cantilever portions of overhanging beams, present special problems.

In Fig. 8-18a is shown a cantilever beam $AC$ whose actual laterally unsupported length is $L$. Vertical loads have caused its free end $A$ to deflect laterally a distance $y$ from a tangent at the fixed support $C$. If this beam $AC$ is now "reflected," as in a mirror, on the opposite side of the support $C$ and if we then consider the entire length $AB$, we find this imaginary beam $AB$ to be identical to the actual simple beam $AB$ in Fig. 8-18b, which has twice the length of the cantilever beam. We may correctly conclude, then, that the "effective length" of a laterally unsupported cantilever beam, to be used in its design, is twice its actual length, or $2L$.

FIG. 8-18 Effective length of cantilever beam. (a) Top view of laterally deflected compression flange of cantilever beam. (b) Top view of laterally deflected compression flange of simple beam.

## ILLUSTRATIVE PROBLEM

**8-41.** The overhanging beam shown in Fig. 8-19a is laterally unsupported except at the two vertical supports A and B. End C is completely free. Select the lightest available wide-flange section capable of carrying the 13-kip concentrated load and in addition a uniformly distributed load of 1,800 lb per ft, which, we may assume, includes the weight of the beam.

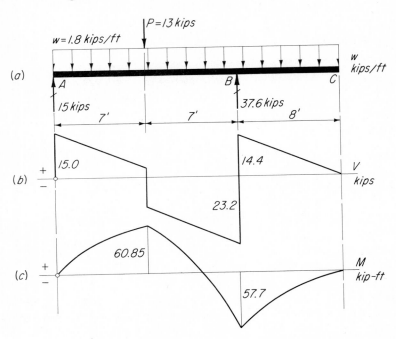

FIG. 8-19 Design of laterally unsupported overhanging beam. (a) Loaded beam. (b) Shear diagram. (c) Moment diagram.

*Solution:* We shall find it convenient here to draw the shear and moment diagrams (Fig. 8-19b and c). The maximum shear is 23.2 kips at B left, and the maximum moment, under the concentrated load, is 60.85 kip-ft. Study of Eq. (8-1) indicates that, for any given section, the allowable stress is inversely proportional to the "effective length" of a span. Consequently the *allowable stress* in the 8-ft cantilever span BC, whose effective length is $2L = (2)(8) = 16$ ft, will be lower than that in the main span AB, whose effective length is 14 ft. But the *actual stress* in span BC will also be lower than that in AB, since the bending moment in span BC is lower than that in span AB. In this case, therefore (though not in all), we must calculate allowable and actual stresses in both spans in order to determine which governs the required size of the beam.

Assuming an allowable stress of 15,000 psi and using the higher main-span moment, the probable required section modulus is

$$\left[Z = \frac{M}{s}\right] \qquad \text{Probable required } Z = \frac{(60.85)(1,000)(12)}{15,000} = 48.7 \text{ in.}^3$$

The lightest suitable wide-flange section found in Table E-6 or E-13 is a W 14 × 34, for which $Z = 48.5$ in.$^3$, $d = 14.0$ in., $b = 6.75$ in., $t = 0.453$ in., and $k = 1.46$ in. In the 14-ft main span, the allowable stress from Eq. (8-1) is

$$\left[s = \frac{12,000,000}{Ld/A_f}\right] \qquad \text{Allowable } s = \frac{12,000,000}{(14)(12)(14)/(6.75)(0.453)} = 15,640 \text{ psi}$$

The actual stress when the moment $M = 60.85$ kip-ft is

$$\left[s = \frac{M}{Z}\right] \qquad \text{Actual } s = \frac{(60.85)(1,000)(12)}{48.5} = 15,050 \text{ psi}$$

Consequently the W 14 × 34 will be adequate *if* the main span governs the selection.

In the 8-ft cantilever span, whose effective length is $(2)(8) = 16$ ft, the allowable stress is

$$\left[s = \frac{12,000,000}{Ld/A_f}\right] \qquad \text{Allowable } s = \frac{12,000,000}{(2)(8)(12)(14)/(6.75)(0.453)} = 13,650 \text{ psi}$$

The actual stress when $M = 57.7$ kip-ft is

$$\left[s = \frac{M}{Z}\right] \qquad \text{Actual } s = \frac{(57.7)(1,000)(12)}{48.5} = 14,260 \text{ psi} \qquad \text{NG}$$

which is 4.47 percent higher than the allowable. Clearly, then, the W 14 × 34 is inadequate.

The next lightest section is a W 12 × 36 for which $Z = 45.9$ in.$^3$, $d = 12.24$ in., $b = 6.565$ in., $t = 0.54$ in., and $k = 1.5$ in. Then in the 14-ft main span the allowable stress is

$$\left[s = \frac{12,000,000}{Ld/A_f}\right] \qquad s = \frac{12,000,000}{(168)(12.24)/3.54} = 20,750 \text{ psi}$$

and the actual stress when $M = 60.85$ kip-ft is

$$\left[s = \frac{M}{Z}\right] \qquad \text{Actual } s = \frac{(60.85)(1,000)(12)}{45.9} = 15,900 \text{ psi} \qquad \text{OK}$$

In the 8-ft cantilever span, the allowable stress is

$$\left[s = \frac{12,000,000}{Ld/A_f}\right] \qquad \text{Allowable } s = \frac{12,000,000}{(2)(8)(12)(12.24)/(6.565)(0.54)} = 18,100 \text{ psi}$$

and the actual stress when $M = 57.7$ kip-ft is

$$\left[s = \frac{M}{Z}\right] \qquad \text{Actual } s = \frac{(57.7)(1,000)(12)}{45.9} = 15,100 \text{ psi} \qquad \text{OK}$$

Hence, use a W 12 × 36 steel beam.    *Ans.*

## PROBLEMS

NOTE: Unless otherwise specifically stated, the weight of the beam shall be considered in the following problems.

**8-42.** A W 14 × 43 section is to be used as a 24-ft simple-span beam. What is the maximum allowable compressive stress if the beam is laterally supported? What is the allowable stress if it is not laterally supported?

**8-43.** A W 12 × 27 section is used as a beam with a simple span of 16 ft. What maximum allowable superimposed uniform load may be placed on this beam if it is not laterally supported? *Ans.* $w = 1,760$ lb per ft

**8-44.** A W 12 × 30 section is used on a simple span of 20 ft. What is the maximum allowable concentrated load $P$ that may be placed at mid-span if the beam is not laterally supported?

**8-45.** A W 8 × 40 section is to be used as a laterally unsupported cantilever beam 8 ft long. What concentrated load $P$ may be placed at the outer end without exceeding the allowable stress? Consider also weight of beam.

**8-46.** A laterally unsupported simple beam 24 ft long is to carry a superimposed uniform load of 1,200 lb per ft. Select the lightest available W section.

**8-47.** Select the lightest available W section with a laterally unsupported simple span of 24 ft to carry a concentrated load $P$ of 20,000 lb at mid-span.
*Ans.* W 12 × 53

**8-48.** The W 10 × 25 beam shown in Fig. 8-20 is laterally unsupported except at supports $A$ and $B$. Calculate the allowable and actual maximum stresses due to the weight of the beam plus a load $P$ of 5,320 lb.

FIG. 8-20 Prob. 8-48.

FIG. 8-21 Probs. 8-49 and 8-62.

**8-49.** The beam shown in Fig. 8-21 is laterally unsupported except at supports $A$ and $B$. Select the lightest available W section when $P_1 = 20,000$ lb and $P_2 = 20,000$ lb.

**8-50.** A laterally unsupported simple-span beam 12 ft long is to carry a concentrated load of 30,000 lb at mid-span. Select lightest available W section.

**8-51.** A simple beam 24 ft long is laterally supported only at its ends and at mid-span in a manner similar to that indicated in Fig. 8-17. Select the lightest W section capable of carrying a uniform load of 1,500 lb per ft.

**8-52.** Solve Prob. 8-41 when the main-span length is 20 ft instead of 14 ft.

## SUMMARY

### Design of Beams. Procedure

1. Calculate the maximum shear $V$ in the beam.
2. Locate the section of maximum moment $M$ in the beam and calculate this moment,

usually by the most suitable method: (a) the free-body-diagram method, (b) the beam-diagram method, or (c) using appropriate moment formula. (The weight of the beam must be appropriately accounted for, here or in step 5.)

3. Ascertain the appropriate allowable stress, convert the unit of moment to pound-inches, and calculate the required section modulus Z.
4. Select a trial beam having a section modulus slightly higher than that required.
5. Calculate the additional moment and the additional required section modulus due to the weight of the trial beam *at the section of maximum load moment.*
6. Ascertain if the section modulus of the trial beam is equal at least to the total required section modulus. If not, select appropriate beam.
7. Determine the safety in shear of the beam selected by calculating the maximum longitudinal shearing stress, if a timber beam, or the average web-shearing stress, if a steel beam, and comparing with the allowable stress.

## REVIEW PROBLEMS

**8-53.** A simply supported timber beam 15 ft long of Douglas fir dense select structural rough must support two concentrated loads of 10 kips each applied at 5 and 10 ft from its left end. Select a suitable and economical beam.

*Ans.* 8 by 16 in. rough

**8-54.** A timber beam of Douglas fir dense No. 1 S4S cantilevers out a distance of 5 ft from its support. It must withstand a concentrated load of 8,000 lb at its outer end. Select a suitable and economical beam.

**8-55.** A pedestrian walk alongside a highway bridge is supported by steel beams spaced 15 ft on centers along the bridge and cantilevered out a length of 6 ft from the side of the bridge. The maximum load is estimated to be 200 psf of walk area, which includes the dead weight of the deck, bracing, and beams. Select a suitable W beam.

**8-56.** A W 16 steel beam with a span of 11 ft is capable of supporting a total uniformly distributed load $W$ of 68,200 lb in addition to its own weight of 396 lb. When the span is 22 ft, $W = 34,100$ lb, and when it is 33 ft, $W = 22,700$ lb. In each case, calculate the percentage of $W$ represented by the weight of the beam and note importance of considering its weight.

*Ans.* 0.58, 2.32, and 5.23 percent

**8-57.** The second floor of a two-story column-free commercial building 34 by 120 ft is to be supported by steel W girders spanning 32 ft and spaced 20 ft on centers. Smaller beams spanning the 20 ft between girders are spaced 8 ft on centers and attach to the girders at the mid-point and at both quarter-points. The floor area supported by each beam is 160 sq ft. (a) Select a suitable W beam section to withstand a uniform floor load, dead and live, of 150 psf. (b) Each girder then supports three 24-kip concentrated loads, one at the center and one at each quarter-point. Select a suitable W girder. *Ans.* (a) W 12 × 27; (b) W 27 × 94

**8-58.** A flagpole 40 ft high whose base is solidly imbedded in concrete is made of four equal-length sections of standard pipe of different diameters. If the pole must resist a horizontal load of 12.5 lb per ft of height without exceeding a unit stress of 16,000 psi, what minimum diameter pipe is required for the bottom section?

**8-59.** A piece of standard steel pipe 66 in. long is to be used as a horizontal lever.

The 1,000-lb vertical load is 6 in. from the fulcrum, and the applied vertical force of 100 lb is 60 in. from the fulcrum, at the opposite end. What nominal diameter of pipe should be used if the unit stress is not to exceed 20,000 psi?

**8-60.** A rustic footbridge crosses a small stream and is supported by two round logs laid parallel, 6 ft on centers, with a simple span of 20 ft. The deck is of 12-in.-wide planks 12 ft long laid across the logs so that they cantilever 3 ft out beyond the logs at both ends. Deck planks and logs must be capable of supporting a load of 125 psf, which includes weight of deck planks but not of logs. Determine (a) the required thickness $t$ of deck planks, using an allowable bending stress of 500 psi (low because planks subject to wear), and (b) the required diameter $d$ of logs, using an allowable bending stress of 1,000 psi. _Ans._ (a) $t = 2.6$ in.; (b) $d = 17.1$ in.

**8-61.** In Prob. 8-41, omit the 13-kip concentrated load and solve.

**8-62.** The beam in Fig. 8-21 is laterally unsupported except at supports $A$ and $B$. Select the lightest available W section when $P_1 = 40,000$ lb and $P_2 = 10,000$ lb.

**8-63.** A simply supported beam 30 ft long is laterally supported only at its ends and at mid-span in a manner similar to that indicated in Fig. 8-17. Select the lightest available W section to carry a concentrated load of 32,500 lb at mid-span.

## REVIEW QUESTIONS

**8-1.** In which type of beam are bending stresses usually critical? In which type are shearing stresses usually critical?

**8-2.** Are shearing stresses often critical (a) in timber beams and (b) in steel beams?

**8-3.** Why should the weight of a beam be considered? Approximately what percentage of the total load is represented by the weight of a simple-span uniformly loaded (a) timber beam and (b) steel beam?

**8-4.** Under what conditions may the estimated weight of a beam be added to the imposed load before design begins? When should it not be so added?

**8-5.** What is meant by the "most economical" beam? What general shape of beam is usually most economical?

**8-6.** Under what conditions may the "most economical" beam not be used?

**8-7.** Define the span of a beam. Discuss spans of timber and steel beams and girders generally used for purpose of design.

**8-8.** By what three methods may the maximum moment in a beam be determined?

**8-9.** Discuss the so-called methods A, B, and C of timber beam design.

**8-10.** What common specification applies to the depth of steel beams?

**8-11.** By what "rule of thumb" may we estimate closely the required depth of (a) a timber floor beam and (b) a steel floor beam?

**8-12.** What is meant by "lateral support" of a beam? Give some typical illustrations.

**8-13.** How may a beam fail when "laterally unsupported" except at the vertical supports?

**8-14.** Upon what factors does the allowable stress in a laterally unsupported beam depend?

# CHAPTER 9

# Deflection of Beams

## 9-1 Relationship between Curvature and Stress

As previously described in Art. 7-1, a condition of pure bending in a beam may be produced by opposing moments applied at its opposite ends. Pure bending will also exist in any part of a beam in which no shear forces are present, such as length $BC$ of the beam in Fig. 9-1. $EF$ and $GH$ are vertical and parallel planes through the straight, unbent beam.

Consider now the central portion of this beam after the loads have produced bending. The planes $EF$ and $GH$, parallel in the unbent beam, now intersect at $O$. Let $R$ be the radius of curvature of the neutral plane, now called the **elastic curve,** and $c$ the distance from the neutral plane to the extreme fiber. If line $IJ$ is now drawn parallel to $EF$ the distance $JH$ is the deformation in the length $KL$, and $JH/KL$ is the unit deformation $\epsilon$. Clearly, angles $KOL$ and $JLH$ are equal and the similarly designated segments $KOL$ and $JLH$ are similar. Therefore

$$\frac{JH}{KL} = \frac{c}{R} = \epsilon \tag{a}$$

Since $\epsilon = s/E$, we obtain

$$\frac{c}{R} = \frac{s}{E} \qquad \text{or} \qquad s = \frac{cE}{R} \tag{9-1}$$

In this equation, $c$ and $R$ are expressed in inches and $s$ and $E$ in psi.

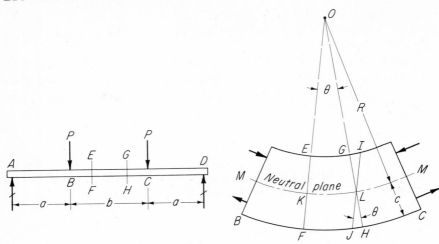

FIG. 9-1 Part *BC* of beam in pure bend-  FIG. 9-2 Section of beam in pure bending.                                       ing.

From Eq. (9-1) we may obtain the maximum unit stress $s$ in any portion of a bent member whose radius of curvature $R$ is known or the radius of curvature $R$ corresponding to any given stress $s$.

### PROBLEMS

**9-1.** A band of sheet steel 0.03 in. thick is to be bent around a circular post. (*a*) What bending stress is produced if the post is 40 in. in diameter? (*b*) What minimum radius $R$ may be used if the elastic-limit stress of 36,000 psi is not to be exceeded ($E = 30,000,000$ psi)?           *Ans.* (*a*) $s = 22,500$ psi; (*b*) $R = 12.5$ in.

**9-2.** The main cables of a suspension bridge are made up of strands of $\frac{3}{16}$-in.-diameter hard-steel wire. Calculate the minimum permissible radius of curvature $R$ of the saddles on which the cables rest at the top of the tower if the unit stress due to bending must not exceed 30,000 psi.

**9-3.** A hard-steel band saw is 0.03 in. thick and runs on pulleys of 36-in. diameter. If the maximum direct tensile stress in the straight portion of the band is 8,000 psi, what is the maximum tensile stress in the bent portion?

**9-4.** A 36-in.-diameter cylinder is to be covered with thin stainless steel which will bend to the 18-in. radius without exceeding the elastic-limit stress of 36,000 psi. What maximum thickness $t$ of steel may be used ($E = 30,000,000$ psi)?

*Ans.* $t = 0.0432$ in.

## 9-2 Relationship between Curvature and Bending Moment

According to the flexure formula, Eq. (7-1), $s = Mc/I$. When this latter quantity is substituted for $s$ in Eq. (9-1), we obtain

$$\frac{Mc}{I} = \frac{cE}{R} \quad \text{or} \quad R = \frac{EI}{M} \tag{9-2}$$

From Eq. (9-2), it is seen that the radius of curvature $R$ at any section in a bent member is equal to the $EI$ of the member divided by the bending moment at that section.

Equations (9-1) and (9-2) were derived to apply to members in pure bending. Since, however, they express only the relations between curvature and stress and between curvature and bending moment, respectively, at any plane section, they are applicable to all bent members.

## PROBLEMS

**9-5.** A timber beam, $E = 1,760,000$ psi, is 6 in. wide and 12 in. deep. If the beam is subjected to a constant bending moment of 144,000 lb-in., calculate the radius of curvature of the beam using (a) Eq. (9-1) and (b) Eq. (9-2).

**9-6.** A rough 2- by 12-in. Douglas fir plank, laid flat, is supported at its ends. A 160-lb man stands at the mid-point of the 16-ft span. Calculate the minimum radius of curvature by Eq. (9-1) and by Eq. (9-2).

**9-7.** Calculate the radius of curvature $R$ of the plank in Prob. 7-12(a) at its left and at points 2, 4, 6, and 8 ft from the left end ($E = 1,760,000$ psi). Give answers in feet. [NOTE: The answers to this problem indicate clearly that the curvature of a simple beam due to a concentrated load at its center varies from zero at either end to a maximum at the center (see Fig. 9-3b).]

   Ans. $R_0 = \infty$; $R_2 = 12,580$ ft; $R_4 = 6,290$ ft; $R_6 = 4,195$ ft; $R_8 = 3,145$ ft

**9-8.** Determine the radius of curvature $R$ of the beam in Prob. 7-1 at the point of maximum stress using (a) Eq. (9-1) and (b) Eq. (9-2). Wood is southern pine.

## 9-3 Deflection of Beams

The deflection of a beam is the deviation of the neutral surface, or elastic curve, of the loaded beam from its original position in the unloaded beam. Since beams are normally horizontal, the deflection is the vertical deviation $\delta$ indicated in Fig. 9-3.

Many reasons exist for determining the deflection of a beam. The most common one is probably the usual limitation that a beam or joist supporting a plastered ceiling must not deflect more than $\frac{1}{360}$ of its span length if cracking of the plaster is to be avoided. Also, in the design of machines and airplanes, the deflection of a member, or of a series of members, is often of vital importance.

Various methods are available for determining beam deflections. Of these, three are presented here: (1) the beam-diagram method, in which slope and

(a)                                              (b)

FIG. 9-3 Beam deflections. (a) Cantilever beam. (b) Simple beam.

deflection diagrams are added to the load, shear, and moment diagrams which were presented in Chap. 6, (2) the moment-area method, in which the two moment-area principles are introduced, and (3) the formula method. In the most commonly occurring problems of the type given herein, the maximum deflection occurs either at the center, as in symmetrically loaded simply supported beams, or at the end, in the case of cantilever beams. Other problems with unsymmetrical loads become more complex, but most of them may be satisfactorily solved by the moment-area method.

### 9-4 Deflection by the Beam-Diagram Method

In Chap. 6, a series of three related diagrams was developed which indicated that certain mathematical relationships existed between the load and shear diagrams and the shear and moment diagrams. The mathematical proof of these relationships, as stated in the first and second laws of beam diagrams, is given in Appendix C. If a fourth and a fifth diagram, drawn in conformity with the first and second laws, are added, ordinates in the fourth diagram will measure the slope of the deflected axis of a beam, and ordinates in the fifth diagram will measure the deflection of that axis from its normal horizontal position.

Since the second law necessitates the determination of areas under curves and since such areas in some cases tend to become rather complex, the beam-diagram method is most easily applied only to certain types of problems, among which are the four common types illustrated in Fig. 9-4.

While the values of shear and moment in a beam are not dependent upon its size or upon the elastic properties of its material, it becomes clear, however, that the slope at any point of the deflected axis, and the deflection of that axis, are definitely dependent upon the size and shape $I$ of the beam and upon the stiffness $E$ of its material. Clearly, the deflection must be inversely proportional to $E$ and $I$.

Values of ordinates in the slope and deflection diagrams will, therefore, be for a beam whose moment of inertia and modulus of elasticity of material both equal unity. The actual slope (in radians) and deflection (usually in inches) may then be obtained by dividing the values of ordinates in these diagrams by $E$ and $I$, since actual slopes and deflections of a beam are *inversely* proportional to its $E$ and $I$. That is, slope $\theta = $ (ordinate value)$/EI$, or ordinate

FIG. 9-4 Deflection by the beam-diagram method. Suitable beam types.

value $= EI\theta$. Also, deflection $\delta = $ (ordinate value)$/EI$, or ordinate value $= EI\delta$.

When the slope and deflection ordinates are thus divided by $E$ and $I$, which express the "elastic properties" of a particular beam, the curves in the slope and deflection diagrams are referred to as the **elastic curves.**

The following examples illustrate the method. For convenience, the laws of beam diagrams are restated:

**First Law of Beam Diagrams:**

$$\begin{bmatrix} \text{Slope of curve at any point} \\ \text{in any diagram} \end{bmatrix} = \begin{bmatrix} \text{length of ordinate at correspond-} \\ \text{ing point in next higher diagram} \end{bmatrix}$$

**Second Law of Beam Diagrams:**

$$\begin{bmatrix} \text{Difference in length of any} \\ \text{two ordinates in any diagram} \end{bmatrix} = \begin{bmatrix} \text{area between corresponding ordi-} \\ \text{nates in next higher diagram} \end{bmatrix}$$

The first law is used to determine the *shape* of a diagram; the second to determine the values of its principal ordinates.

In order to avoid extremely high, and therefore cumbersome, ordinate values, it is advisable that all loads be in kips or kips per foot and that beam spans be in feet. Then shear ($V$) values will be in kips, moment ($M$) values in kip-feet, slope ($EI\theta$) values in kip-feet$^2$, and deflection ($EI\delta$) values in kip-feet$^3$.

## ILLUSTRATIVE PROBLEMS

**9-9.** The beam shown in Fig. 9-5 extends 12 ft horizontally out of a solid concrete pier and supports a uniformly distributed load of 0.5 kip-ft. Determine the maximum deflection $\delta_A$ of this beam at its free end $A$ due to ($a$) the load only if a W 10 × 25 steel beam is used, ($b$) both load and weight of steel beam, by proportion, ($c$) load only if the beam is a 10- by 16-in. timber of southern pine S4S, and ($d$) both load and weight of timber beam, by proportion. ($e$) If the maximum permissible deflection is 0.4 in., what size beams should be used?

Fɪɢ. 9-5 Maximum deflection by the beam-diagram method.

*Solution:* The load, shear, and moment diagrams are first drawn in the usual manner, using the laws of beam diagrams restated above. We note that at $A$ the shear and the moment both are zero. Before the slope and deflection diagrams are drawn, it must be clearly seen and understood from the physical conditions of the problem that the slope of the axis of the beam is zero at $B$, where it extends *horizontally* out of the pier and is maximum at $A$, and that the deflection is zero at $B$ and is maximum at the free end $A$.

FIG. 9-6 Beam diagrams.

The slope of the curve in the slope diagram is negative and increasing, as are the moment ordinates above. Hence the slope diagram must begin with some positive value of ordinate at $A$ in order to end up with zero ordinate at $B$. The slope of the curve in the deflection diagram is positive and decreasing, as are the ordinates in the $EI\theta$ diagram above. But since $\delta$ is zero at $B$, the deflection diagram must start with a negative value of ordinate at $A$.

The principal ordinates may now be evaluated by the second law as follows: The shear ordinate $V$ at $B$ is $(12)(0.5) = 6$ kips, the area under the load curve between $A$ and $B$, and is negative; the moment ordinate $M$ at $B$ is $(\frac{1}{2})(12)(6) = 36$ kip-ft, the area under the shear curve above, and is negative; the $EI\theta$ ordinate at $A = (\frac{1}{3})(12)(36) = 144$ kip-ft$^2$, the area under the moment curve, and is positive; and the $EI\delta$ ordinate at $A$ is $(\frac{3}{4})(12)(144) = 1,296$ kip-ft$^3$, the area under the $EI\theta$ curve (see Fig. 9-11e), and is negative. Before these areas are calculated, it should be noted that the $M$ and $EI\theta$ curves *begin* with zero slope at $A$ because in each case the ordinate at $A$ in the next higher diagram is zero and that the $EI\delta$ curve *ends* with zero slope at $B$, since the $EI\theta$ ordinate at $B$ is zero.

To convert $EI\delta$ from kip-feet$^3$ to pound-inches$^3$, as must be done if $\delta$ is to be in inches, we first multiply the $EI\delta$ value of $-1,296$ by 1,000, thus converting kip-feet$^3$ to pound-feet$^3$ and next by 1,728 ($= 12^3$), thereby converting pound-feet$^3$ to pound-inches$^3$.

(a) For the W $10 \times 25$ steel beam, $EI\delta_A = -1,296$ kip-ft$^3$, $E = 29,000,000$ psi, and $I = 133.2$ in.$^4$. Then, $\delta_A = -1,296/EI$, or

$$\delta_A = \frac{(-1,296)(1,000)(1,728)}{(29,000,000)(133.2)} = -0.580 \text{ in.} \qquad Ans.$$

(*b*) By proportion, the deflection at *A* due both to load, 500 lb per ft, and weight of beam, 25 lb per ft, is

$$\delta_A = -0.580 \frac{500 + 25}{500} = -0.609 \text{ in.} \qquad Ans.$$

(*c*) Using a 10- by 16-in. timber beam S4S, for which $I = 2,948$ in.[4] and $E = 1,760,000$ psi,

$$\delta_A = \frac{(1,296)(1,000)(1,728)}{(1,760,000)(2,948)} = -0.432 \text{ in.} \qquad Ans.$$

(*d*) Considering also the weight of 40.9 lb per ft of this beam, the total deflection by proportion is

$$\delta_A = -0.432 \frac{500 + 40.9}{500} = -0.467 \text{ in.} \qquad Ans.$$

(*e*) If the maximum deflection of the steel beam is limited to $-0.4$ in. and $EI\delta_A = -1,296$, its required moment of inertia is $I = -1,296/E\delta_A$, or

$$\text{Required } I = \frac{(-1,296)(1,000)(1,728)}{(29,000,000)(-0.4)} = 193 \text{ in.}[4]$$

Try W 12 × 27 ($I = 204.1$ in.[4]).
For both load and weight of beam

$$\text{Required } I = 193 \frac{500 + 27}{500} = 203 \text{ in.}[4] \qquad \text{OK}$$

$$\text{Use W 12} \times 27 \text{ steel beam} \qquad Ans.$$

When $\delta_A = -0.4$ for the timber beam,

$$\text{Required } I = \frac{(-1,296)(1,000)(1,728)}{(1,760,000)(-0.4)} = 3,180 \text{ in.}[4]$$

Try 10- by 18-in. southern pine S4S ($I = 4,243$ in.[4]; weight = 46.1 lb per ft). For both load and weight of beam

$$\text{Required } I = 3,180 \frac{500 + 46.1}{500} = 3,475 \text{ in.}[4] \qquad \text{OK}$$

$$\text{Use 10- by 18-in. southern pine S4S } (I = 4,243 \text{ in.}[4]) \qquad Ans.$$

NOTE: The method of proportion used in this example holds good only for beams with uniformly distributed loads over entire length.

**9-10.** A simple beam 20 ft long supports a single concentrated load of 12 kips at its center, as shown in Fig. 9-7. Calculate the maximum deflection at the center

FIG. 9-7 Maximum deflection by the beam-diagram method.

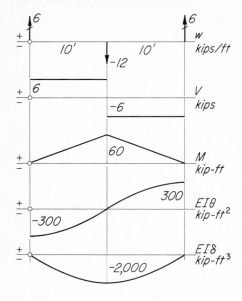

FIG. 9-8 Beam diagrams.

of this beam (a) if it is a W 14 × 30 steel beam and (b) if it is a 12- by 18-in. timber beam, southern pine S4S.

*Solution:* The load, shear, and moment diagrams are drawn in the usual manner, as shown in Fig. 9-8. Next are the slope $(EI\theta)$ and deflection $(EI\delta)$ diagrams. Because of symmetry of loading, it is clear that the maximum deflection will be at the center of the beam. At this point also, the slope of the beam axis will be zero, since a tangent to the axis at that point will be horizontal. That is, the $EI\theta$ ordinate at $C$ will be zero. The *slope* of the $EI\theta$ curve between $A$ and $C$ is increasing and positive, as are the ordinates in the moment diagram above. Since the curve must end with zero ordinate at the center $C$, it must necessarily begin at $A$ with a negative ordinate of $EI\theta_A = (\frac{1}{2})(10)(60) = 300$, which is the area under the moment curve between $A$ and $C$. The deflections of the beam at supports $A$ and $B$ are clearly zero. Hence $EI\delta_A$ and $EI\delta_B$ are zero. The *slope* of the deflection curve between $A$ and $C$ is decreasing and negative, as are the ordinates in the $EI\theta$ diagram above. The maximum deflection ordinate at the center $C$ of the beam is $EI\delta_C = (\frac{2}{3})(10)(300) = 2,000$ and is negative.

The deflection $\delta_C$ of the W 14 × 30 steel beam, for which $I = 289.6$ in.$^4$ and $E = 29,000,000$ psi, then is

$$\delta_C = \frac{(-2,000)(1,000)(1,728)}{(29,000,000)(289.6)} = -0.412 \text{ in.} \qquad Ans.$$

The deflection $\delta_C$ of the 12- by 18-in. timber beam, for which $E = 1,760,000$ psi and $I = 5,136$ in.$^4$, is

$$\delta_C = \frac{(-2,000)(1,000)(1,728)}{(1,760,000)(5,136)} = -0.383 \text{ in.} \qquad Ans.$$

**PROBLEMS**

NOTE: Solve the following problems by the beam-diagram method. Disregard weight of beam unless otherwise stated.

**9-11.** Calculate the maximum deflection at the free end of a cantilever beam similar to that shown in Fig. 9-4a when $P = 5$ kips and $L = 6$ ft (a) for an 8- by 16-in. timber beam of Douglas fir S4S and (b) for a W 10 × 21 steel beam.

**9-12.** Compute the maximum deflection at the free end of a cantilever beam similar to that in Fig. 9-4a when $P = 4$ kips and $L = 12$ ft (a) for an 8- by 18-in. beam of southern pine S4S and (b) for a W 12 × 27 steel beam. If this deflection is limited to 0.5 in., select the appropriate size (c) of timber beam, letting $b/h = \frac{1}{2}$, and (d) of steel W beam.

Ans. (a) $\delta = -0.675$ in.; (b) $\delta = -0.672$ in.; (c) 10- by 20-in.; (d) W 14 × 30

**9-13.** Calculate the maximum deflection at the top of a 20-ft-high 6-in. standard pipe pole due to an 800-lb horizontal load applied at the top. If the deflection must not exceed 0.5 in., what diameter of pipe should be used?

**9-14.** A cantilever beam similar to that shown in Fig. 9-4b has a length $L$ of 8 ft and its load $w$ is 0.6 kip per ft. Calculate the maximum deflection at the free end if the beam is (a) an 8- by 12-in. southern pine S4S or (b) a steel 8 I 18.4.

**9-15.** Solve all parts of Prob. 9-9 for a beam whose length $L$ is 9 ft and load $w$ is 0.5 kip per ft if the steel beam in (a) is an W 8 × 17, the timber beam in (c) is an 8- by 14-in. timber S4S, and the maximum permissible deflection in (e) is 0.25 in.        Ans. (a) $\delta = -0.433$ in.; (b) $\delta = -0.448$ in.; (c) $\delta = -0.262$ in.;
(d) $\delta = -0.277$ in.; (e) W 10 × 21; 6- by 16-in. S4S

**9-16.** A cantilever beam similar to those shown in Fig. 9-4a and b is 6 ft long and carries a uniformly distributed load $w$ of 0.6 kip per ft in addition to a concentrated load $P$ of 2 kips at its left end. Calculate the total maximum deflection at the free end due to both loads if a W 8 × 17 steel beam is used. (HINT: Solve separately for $\delta$ due to each load.)

**9-17.** An 8- by 12-in. timber beam of Douglas fir S4S 12 ft long is supported at its outer ends and carries a concentrated load $P$ of 10 kips at its center, as illustrated in Fig. 9-4c. Calculate the maximum deflection $\delta$ at the center.

**9-18.** Calculate the maximum deflection $\delta$ at the center of the shaft described in Prob. 8-35. If $\delta$ must not exceed 0.035 in., what diameter of shaft should be used?

Ans. $\delta = -0.0994$ in.; use 4-in. diameter

**9-19.** A steel girder, W 16 × 45, is 24 ft long and is supported at its outer ends. It carries two concentrated loads of 12 kips each applied at the third-points ($a = 8$ ft), similarly to that illustrated in Fig. 9-4d. Calculate the maximum deflection $\delta$ at its center.

**9-20.** The loads $P$ in Fig. 9-1 are 12 kips each. The dimensions $a$ and $b$ are 9 ft and 7 ft, respectively. Calculate the deflection under one of the loads if the beam is a steel W 24 × 76.

## 9-5 Deflection by the Moment-Area Method

The determination of the deflections of unsymmetrically loaded simple beams and of overhanging beams by the beam-diagram method becomes rather complex. The closely related **moment-area method,** developed and illustrated in Arts. 9-6 to 9-11, uses the principle embodied in the second law of beam diagrams for the derivation of the first moment-area principle, which, then, in turn is used to derive the second moment-area principle.

### 9-6 The First Moment-Area Principle

The second law of beam diagrams, as stated in Art. 9-4, is "The difference in length of any two ordinates in any diagram equals the area between corresponding ordinates in the next higher diagram."

In the slope and deflection diagrams of the beam-diagram series, the lengths of ordinates represent, respectively, $EI\theta$ and $EI\delta$. Actual slopes $\theta$ and deflections $\delta$ are then obtained by dividing these ordinate values by $EI$.

In Fig. 9-9 are shown portions of a moment diagram (upper) and the corresponding slope ($EI\theta$) diagram (see Art. 9-4). In this slope diagram the length of ordinate at any point, divided by $EI$, expresses the slope-angle $\theta$ (in radians) between a tangent to the elastic curve of the loaded beam at that point ($B$, for example), and the original undeflected horizontal axis of the beam, as illustrated in Fig. 9-10$b$. Therefore, the *difference* in length between slope ordinates $A$ and $B$ (in Fig. 9-9$b$), divided by $EI$, must equal the *difference* $\phi$ in the slopes of the tangents at $A$ and $B$, in Fig. 9-10$b$. But this difference, according to the second law, equals the area between corresponding ordinates in the moment diagram. We may therefore state

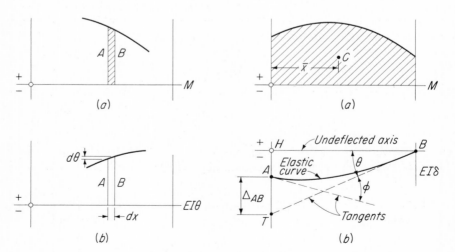

FIG. 9-9 First moment-area principle. (*a*) Moment diagram. (*b*) Slope diagram.

FIG. 9-10 Second moment-area principle. (*a*) Moment diagram. (*b*) Deflection diagram.

**The First Moment-Area Principle:**

$$\left[ \begin{array}{l} \text{The angle } \phi \text{ between the tan-} \\ \text{gents to the elastic curve of a} \\ \text{beam at any two points } A \text{ and } B \end{array} \right] = \left[ \begin{array}{l} \text{the area in the moment dia-} \\ \text{gram, lying between verticals} \\ \text{through } A \text{ and } B, \text{ divided by } EI \end{array} \right]$$

Expressing the first principle in symbols gives

$$\phi_{AB} = \frac{A_{AB}}{EI} \tag{9-3}$$

Clearly, from Fig. 9-10$b$, when either one of the two tangents is horizontal, the angles $\phi$ and $\theta$ are equal.

**Unit of slope.** In the vast majority of practical problems the actual beam deflections, and hence the slope-angles of tangents to the elastic curve of the beam, are very small, seldom exceeding $1°$. A well-known mathematical fact is that, for such small angles, the value of the angle, measured in radians, is very nearly numerically equal to the value of its tangent. The unit of the slope-angles $\phi$ and $\theta$ is radians. Conversion to degrees is accomplished by multiplying by $360/2\pi$, the number of degrees in one radian.

The areas under curves shown in Fig. 9-11 are commonly encountered in the application to problems which involve the two moment-area principles.

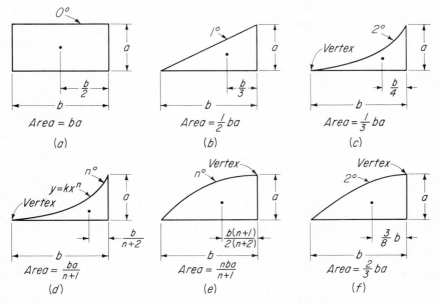

FIG. 9-11 Areas under curves. Location of centroids. ($a$) Rectangle. ($b$) Triangle. ($c$) Second-degree parabola. ($d$) $n$th-degree parabola. ($e$) $n$th-degree parabola. ($f$) Second-degree parabola.

## ILLUSTRATIVE PROBLEM

**9-21.** The beam shown in Fig. 9-12 is a W 8 × 24 steel section. Calculate the angle $\phi_{AB}$ (in radians and degrees) between the tangents to the elastic curve of this beam at points $A$ and $B$ (left end and center).

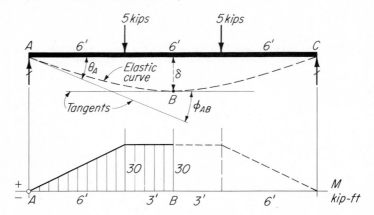

FIG. 9-12 Calculation of slope-angle of beam. Moment diagram.

*Solution:* The moment of inertia $I$ of this beam is 82.5 in.$^4$ (from Table E-6). $E = 29,000,000$ psi, the usual value for structural steel. Since the tangent at $B$ is horizontal, $\phi_{AB} = \theta_A$, the slope of the tangent at $A$ with respect to the original horizontal axis of the beam. The bending moment at $B$ is 30 kip-ft. Hence the moment-area divided by $EI$ is

$$\phi_{AB} = \theta_A = \frac{[(0.5)(6)(30) + (3)(30)](1,000)(144)}{(29,000,000)(82.5)} = 0.010\ 83 \text{ radian}$$

$$= \frac{(0.010\ 83)(360)}{2\pi} = 0.62° \quad Ans.$$

The unit of the moment-area is (kip-ft)(ft) = kip-ft$^2$. The necessary conversion from kip-feet$^2$ to pound-inches$^2$ is accomplished by multiplying by 1,000 and 144.

## PROBLEMS

**9-22.** Solve Prob. 9-21 when each of the two loads is 500 lb and the beam is a 6- by 6-in. rough (full-size) timber of Douglas fir ($E = 1,760,000$ psi).

**9-23.** A cantilever beam, W 8 × 40, 10 ft long extends horizontally out of a completely fixed support at its right end $B$ and supports a concentrated load of 6 kips at its left end $A$. Calculate the slope-angle $\theta_A$, in radians and degrees, of its elastic curve at the left end $A$.          *Ans.* $\theta_A = 0.010\ 17$ radian $= 0.583°$

**9-24.** A steel beam, W 10 × 25, cantilevers 10 ft horizontally out of a completely fixed support at its right end $B$. Calculate the slope-angle $\theta_A$ of its elastic curve at the left end $A$ caused by a uniformly distributed load of 3 kip per ft. Check the results by the appropriate formula found in Fig. 9-30.

**9-25.** A man weighing 200 lb stands at the center of a 2- by 12-in. S4S white oak plank laid flat and supported at its outer ends. Calculate the slope-angle $\theta_A$ at the left end $A$ of the plank. (HINT: A tangent to the elastic curve at the center $B$ is horizontal.) Plank is 16 ft long.

**9-26.** A simple beam of length $L$ supports a single concentrated load $P$ at its center. Derive the formula giving the slope-angle $\theta_A$ at its left end $A$. Check this against the similar formula shown in Fig. 9-31.

**9-27.** A 2- by 8-in. (full-size) Douglas fir plank 80 in. long and laid flat is supported at its ends and carries a uniform load of 12 lb per in. Calculate the slope-angle $\theta_A$ at its left end $A$. *Ans.* $\theta_A = 0.0273$ radian $= 1.565°$

**9-28.** A simple beam of length $L$ supports two equal concentrated loads $P$, one located distance $a$ from its right end. Derive the formula giving the slope-angle $\theta_A$ at the left end $A$ of the beam.

## 9-7 The Second Moment-Area Principle

We may now use the first moment-area principle as a means of developing the second moment-area principle. Together, these two principles enable us to calculate actual beam deflections $\delta$.

In Fig. 9-13 is shown a portion, between two points $A$ and $B$, of the moment diagram for a loaded beam in which all ordinates, for simplicity of development, have been divided by $EI$. This diagram is usually referred to as the $M/EI$ diagram. Below is shown the corresponding portion of the deflected elastic curve of the beam. These two diagrams are similar to those shown

FIG. 9-13 Derivation of the second moment-area principle.

in Fig. 9-10. We shall now show that the total displacement $\Delta_{AB}$ of point $A$ on the elastic curve from a tangent to the curve at $B$ equals the *moment* about $A$ of the area of the $M/EI$ diagram between $A$ and $B$.

First we select three very closely adjacent ordinates $C$, $D$, and $E$ in the $M/EI$ diagram. The $M/EI$ area between $C$ and $D$ is $A_1$ and that between $D$ and $E$ is $A_2$. Next we project ordinates $C$, $D$, and $E$ downward to intersect the elastic curve and then draw tangents to the curve at the three points of intersection, extending these tangents left to a vertical through $A$, thus obtaining the intercepts $\Delta_{CD}$ and $\Delta_{DE}$. The slopes of the elastic curve and of these tangents are greatly exaggerated in Fig. 9-13, merely for greater clarity of development. Actually, in practical beam problems, these slopes seldom exceed $1°$ from the original horizontal position of the beam.

Under these given circumstances and also under the condition that the distances from $C$ to $D$ and from $D$ to $E$ are infinitesimally small, it is clear that the slope-lengths of the tangents at $C$ and $D$ are very nearly equal to the horizontal distance $\bar{x}_1$. As previously explained, the numerical values of the very small slope-angles $\phi$ (in radians) and of their tangents are sufficiently close to be considered equal. Consequently $\tan \phi = \phi = \Delta/\bar{x}$ and $\Delta = \phi\bar{x}$. But, according to the first moment-area principle, $\phi_{CD} = A_1$ and $\phi_{DE} = A_2$. Therefore $\Delta_{CD} = A_1\bar{x}_1$, the *moment* about $A$ of the area $A_1$ in the $M/EI$ diagram between $C$ and $D$. Similarly, $\Delta_{DE} = A_2\bar{x}_2$. If *all* the intercepts between $A$ and $T$ are now summed up, we obtain $\Delta_{AB}$, the displacement of point $A$ from a tangent at $B$. But $\Delta_{AB}$ would then equal the moment about $A$ of the *entire* area of the $M/EI$ diagram between $A$ and $B$. We may therefore state

**The Second Moment-Area Principle:**

$$
\begin{bmatrix}
\text{The vertical displacement } \Delta_{AB} \\
\text{of any point } A \text{ on the elastic} \\
\text{curve of a beam from a tangent} \\
\text{to the curve at any other point } B
\end{bmatrix}
=
\begin{bmatrix}
\text{the moment about } A \text{ of the area} \\
\text{in the moment diagram lying} \\
\text{between verticals through } A \\
\text{and } B \text{, divided by } EI
\end{bmatrix}
$$

Expressing the second principle in symbols gives

$$\Delta_{AB} = \frac{(A_{AB})(\bar{x}_A)}{EI} \tag{9-4}$$

Clearly, $\Delta_{AB}$ is *not* the actual deflection $\delta$ of point $A$ from its original position. Methods of calculating actual deflections $\delta$ are outlined and illustrated in the following articles.

## 9-8 Deflection of Cantilever Beams

The vertical deflection $\delta$ of any point on a cantilever beam is readily calculated by use of the second moment-area principle, as is illustrated in the following illustrative problem. Unless otherwise stated and shown, all cantilever beams are presumed to be, and to remain, horizontal at the point of fixity. A tangent to the elastic curve of the beam at the point of fixity will then also be horizontal, thus greatly simplifying the solution of this type of problem.

### ILLUSTRATIVE PROBLEMS

**9-29.** The cantilever beam shown in Fig. 9-14 is a 2- by 12-in. rough wood plank laid flat. It is completely fixed at $B$. $E = 1,760,000$ psi; $I = 8$ in.[4]. Calculate its maximum deflection caused by the load shown.

FIG. 9-14 Deflection of cantilever beam. Moment diagram.

*Solution:* The upper diagram in Fig. 9-14 shows the originally horizontal beam and also its deflected elastic curve. Clearly, the maximum deflection $\delta$ occurs at $A$. Also, a tangent to the elastic curve at $B$ is parallel to the horizontal axis of the undeflected beam. Therefore, $\delta = \Delta_{AB}$, the vertical displacement of $A$ from a tangent to the elastic curve at $B$. This vertical displacement $\Delta_{AB}$ equals the moment *about* $A$ of the moment-area between $A$ and $B$. The moment curve is a second-degree parabola, the moment at $B$ is $(wL)(L/2) = (150)(4)(2) = 1,200$ lb-ft, the moment-area $MA$ equals $(\frac{1}{3})(4)(1,200) = 1,600$ lb-ft$^2$, and the distance $\bar{x}_A$ from $A$ to the centroid of this moment-area is $(\frac{3}{4})(4) = 3$ ft (see Fig. 9-11c). Then $\delta_A$ equals the moment *about* $A$ of this moment-area divided by $EI$ and is

$$\delta_A = \Delta_{AB} = \frac{(3)(1,600)(1,728)}{(1,760,000)(8)} = 0.59 \text{ in.} \qquad Ans.$$

The *moment* of the moment-area is expressed in pound-feet$^3$. Its necessary conversion to pound-inches$^3$ is accomplished by multiplying by 1,728, the cube of 12.

**9-30.** Calculate the maximum deflection $\delta$ (at $A$) of the cantilever beam shown in Fig. 9-15. The beam is a W 16 × 71, for which $E$ is 29,000,000 psi and $I$ is 936.9 in.[4].

*Solution:* We shall find it simplest here to divide the moment diagram into two parts, one for the concentrated load of 3 kips at $A$ forming a triangle and the other

FIG. 9-15 Deflection of cantilever beam. Moment diagram.

for the distributed load forming a parabolic area tangent to the triangle at $A$. Here also, $\delta_A = \Delta_{AB}$, the sum of the moments about $A$ of the moment-areas between $A$ and $B$ divided by $EI$. Therefore

$$\delta_A = \Delta_{AB} = \frac{[(216)(8) + (576)(9)](1{,}000)(1{,}728)}{(29{,}000{,}000)(936.9)} = 0.439 \text{ in.} \qquad Ans.$$

Since the moment of the moment-area here is expressed in kip-feet³, the necessary conversion to pound-inches³ is made by multiplying by 1,000 and 1,728.

## PROBLEMS

**9-31.** In Prob. 9-29 (Fig. 9-14) let $L = 12$ ft and $w = 29$ lb per in. Calculate the maximum deflection $\delta_A$ at the left end $A$ if the beam is W 8 × 24 structural-steel section.

**9-32.** Using the moment-area method, check the deflections obtained in Prob. 9-9.

**9-33.** A 6- by 6-in. rough (full-size) Douglas fir timber beam cantilevers a distance of 6 ft horizontally out of a completely fixed support at its right end $B$. Calculate the deflection $\delta_A$ at its left end $A$ that would be caused by a concentrated load there of 1,000 lb. *Ans.* $\delta_A = 0.656$ in.

**9-34.** The beam shown in Fig. 9-16 is a 3-in.-diameter standard steel pipe. Calculate the maximum deflection $\delta_A$ at the left end $A$. Use $E = 20{,}000{,}000$ psi.

FIG. 9-16 Prob. 9-34.     FIG. 9-17 Prob. 9-35.     FIG. 9-18 Prob. 9-36.

**9-35.** The cantilever beam in Fig. 9-17 is a 2-in.-diameter solid wood rod of Douglas fir completely fixed at $B$. Calculate ($a$) the deflection $\delta_C$ of point $C$ and ($b$) the deflection $\delta_A$ of point $A$. ($c$) As a check on $\delta_A$, calculate $\theta_C$ and multiply $\theta_C$ by 48 in. to obtain the difference in deflections between $A$ and $C$.

*Ans.* $\delta_C = 0.45$ in.; $\delta_A = 1.35$ in.

**9-36.** In Fig. 9-18 is shown a 1- by 4-in. rough board of southern cypress 24 in. long fixed at its right end $B$ (1-in. dimension is vertical). Calculate the maximum deflection $\delta_A$ at $A$.

## 9-9 Bending-Moment Diagrams by Parts

In the application of the two moment-area principles, we are concerned with the *area* in the bending-moment diagram between verticals drawn through two points, with the location of the centroid of that area, and with the *moment of that area* about one of these points. It is advantageous, therefore, that we draw these moment diagrams in their simplest possible form. In Fig. 9-19, for example, are shown the three different moment diagrams that might have been drawn for the beam in Prob. 9-56, shown at top. All would have given identical results in equally simple manner. In Fig. 9-20 are shown three possible moment diagrams for the beam in Prob. 9-57. Of these, diagram $c$

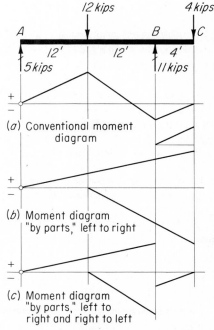

FIG. 9-19 Moment diagrams "by parts."

FIG. 9-20 Moment diagrams "by parts."

will give the desired results in the simplest manner (see procedure in Art. 9-11).

In a moment diagram "by parts," a separate moment diagram is drawn for each force or load showing its moment-effect along the beam, positive (upward-acting) forces and loads giving positive moment ordinates, and vice versa. As illustrated in Fig. 9-19 and 9-20 and also in Figs. 9-21 and 9-22, these separate diagrams may be drawn from left to right or from right to left, or they may be drawn from left to right up to a certain point and then from right to left back to that same point, as illustrated in Fig. 9-20c. The best form to use for any particular problem depends on the type of load and on its position along the beam. Hence only a general *procedure*, such as is outlined in Art. 9-11, can be used as a guide.

## PROBLEMS

**9-37.** Calculate the values of the principal ordinates in the three moment diagrams shown in Fig. 9-19. Show that the algebraic sums of the moment-areas of all three diagrams are equal, and also that the algebraic sums of the moments about $A$ of these moment-areas are equal.

**9-38.** Evaluate the principal ordinates in the three moment diagrams shown in Fig. 9-20. Show that the moment-areas of all three diagrams are equal and also that the moments about $A$ of these moment-areas are also equal.

**9-39.** Draw three separate moment diagrams, similar to those in Fig. 9-20, for the beam in Fig. 8-21. Evaluate the principal ordinates if $P_1 = 8$ kips and $P_2 = 7$ kips.

**9-40.** For the loaded beam in Fig. 9-24, draw the conventional moment diagram and the left-to-right and the right-to-left moment diagrams "by parts," and show that the algebraic sums of the moment-areas of all are equal and that the moment at $B$ is $-24$ kip-ft in each diagram.

**9-41.** For the beam shown in Fig. 9-26 draw the conventional moment diagram and the left-to-right and right-to-left moment diagrams "by parts." Show that the moment-areas of the three diagrams are equal and that the moments about $A$ of these areas are also equal.

## 9-10 Deflection of Symmetrically Loaded Simple Beams

A symmetrically loaded simple beam has complete symmetry of beam, load, and reactions about a vertical center line. Clearly, then, the maximum deflection of such a beam always occurs at the center. A tangent to the elastic curve at the point of maximum deflection will always be horizontal. This fact greatly simplifies the solution of this type of problem, as is illustrated in Probs. 9-42 and 9-43.

## ILLUSTRATIVE PROBLEMS

**9-42.** A symmetrically loaded simple beam, W 8 × 24, is shown in Fig. 9-21. Calculate (a) the maximum deflection $\delta$ at the center of the beam and (b) the deflection at a random point $D$, 3 ft from the left end.

FIG. 9-21 Deflection of a symmetrically loaded simple beam. (a) Beam and elastic curve. (b) Moment-area between $A$ and $B$, moment diagram. (c) Moment-area between $D$ and $B$ moment diagram.

*Solution:* (a) When a tangent is drawn to the elastic curve at the center $B$, the point of maximum deflection, we see clearly that this maximum deflection $\delta_B$ is equal to $\Delta_{AB}$, the displacement of point $A$ from the tangent at $B$. According to the second moment-area principle, this displacement $\Delta_{AB}$ equals the moment *about* $A$ of the moment-area between $A$ and $B$, divided by $EI$. A conventional moment diagram is most suitable here. Therefore, since $E = 29,000,000$ psi and $I = 82.5$ in.$^4$,

$$[M_A] \qquad \Delta_{AB} = \delta_B = \frac{[(90)(4) + (90)(7.5)](1,000)(1,728)}{(29,000,000)(82.5)} = 0.749 \text{ in.} \qquad Ans.$$

(b) To obtain the actual deflection $\delta_D$ at $D$, we must first calculate $\Delta_{AB}$ and $\Delta_{DB}$. Then $\delta_D = \Delta_{AB} - \Delta_{DB}$, as is clearly seen in Fig. 9-21. In part (a), $\Delta_{AB}$ was found to equal 0.749 in. We may now obtain $\Delta_{DB}$ by taking moments *about* $D$ of the moment-area between $D$ and $B$, shown in Fig. 9-21c, and dividing the result by $EI$. That is,

$$[M_D] \qquad \Delta_{DB} = \frac{[(\tfrac{1}{2})(3)(15)(2) + (3)(15)(1.5) + (3)(30)(4.5)](1,000)(1,728)}{(29,000,000)(82.5)} = 0.374 \text{ in.}$$

Then

$$\delta_D = 0.749 - 0.374 = 0.375 \text{ in.} \qquad Ans.$$

FIG. 9-22 Deflection of a symmetrically loaded simple beam. Moment diagram "by parts." Moment-area between $A$ and $B$.

In this example, the conventional moment diagram was used because a moment diagram "by parts" had no particular advantage. In Prob. 9-43 it will have considerable advantage and will be used.

**9-43.** Calculate the maximum deflection $\delta$ of the 10- by 16-in. Douglas fir timber beam shown in Fig. 9-22.

*Solution:* The maximum deflection $\delta$ occurs at the center of the beam. A tangent to the elastic curve at the center $B$ will be horizontal. Consequently, the displacement $\Delta_{AB}$ at $A$ equals the deflection $\delta_B$ at $B$. We shall find it simplest here to draw the moment diagram "by parts" from left to right *away from* the point $A$. The upward-acting (positive) reaction gives a positive triangular moment-area between $A$ and $B$. The downward-acting (negative) uniform load gives a negative parabolic moment-area between $D$ and $B$. Then $\Delta_{AB} = \delta_B$ and is also equal to the algebraic sum of the moments *about* $A$ of the moment-areas between $A$ and $B$. Since $E = 1,760,000$ psi and $I = 2,948$ in.[4], we obtain

$$[M_A] \qquad \Delta_{AB} = \delta_B = \frac{[(294)(4.67) - (32)(6)](1,000)(1,728)}{(1,760,000)(2,948)} = 0.393 \text{ in.} \qquad Ans.$$

## PROBLEMS

**9-44.** A 3- by 12-in. rough Douglas fir plank 18 ft long is laid flat on supports at its outer ends. Calculate the maximum deflection $\delta$ caused by a man weighing 176 lb standing at the center of the plank.

**9-45.** In Prob. 9-44 calculate the deflection $\delta$ at a point 3 ft from the left end.
*Ans.* $\delta = 0.375$ in.

**9-46.** A structural-steel beam, W 12 $\times$ 36, 24 ft long is supported at its ends and carries a uniformly distributed load of 1 kip per ft. Calculate the maximum deflection $\delta$ of this beam.

**9-47.** A simple beam 24 ft long of structural steel, W 12 $\times$ 50, supports three 8-kip concentrated loads applied, respectively, at 6, 12, and 18 ft from the left end. Calculate (*a*) the maximum deflection $\delta_{max}$ at the center and (*b*) the deflection $\delta_6$ at a point 6 ft from the left end.    *Ans.* (*a*) $\delta_{max} = 0.826$ in.; (*b*) $\delta_6 = 0.577$ in.

**9-48.** A 2- by 12-in. rough (full-size) Douglas fir plank 16 ft long laid flat is supported at its outer ends. It carries a uniform load of 150 lb per ft on the extreme left 4 ft and a similar load on the extreme right 4 ft. Calculate the maximum deflection $\delta$.

**9-49.** Calculate the deflection $\delta_3$ of the beam in Prob. 9-43 at a point 3 ft from the left end.

**9-50.** Determine the maximum value of $EI\delta$ in Prob. 9-19.

**9-51.** Calculate the maximum $EI\delta$ in Prob. 9-21.

**9-52.** A steel beam, W 21 $\times$ 73, spans 40 ft between the exterior walls of a building. It supports a uniform load of 1,250 lb per ft, which includes its own weight. (*a*) Calculate the maximum deflection $\delta$. (*b*) Select the most economical W section for which $\delta$ will not exceed $\frac{1}{360}$ of the span. (*c*) By proportion, what is $\delta$ of the beam selected?    *Ans.* (*a*) $\delta = -1.55$ in.; (*b*) W 24 $\times$ 76; (*c*) $\delta = -1.18$ in.

**9-53.** Solve Prob. 9-52 for a W 12 $\times$ 50 steel beam whose span is 24 ft and whose load is 1,450 lb per ft.

**9-54.** Calculate the maximum deflection of a 6- by 14-in. timber beam of southern pine S4S whose simple span $L$ is 16 ft and uniformly distributed load $w$ is 500 lb per ft, which includes weight of beam.    *Ans.* $\delta = -0.372$ in.

**9-55.** In Prob. 8-23, adjust the design load on beam $B1$ to include the actual weight of beam. Then (*a*) calculate the maximum deflection $\delta$ of the beam. (*b*) If the permissible deflection is $\frac{1}{360}$ of the span, what other W section would be most economical? (*c*) By proportion, calculate the deflection of the beam selected.

## 9-11 Deflection of Unsymmetrically Loaded Beams. Overhanging Beams

Lack of symmetry of either load or reactions (overhang) makes the determination of deflection a somewhat more extensive problem. Determination of *maximum* deflection caused by distributed loads usually requires the solution of a cubic, or higher, equation. Such problems, while not too difficult to solve, are beyond the intended scope of this text. The following procedures and examples will indicate the method of solution in problems with concentrated loads.

### Procedure

1. Draw the beam, its probable elastic curve, and the moment diagram "by parts" (usually) *from* the support farthest from the load (concentrated or resultant). Moments may be in kip-feet, pound-feet, or pound-inches.

2. Draw a tangent to the elastic curve of the beam at the support farthest from load or from the resultant of loads. Using the second principle, calculate desired displacement $\Delta$ from this tangent. To obtain $\Delta$ in inches, the moment must be in pound-inches or must be converted to pound-inches.
3. Calculate the slope-angle $\theta$ of this tangent: $\theta = \tan \theta = \Delta/L$, where $\Delta$ and $L$ must be in inches.
4. Using the first principle, solve for distance $m$ from point of maximum deflection $\delta$ to the support *farthest* (usually) from load, which support should also be point of beginning of the moment diagram. The moment *must* be in pound-inches.
5. Draw a second tangent to the elastic curve at point of maximum deflection (this tangent will be horizontal). Then calculate $\delta_{max} = \Delta$, using the second principle. The moment of the area of the moment diagram *must* be converted to pound-inches in this calculation.

NOTE: No set of rules can be established to cover all cases. In general, however, the origin of the moment diagram "by parts" and the *first* tangent to the elastic curve should be at the same end of the beam. If one scheme does not work out to be simple, try another.

## ILLUSTRATIVE PROBLEMS

**9-56.** Calculate the maximum deflection $\delta$ of the unsymmetrically loaded beam shown in Fig. 9-23. The beam is a 6- by 6-in. (full-size) timber; $E = 1,760,000$ psi and $I = 108$ in.[4].

*Solution:* The probable elastic curve of the beam having been drawn, a tangent is drawn to this curve at $B$, farthest from load. Point $B$ is also the starting point of the moment diagram "by parts." Positive (upward-acting) forces give positive moment-areas and vice versa. $\Delta_{AB}$ is the displacement of $A$ from a tangent at $B$ and is equal to the moment *about* $A$ of the moment-area between $A$ and $B$. That is, using the second principle, we have

$$[M_A] \qquad \Delta_{AB} = \frac{[(67,500)(5) - (27,000)(2)](1,728)}{(1,760,000)(108)} = 2.58 \text{ in.}$$

Since the slope-angle $\theta_B$ (in radians) is very small, its tangent equals the angle. Therefore

$$\theta_B = \tan \theta_B = \frac{\Delta_{AB}}{L} = \frac{2.58}{(15)(12)} = 0.014\,35$$

We may now solve for the distance $m$ from $B$ to the point of maximum deflection $C$. At $C$ the moment equals $600m$. The moment-area between $B$ and $C$ is

$$(\tfrac{1}{2})(m)(600m) = 300m^2$$

which, according to the first principle, equals $EI\theta_B$. Hence, since moment-area $BC = EI\theta_B$,

$$300m^2 = (1,760,000)(108)(0.014\,35) = 2,720,000$$

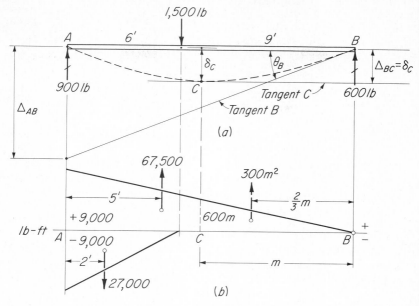

FIG. 9-23 Maximum deflection of unsymmetrically loaded simple beam. Moment diagram "by parts."

from which

$$m^2 = \frac{2,720,000}{300} = 9,066 \qquad \text{and} \qquad m = 95.2 \text{ in.} = 7.93 \text{ ft}$$

A second tangent, drawn to the elastic curve at $C$, the point of maximum deflection, will be horizontal. The displacement $\Delta_{BC}$ of point $B$ from this tangent, according to the second principle, equals the moment about $B$ of the moment-area between $B$ and $C$. But $\Delta_{BC}$ equals the required maximum deflection $\delta_C$. Therefore

$$\delta_C = \Delta_{BC} = \frac{(300)(9,066)(\frac{2}{3})(95.2)}{(1,760,000)(108)} = 0.91 \text{ in.} \qquad Ans.$$

**9-57.** A steel beam, W 8 × 48, overhanging the right support carries two concentrated loads, as shown in Fig. 9-24 ($E = 29,000,000$ psi; $I = 183.7$ in.⁴). Calculate (a) the maximum deflection $\delta_D$ between supports $A$ and $B$ and (b) deflection $\delta_C$ of the right end $C$, both from the original horizontal position of the beam.

*Solution: (a)* The probable elastic curve as drawn is undoubtedly correct, at least between $A$ and $D$. In this problem the distance $m$ is most easily obtained if the moment diagram "by parts" is drawn *from A to B.* The diagram could be correctly *continued* to $C$, but the part on the overhang may also be drawn from $C$ to $B$, giving a much simpler form.[1] We shall now draw a tangent to the elastic curve at $A$ so

[1] If a "conventional" moment diagram is also drawn, a little study will show that the moment-area between any two points, and also the moment about any point of such moment-area, is the same for the conventional diagram and for the diagram "by parts" in Fig. 9-24.

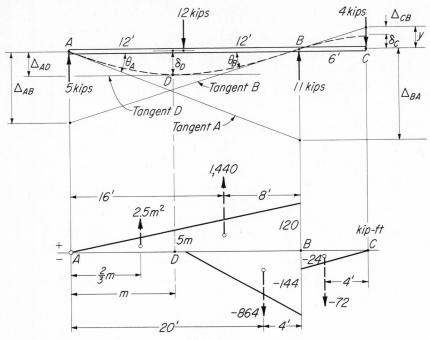

FIG. 9-24 Maximum deflection of overhanging beam. Moment diagram "by parts."

that we may solve for $\Delta_{BA}$, $\theta_A$, and $m$, in that order. Then

$$[M_B] \qquad \Delta_{BA} = \frac{[(1,440)(8) - (864)(4)](1,000)(1,728)}{(29,000,000)(183.7)} = 2.613 \text{ in.}$$

Also, now,

$$\theta_A = \tan \theta_A = \frac{\Delta_{BA}}{L} = \frac{2.613}{(24)(12)} = 0.009\ 08$$

and, since moment-area $AD = EI\theta_A$,

$$(2.5m^2)(1,000) = (29,000,000)(183.7)(0.009\ 08) = 48,300,000$$

from which

$$m^2 = \frac{48,300,000}{2,500} = 19,300 \qquad \text{and} \qquad m = 139 \text{ in.} = 11.58 \text{ ft}$$

Had $m$ exceeded 12 ft, the solution would have been incorrect, because the moment-area between $A$ and $D$ would not then have equaled $2.5m^2$.

In order now to solve for $\delta_D$, we draw a second tangent $D$ to the elastic curve at $D$. Clearly, now, $\delta_D = \Delta_{AD}$, the displacement of $A$ from tangent $D$. Then, using the second principle, we have

$$[M_A] \qquad \Delta_{AD} = \delta_D = \frac{(2.5)(1,000)(19,300)(\frac{2}{3})(139)}{(29,000,000)(183.7)} = 0.838 \text{ in.} \qquad \textit{Ans.}$$

(b) To obtain the deflection $\delta_C$, we must draw a third tangent $B$ to the elastic curve at $B$, after which we calculate $\Delta_{AB}$ at $A$ and, by proportion, $y$ at $C$. The slope of the tangent at $B$ will be positive (as it is in *this* problem) if the moment about $A$ of the moment-area between $A$ and $B$ is positive and will be negative if this moment is negative. Then

$$[M_A] \qquad \Delta_{AB} = \frac{[(1,440)(16) - (864)(20)](1,000)(1,728)}{(29,000,000)(183.7)} = 1.87 \text{ in.}$$

We now obtain $y$ by proportion thus:

$$\frac{y}{6} = \frac{\Delta_{AB}}{24} \qquad \text{or} \qquad y = \frac{6}{24}1.87 = 0.467 \text{ in.}$$

The displacement $\Delta_{CB}$ of point $C$ from the tangent at $B$ is

$$[M_C] \qquad \Delta_{CB} = \frac{(-72)(4)(1,000)(1,728)}{(29,000,000)(183.7)} = -0.094 \text{ in.}$$

The deflection $\delta_C$ at $C$ then is

$$\delta_C = y - \Delta_{CB} = 0.467 \text{ in.} - 0.094 \text{ in.} = 0.373 \text{ in.} \qquad Ans.$$

Since $\Delta_{CB}$ is less than $y$, the deflection $\delta_C$ is *upward* from the original horizontal position of the beam.

### PROBLEMS

**9-58.** A simple beam, W 16 × 45, 30 ft long supports a single concentrated load of 15 kips at a point 12 ft from the left end. Calculate the maximum deflection of this beam ($E = 29,000,000$ psi; $I = 583.3$ in.⁴).

**9-59.** In Fig. 9-25 is shown a wood board 8 in. wide by 1 in. thick, laid flat. Calculate the maximum deflection $\delta$ caused by loads as shown. Use $E = 1,760,000$ psi.

FIG. 9-25 Prob. 9-59.                    FIG. 9-26 Prob. 9-60.

**9-60.** The beam in Fig. 9-26 is a 2- by 6-in. rough wood plank, $E = 1,760,000$ psi, laid flat. Calculate the deflection $\delta$ at the center (not maximum).

**9-61.** The beam shown in Fig. 9-27 is a full-size 8- by 8-in. timber, for which $E = 1,760,000$ psi. Calculate the deflection $\delta_C$ of its extreme right end $C$.

Ans. $\delta_C = 0.207$ in.

**9-62.** Calculate the deflection of end $C$ of the beam shown in Fig. 9-28. The beam is a W 8 × 28 structural-steel section.

**9-63.** Determine the maximum deflection between supports $A$ and $B$ of the W 8 × 28 structural-steel beam shown in Fig. 9-28.         Ans. $\delta = 0.338$ in.

FIG. 9-27  Prob. 9-61.

FIG. 9-28  Probs. 9-62 and 9-63.

### 9-12 Derivation of Deflection Formulas

In the preceding articles, deflections of beams supporting specified loads were determined by the beam-diagram method and the moment-area method. These methods will now be used to derive a series of simple formulas giving the deflections of various types of beams in terms of their lengths $L$ and loads $P$ or $w$, as the case may be. Formulas for the cases $a$, $b$, and $c$ shown in Fig. 9-4 will be derived by the beam-diagram method. The formula for a simple beam with a uniformly distributed load and that for case $d$ in Fig. 9-4 will be derived by the moment-area method.

### 9-13 Cantilever Beam with Concentrated Load at Free End

The load, shear, and moment diagrams are drawn in the usual manner, as shown in Fig. 9-29, in accordance with the laws of beam diagrams. The shear ordinates have a constant value of $P$ between $A$ and $B$. The moment at the free end $A$ is zero. Hence the moment at $B$ is $PL$, the area under the shear curve between $A$ and $B$.

From the physical conditions of the problem, it is apparent that the beam must remain horizontal at $B$, indicating that the $EI\theta$ ordinate at $B$ is zero. The shape of the curve in the slope diagram must be increasing negative, as are the moment ordinates above. Hence the initial $EI\theta$ ordinate is positive and has a value, according to the second law, of $(\frac{1}{2})(L)(PL) = PL^2/2$, or $PL^2/2EI$ after being divided by $EI$.

Clearly, the deflection at $B$ is zero; that is, $\delta_B = 0$. Hence the maximum deflection ordinate at $A$ is $(\frac{2}{3})(L)(PL^2/2EI) = PL^3/3EI$, the area under the slope curve between $A$ and $B$.

Therefore the maximum deflection at the free end of a cantilever beam due to a load $P$ at that end is

$$\delta = \frac{PL^3}{3EI} \qquad (9\text{-}5)$$

### 9-14 Cantilever Beam with Uniformly Distributed Load

From the physical conditions of this beam we find that the shear $V$ and moment $M$ at $A$ are both zero and that at $B$ the slope $\theta$ and deflection $\delta$ are zero.

The load, shear, and moment diagrams are drawn, as illustrated in Fig. 9-30. The slope of the shear curve is constant and negative, as are the load ordinates

FIG. 9-29 Derivation of deflection formula.

FIG. 9-30 Derivation of deflection formula.

above. The shear at $B$ equals $wL$, the area under the load curve between $A$ and $B$.

The slope of the moment curve is increasing and negative, as are the shear ordinates above. The moment at $B$ is $(\frac{1}{2})(L)(wL) = wL^2/2$, the area under the shear curve above.

Clearly, the beam remains horizontal at $B$ and the $\theta$ ordinate at $B$ is therefore zero. The slope of the $\theta$ curve is increasing and negative, as are the moment ordinates above. Hence, since $\theta_B = 0$, the curve starts above the axis and the initial $\theta$ ordinate is $(\frac{1}{3})(L)(wL^2/2) = wL^3/6$, or $wL^3/6EI$ after being divided by $EI$.

Beginning at $A$, the slope of the deflection curve is positive and decreasing, as are the $\theta$ ordinates above. Because $\delta_B$ is zero, the curve must start below the axis, and the ordinate $\delta_A$ then equals $(\frac{3}{4})(L)(wL^3/6EI) = wL^4/8EI$ or $WL^3/8EI$, since $W = wL$.

Hence, the maximum deflection at the free end of a cantilever beam due to a uniformly distributed load is

$$\delta = \frac{WL^3}{8EI} \tag{9-6}$$

### 9-15 Simple Beam with Concentrated Load at Center

Having drawn the load, shear, and moment diagrams, as in Fig. 9-31, we find the maximum moment to be $(P/2)(L/2) = PL/4$, the area under the shear curve between $A$ and $B$.

Because of symmetry of beam and of loading, it is clear that the deflection $\delta$ will be maximum at the center $B$ of the beam, at which point, therefore, the slope of the beam will necessarily be zero. Also, clearly, the deflections at supports $A$ and $C$ will be zero.

Between $A$ and $B$ the slope of the $\theta$ curve is positive and increasing, and between $B$ and $C$ it is positive and decreasing, as are the moment ordinates.

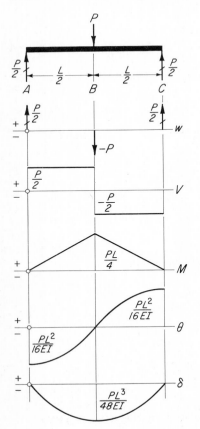

FIG. 9-31 Derivation of deflection formula.

Since $\theta_B = 0$, the curve must start at $A$ with a negative ordinate (according to the second law)

$$\theta_A = \frac{1}{2} \cdot \frac{L}{2} \cdot \frac{PL}{4} \cdot \frac{1}{EI} = \frac{PL^2}{16EI}$$

Starting with zero ordinate at $A$, the slope of the deflection curve is negative and decreasing from $A$ to $B$ and is positive and increasing from $B$ to $C$, as are the $\theta$ ordinates above. The maximum deflection at the center $B$, according to the second law, then is

$$\frac{2}{3} \cdot \frac{L}{2} \cdot \frac{PL^2}{16EI}$$

or the maximum deflection at the center of a simple beam due to a concentrated load $P$ at the center is

$$\delta = \frac{PL^3}{48EI} \tag{9-7}$$

## 9-16 Simple Beam with Uniformly Distributed Load

In this derivation we shall use the moment-area method, which is the simplest for this case.

The beam and its deflected elastic curve are shown in Fig. 9-32. Clearly,

FIG. 9-32 Derivation of deflection formula. Moment diagram "by parts."

the maximum deflection occurs at the center $C$, and a tangent to the elastic curve at $C$ will be horizontal, that is, parallel to the undeflected axis of the beam. Hence the maximum deflection $\delta_C$ equals $\Delta_{AC}$, which in turn equals the moment about $A$ of the moment-area between $A$ and $C$, divided by $EI$. That is,

$$[M_A] \qquad \delta_{max} = \Delta_{AC} = \left[ \left( \frac{wL^3}{16} \frac{L}{3} \right) - \left( \frac{wL^3}{48} \frac{3L}{8} \right) \right] \frac{1}{EI} = \frac{5wL^4}{384EI} \qquad (9\text{-}8)$$

### 9-17 Simple Beam with Two Equal Concentrated Loads Symmetrically Placed

Here also we shall use the moment-area method because it gives us the required formula in the simplest possible manner.

The loaded beam and its deflected elastic curve are shown in Fig. 9-33. Because of symmetry of forces, the maximum deflection $\delta$ occurs at the center. A tangent at this point $C$ will be horizontal. Clearly, therefore, $\delta_{max} = \Delta_{AC}$, which in turn equals the moment about $A$ of the moment-area between $A$ and $C$, divided by $EI$. Hence

$$[M_A] \qquad \delta_{max} = \Delta_{AC} = \left\{ \left( \frac{Pa^2}{2} \frac{2a}{3} \right) + \left[ \frac{Pa(L - 2a)}{2} \right] \left( a + \frac{L - 2a}{4} \right) \right\} \frac{1}{EI}$$

which when simplified becomes

$$\frac{Pa}{24EI}(3L^2 - 4a^2) \qquad (9\text{-}9)$$

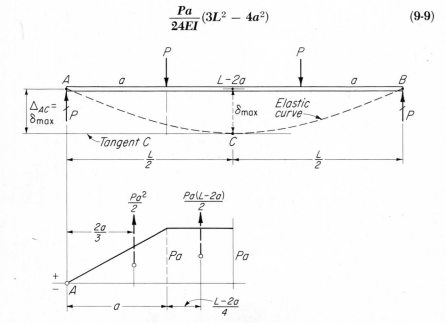

FIG. 9-33 Derivation of deflection formula. Conventional moment diagram.

## 9-18 Summary of Deflection Formulas

In the preceding articles, the deflection formulas for five commonly used beams and beam loadings were derived. A score or more additional formulas for other types of beams and loadings and for partial loadings could be derived, some in similar manner and others with the aid of calculus. In practice, the vast majority of deflection problems encountered by engineers and architects are solved by the use of such formulas. In Table 9-1 are listed the five derived

**Table 9-1  BEAM-DEFLECTION FORMULAS**

$P (lb)$     $L (in)$     $E$ - Modulus of Elasticity $(psi)$
$I$ - Moment of Inertia $(in^4)$

| Case no. | Beam and load | Maximum moment | Maximum deflection |
|---|---|---|---|
| 1 ✓ | | $M = -PL$ | $\delta = -\dfrac{PL^3}{3EI}$ |
| 2 | | $M = -Pa$ | $\delta = -\dfrac{Pa^2}{6EI}(3L - a)$<br><br>($a$ and $L$ must be in inches) |
| 3 ✓ | | $M = -\dfrac{wL^2}{2}$<br><br>$= -\dfrac{WL}{2}$ | $\delta = -\dfrac{wL^4}{8EI}$<br><br>$= -\dfrac{WL^3}{8EI}$ |
| 4 | | $M = -\dfrac{wL^2}{6}$<br><br>$= -\dfrac{WL}{3}$ | $\delta = -\dfrac{wL^4}{30EI}$<br><br>$= -\dfrac{WL^3}{15EI}$ |
| 5 *Ignore.* | | $M = M_L$ | $\delta = -\dfrac{0.0642 ML^2}{EI}$<br><br>at $x = 0.423L$<br><br>($M$ must be in pound-inches) |
| 6 ✓ | | $M = \dfrac{PL}{4}$ | $\delta = -\dfrac{PL^3}{48EI}$ |

**Table 9-1  BEAM-DEFLECTION FORMULAS** *(continued)*

| Case no. | Beam and load | Maximum moment | Maximum deflection |
|---|---|---|---|
| 7 | | $M = \dfrac{Pab}{L}$ | $\delta = -\dfrac{Pab(a + 2b)\sqrt{3a(a + 2b)}}{27EIL}$<br><br>at $x = \sqrt{\dfrac{a(a + 2b)}{3}}$<br><br>when $x = <a$ and $a = >b$<br><br>($a$, $b$, and $L$ must be in inches) |
| 8 | | $M = Pa$ | $\delta = -\dfrac{Pa}{24EI}(3L^2 - 4a^2)$<br><br>($a$ and $L$ must be in inches) |
| 9 | | $M = \dfrac{wL^2}{8}$<br><br>$= \dfrac{WL}{8}$ | $\delta = -\dfrac{5wL^4}{384EI}$<br><br>$= -\dfrac{5WL^3}{384EI}$ |
| 10 | | $M = \dfrac{wL^2}{9\sqrt{3}}$<br><br>$= 0.1283WL$<br><br>at $x = 0.577L$ | $\delta = -\dfrac{0.01304WL^3}{EI}$<br><br>at $x = 0.519L$ |
| 11 | | $M = \dfrac{wL^2}{12}$<br><br>$= \dfrac{WL}{6}$ | $\delta = -\dfrac{wL^4}{120EI}$<br><br>$= -\dfrac{WL^3}{60EI}$ |

Important notes: To obtain deflection $\delta$ in inches:

In all cases with distributed loads, use the formulas containing $W$. (If formulas containing $w$ are used, $w$ must be in pounds per inch.)

$P$ and $W$ must be in pounds.

$E$ must be in pounds per square inch.

$I$ must be in inches$^4$.

If $L$ is given in feet, multiply $L^2$ by $(12)^2 = 144$; multiply $L^3$ by $(12)^3 = 1,728$, or $(144)(12)$; multiply $L^4$ by 20,736, or $(144)(144)$.

deflection formulas plus an additional six, together with the formulas giving the maximum moment in each case.

The calculations of deflections due to irregular or partial loads become rather complex and are rarely performed in practice. For the practical designer, therefore, the following facts are useful: The deflection at the mid-span of a simple beam due to a concentrated load applied anywhere on the beam differs from the maximum deflection by less than 3 percent. Similarly, the mid-span deflection caused by a partial distributed load, covering not less than half of a simple beam, is less than 5 percent less than the maximum due to a load of equivalent total magnitude distributed uniformly over the entire span.

## PROBLEMS See p. 184

NOTE: Use the appropriate deflection formula listed in Table 9-1.

**9-64.** Calculate the maximum deflection called for in parts (a) and (c) of Prob. 9-9.

**9-65.** Check, by formulas, the maximum deflections of the two beams described in Prob. 9-10.

**9-66.** Compute the deflections called for in Prob. 9-11.
$$Ans. \ (a) \ \delta = -0.152 \ in.; \ (b) \ \delta = -0.202 \ in.$$

**9-67.** Solve Prob. 9-12 by using the appropriate formula listed in Table 9-1.

**9-68.** Using the appropriate formula from Table 9-1, solve Prob. 9-13.
$$Ans. \ \delta = 4.52 \ in.; \ use \ 12\text{-}in. \ pipe$$

**9-69.** Calculate the deflections called for in Prob. 9-14.

**9-70.** Compute by formulas the deflections called for in Prob. 9-16.
$$Ans. \ \delta = -0.148 \ in.; \ -0.37 \ in.; \ -0.518 \ in.$$

**9-71.** Using the appropriate formula, calculate the deflection called for in Prob. 9-17.
$$Ans. \ \delta = -0.371 \ in.$$

**9-72.** Solve Prob. 9-18 by using the appropriate formula listed in Table 9-1.

**9-73.** Calculate by formula the deflection called for in Prob. 9-19.

**9-74.** Using the appropriate formulas from Table 9-1, calculate the deflection called for in Prob. 9-20.

**9-75.** Calculate, using the proper formulas, the maximum deflection $\delta$ of girder $G1$ in Prob. 8-23 that results from the three concentrated loads and the actual weight of the beam.

[NOTE: Calculate the deflection in three separate parts: (1) $\delta$ due to weight of beam, (2) $\delta$ due to the middle concentrated load, and (3) $\delta$ due to the two outer concentrated loads.]
$$Ans. \ \delta = -0.019 \ in. \ -0.306 \ in. \ -0.408 \ in. = -0.731 \ in.$$

## SUMMARY
### (By article number)

**9-1.** The **relationship between radius of curvature** $R$ **and stress** $s$ in a member subjected to pure bending is

$$s = \frac{cE}{R} \tag{9-1}$$

in which $c$ is the distance from the neutral plane to the extreme fiber and $E$ is the modulus of elasticity of the material.

**9-2.** The **relationship between radius of curvature** $R$ **and bending moment** $M$ is

$$R = \frac{EI}{M} \tag{9-2}$$

in which $E$ is the modulus of elasticity and $I$ is the moment of inertia of the section.

**9-4. Deflections by the beam-diagram method** are readily obtained only for cantilever beams and for simple beams with concentrated loads symmetrically placed. The five beam diagrams are drawn in accordance with the first and second laws of beam diagrams. Ordinates in the fourth (slope) and fifth (deflection) diagrams must be divided by $E$ and $I$.

**9-5.** The **moment-area method** is slightly more complex but is also more useful in that it is readily applicable to beams with unsymmetrical loads. The two moment-area principles should be memorized.

**9-12. Deflection formulas** are most commonly used in practice. Table 9-1 gives the formulas for 11 typical cases. In more difficult problems, approximate results, usually sufficiently close, may readily be obtained by calculating the mid-span deflection in the manner outlined in the second paragraph of Art. 9-18.

## REVIEW PROBLEMS

**9-76.** A horizontal cantilever beam 9 ft long supports a uniformly distributed load of 800 lb per ft, which includes its own weight. Using the beam-diagram method, calculate the deflection $\delta$ at the free end if the beam is (a) a W 10 × 21 steel section and (b) a 6- by 18-in. Douglas fir timber S4S.

*Ans.* (a) $\delta = -0.367$ in.; (b) $\delta = -0.262$ in.

**9-77.** Assume that the beam described in Prob. 9-76 is of steel W 12 × 27. Calculate the actual maximum deflection $\delta$ (a) by proportion, using the answer to Prob. 9-76, and (b) using the appropriate formula.

**9-78.** A simple beam 18 ft long supports two equal concentrated loads of 8 kips each at the third-points of the beam. Calculate the maximum value of $EI\delta$ using (a) the beam-diagram method, and (b) the moment-area method.

*Ans.* $EI\delta = -1,656$ kip-ft$^3$

**9-79.** Using the moment-area method, calculate the maximum deflection $\delta$ of girder G1 in Prob. 8-3. Adjust the load to include actual weight of beam. Check $\delta$ by the appropriate formulas.

**9-80.** A cantilever beam, W 8 × 17, extends 6 ft horizontally out from a completely solid abutment. It supports a single concentrated load of 2,900 lb at its free end. Using the beam-diagram method, calculate (a) the slope-angle $\theta$ at the free end in radians and degrees and (b) the maximum deflection $\delta$. Neglect weight of beam.

**9-81.** Consider a simple beam of length $L$ and with uniform load $w$ per foot. Using the moment-area method, show that the maximum slope-angle at either end is $wL^3/24EI$ and that the maximum deflection at the center is $5wL^4/384EI$.

**9-82.** Calculate the deflection $\delta$ of the beam mentioned in Prob. 9-42 and illustrated in Fig. 9-21 at the point of application $E$ of the left 5-kip load. Use the moment-area method.

**9-83.** Using the moment-area method, calculate the maximum deflection δ of the W 24 × 76 structural-steel beam shown in Fig. 9-34.  *Ans.* δ = 0.419 in.

FIG. 9-34 Prob. 9-83.

FIG. 9-35 Prob. 9-84.

**9-84.** The cantilever beam shown in Fig. 9-35 is a 2- by 6-in. wood plank (full size). Using the moment-area method, calculate the deflections at $C$ and $A$. Check the difference in the deflections at $C$ and $A$ by calculating $\theta_A$ and evaluating this difference in terms of $\theta_A$ and the distance $AC$. Use $E = 1,760,000$ psi.

*Ans.* $\delta_C = 0.942$ in.; $\delta_A = 1.46$ in.

**9-85.** Calculate the center deflection of the beam of Prob. 8-2 by the moment-area method. Neglect weight of beam.

**9-86.** Refer to Prob. 8-41. Calculate the slope $\theta_B$ over the support $B$. Use the second moment-area principle and solve "by parts." Also calculate the deflection at $C$.

## REVIEW QUESTIONS

**9-1.** What is meant by pure bending?

**9-2.** What is the elastic curve?

**9-3.** If a section of a bent member of prismatic cross section and of homogeneous material is in pure bending, is the radius of curvature of the elastic curve necessarily constant?

**9-4.** Express in equation form the mathematical relationship between unit stress $s$ and radius of curvature $R$.

**9-5.** Give the mathematical relationship between radius of curvature $R$ and bending moment $M$.

**9-6.** What is meant by the deflection of a beam?

**9-7.** What objectionable results can be caused by the excessive deflection (a) of a beam in a building and (b) of a part in a precision machine?

**9-8.** What is the commonly accepted maximum deflection of beams and joists to avoid cracking of plastered ceilings?

**9-9.** Describe the beam-diagram method of determining deflections.

**9-10.** Explain the first moment-area principle.

**9-11.** If a moment-area is expressed in kip-ft², how may a conversion to pound-inches² be accomplished, in order to obtain $\phi$ or $\theta$ in radians?

**9-12.** For very small angles (say, less than 3°), what relationship exists between the angle (in radians) and its tangent?

**9-13.** Give the essence of the second moment-area principle.

**9-14.** When is the displacement $\Delta$ of a point equal to the actual deflection $\delta$? Cite some cases.

**9-15.** When the moment of a moment-area is expressed in kip-feet$^3$, how may a conversion to pound-inches$^3$ be accomplished in order to obtain $\delta$ in inches?

**9-16.** Explain how a moment diagram "by parts" is constructed.

**9-17.** If in a beam-deflection problem $W$ is in pounds and $L$ in feet, what will be the units of $EI\delta$?

**9-18.** In the equation $\delta = (EI\delta)/EI$, in what units must each quantity be expressed to give $\delta$ in inches?

**9-19.** Give the units in which each of the following quantities in formulas 1 to 11, Table 9-1, must be expressed in order to give the deflections $\delta$ in inches: $P$, $W$, $w$, $L$, $a$ and $b$, $E$, and $I$.

# CHAPTER 10

# Statically Indeterminate Beams

## 10-1 Statically Indeterminate Beams

Any loaded beam for which the simple laws of statics ($\Sigma F = 0$ and $\Sigma M = 0$) are insufficient for the determination of *all* support reactions is said to be **statically indeterminate.** Such a condition exists whenever the number of unknown reactions (forces or moments) exceeds the number of available *independent* statics equations.

When a beam extends over several supports, as illustrated in Fig. 10-1a, it is said to be a **continuous beam.** The four unknown support reactions make this beam statically indeterminate, since only two *independent* statics equations, $\Sigma F = 0$ and $\Sigma M = 0$, are available. A similar situation exists in the "supported cantilever" beam in Fig. 10-1b. Here there are three unknowns, the support reactions $R_1$ and $V_2$ and the support reacting moment $M_2$, and only two independent equations can be written.

To solve problems of this type, therefore, we must introduce certain other relationships that are known to exist. According to the second moment-area principle, stated in Chap. 9, and using a slight modification of the subscripts, we have the following relationship:

**Second Moment-Area Principle:**

$$
\begin{bmatrix}
\text{The vertical displacement of any} \\
\text{point 1 on the elastic curve of a} \\
\text{beam from a tangent to the curve} \\
\text{at any other point 2}
\end{bmatrix}
=
\begin{bmatrix}
\text{the moment about } A \text{ of the} \\
\text{area in the moment diagram} \\
\text{lying between verticals through} \\
\text{1 and 2, divided by } EI
\end{bmatrix}
$$

We shall find this relationship useful in the following derivation.

FIG. 10-1 Statically indeterminate beams.

## 10-2 Derivation of the Three-Moment Equation

We shall now show that the second moment-area principle may be used to derive an equation expressing a relationship existing among the bending moments $M$ at any three consecutively adjacent supports of a continuous beam, such as supports 1, 2, and 3 of the beam shown in Fig. 10-2a.

Let us assume, for example, that $L_1$ and $L_2$ are any two adjacent spans cut out of a multiple-span continuous beam and that the loads shown merely represent any loads that may be applied on these two spans. If each span were a simple beam, the moments produced by the loads would be *positive*, and the simple-beam moment diagrams could then be drawn on the positive side of the axis, as shown in Fig. 10-2b. In Fig. 10-2c is shown the probable deflected "elastic curve" of this two-span section. The curvature over each of the three supports indicates tension in the top fibers, which in turn indicates negative moments at supports 1, 2, and 3. We may designate these moments $M_1$, $M_2$, and $M_3$ and may indicate them as ordinates on the *negative* side of the axis in Fig. 10-2b. When now we join the lower ends of these ordinates with straight lines, we have the complete moment diagrams for spans $L_1$ and $L_2$. For simplicity, the trapezoidal areas are divided into two triangles.

While it is true that the bending moments at interior supports of continuous beams are negative in the vast majority of problems, it is, however, not always so. We shall, therefore, in this derivation, consider them to be positive. By so doing, each moment solved for by the three-moment equation will bear its true sign, positive or negative.

When this beam is considered to be weightless and is not otherwise loaded, a tangent to its elastic curve at support 2 would coincide with the undeflected horizontal axis of the beam and would, of course, be a horizontal line passing through support points 1, 2, and 3. But a tangent to the elastic curve at 2 of the *loaded* beam would slope, as shown in Fig. 10-2c, and the displacements at 1 and 3 of this tangent from the horizontal would be $\Delta_1$ and $\Delta_3$, positive and negative, respectively.

If we now take moments about point 1 of the moment-diagram areas in span $L_1$, we would, according to the second moment-area principle, obtain $EI\Delta_1$. Similarly, moments about 3 of the moment-diagram areas in span $L_2$ would yield $EI\Delta_3$. All moment-areas are assumed to be positive, or upward-acting. If $A_1$ and $A_2$ are the simple-beam moment areas in spans $L_1$ and $L_2$, and $\bar{a}_1$ and $\bar{b}_2$ are the distances from 1 and 3, respectively, to the centroids of those areas, we obtain, by moments about point 1,

$[M_1]$    $$EI\Delta_1 = A_1\bar{a}_1 + M_1\left(\frac{L_1}{2}\frac{L_1}{3}\right) + M_2\left(\frac{L_1}{2}\frac{2L_1}{3}\right)$$

or    $$\Delta_1 = \left(A_1\bar{a}_1 + \frac{M_1L_1{}^2}{6} + \frac{M_2L_1{}^2}{3}\right)\frac{1}{EI} \qquad (a)$$

FIG. 10-2 Derivation of the three-moment equation. (a) Any two spans of continuous beam. (b) Moment diagram. (c) Deflection diagram.

Similarly, by moments about point 3,

$$[M_3] \qquad EI\Delta_3 = A_2\bar{b}_2 + M_2\left(\frac{L_2}{2}\frac{2L_2}{3}\right) + M_3\left(\frac{L_2}{2}\frac{L_2}{3}\right)$$

or
$$\Delta_3 = \left(A_2\bar{b}_2 + \frac{M_2L_2{}^2}{3} + \frac{M_3L_2{}^2}{6}\right)\frac{1}{EI} \qquad (b)$$

A simple relationship may now be established: From Fig. 10-2c by similar triangles, if $\Delta_1$ is positive and $\Delta_3$ is negative,

$$\frac{\Delta_1}{L_1} = -\frac{\Delta_3}{L_2} \qquad (c)$$

If we now divide each term in Eq. (a) by $L_1$ and each term in Eq. (b) by $L_2$ and then equate the right sides of Eqs. (a) and (b) in accordance with the relationship established in Eq. (c), the $\Delta$ and $1/EI$ terms drop out and we obtain

$$\frac{A_1\bar{a}_1}{L_1} + \frac{M_1L_1}{6} + \frac{M_2L_1}{3} = -\left(\frac{A_2\bar{b}_2}{L_2} + \frac{M_2L_2}{3} + \frac{M_3L_2}{6}\right) \qquad (d)$$

If next we multiply each term in Eq. (d) by 6 and rearrange the terms, we obtain

$$M_1L_1 + 2M_2(L_1 + L_2) + M_3L_2 = -\frac{6A_1\bar{a}_1}{L_1} - \frac{6A_2\bar{b}_2}{L_2} \qquad (10\text{-}1)$$

Equation (10-1) is called the **three-moment equation.** It expresses, in terms of the loads on any two adjacent spans (such as 1–2 and 2–3), the relationship among the bending moments at the three consecutive supports (1, 2, and 3) of those two spans. The presumptions are (1) that the beam is initially straight before loading, (2) that its $E$ and $I$ are constant throughout its entire length, and (3) that the supports remain in a straight line after loading.

The three-moment equation (10-1) enables us to calculate the bending moments at *all* supports of continuous and other statically indeterminate beams. When these moments have been evaluated, the support reactions are easily calculated.

### 10-3 Load Terms for the Three-Moment Equation
The two terms on the right side of Eq. (10-1) are called **load terms.** For any given type of load, the value of a load term may be expressed in terms of the load and span.

Each type of load, except those that are symmetrically distributed about the vertical center line of a span, will have two slightly differing load terms, depending on whether the load is on the *left* or the *right* span of the two-span section under consideration.

For example, to determine the two load terms for a single off-center concentrated load $P$, as illustrated in Fig. 10-3, we draw the moment diagrams shown in Fig. 10-3$b$ and $f$. To obtain the left-span load term, the $A\bar{a}$ portion of the term $6A\bar{a}/L$ is obtained by calculating the moment of the moment-diagram area about point 1. For the right-span load term, the $A\bar{b}$ portion of $6A\bar{b}/L$ is obtained by moments about point 3.

This moment of the moment-area may be obtained in one of several ways. Perhaps the simplest of all, for the left span, is to construct the two triangles 1–2–$n$ and 1–$m$–$n$, as in Fig. 10-3$c$. Clearly, now, according to the principles of moments of areas outlined in Chap. 5, the moment about point 1 of the

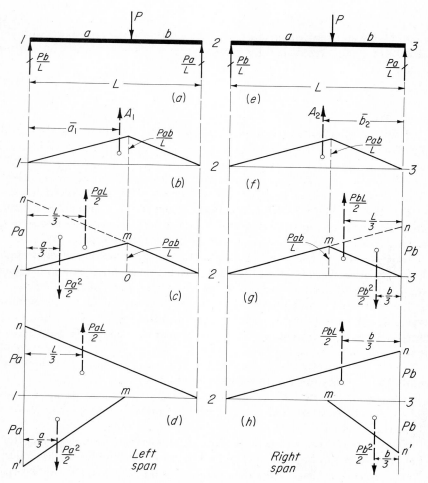

FIG. 10-3 Determination of the load terms.

moment-area 1–2–$m$ equals the moment of area 1–2–$n$ less the moment of area 1–$m$–$n$. Triangles $n$–1–2 and $m$–$o$–2 are similar. Hence, by proportion,

$$\frac{n-1}{L} = \frac{Pab/L}{b} \quad \text{or} \quad \frac{n-1}{L} = \frac{Pa}{L} \quad \text{and} \quad n-1 = Pa$$

Then by moments about point 1

$$[M_1] \qquad A_1\bar{a}_1 = \left(\frac{PaL}{2}\frac{L}{3}\right) - \left(\frac{Pa^2}{2}\frac{a}{3}\right) = \frac{Pa}{6}(L^2 - a^2) \qquad (a)$$

To obtain the complete left-span load term for this case, the above expression must be multiplied by $6/L$ [see Eq. (10-1)]. We then have

**Case 1, Left-Span Load Term:** $\qquad \dfrac{6A\bar{a}}{L} = \dfrac{Pa}{L}(L^2 - a^2) \qquad$ **(10-2)**

In similar manner, the right-span load term for case 1 is

**Case 1, Right-Span Load Term:** $\qquad \dfrac{6A\bar{b}}{L} = \dfrac{Pb}{L}(L^2 - b^2) \qquad$ **(10-3)**

But note here that the two triangles making up the moment diagram in Fig. 10-3$c$ could also be shown separately, as in Fig. 10-3$d$, without changing the value of $A\bar{a}$. In this illustration, the ordinate at any point from line 1–2 to curve $n$–2 represents the moment about point 1 of the reaction $Pa/L$; similarly, the ordinate at any point from line 1–$m$ to curve $n'$–$m$ represents the moment at that point of load $P$. From this it is clear that **a moment diagram may be constructed by parts.** This concept was used in deriving the three-moment equation (Fig. 10-2$b$) and will be of great advantage in determining the $A\bar{a}$ and $A\bar{b}$ portions of the other load terms listed in Table 10-1.

Because of symmetry of load, the load terms for a uniformly distributed load (Fig. 10-4$a$) are identical for left and right spans. In Fig. 10-4$b$ is shown the moment diagram (by parts) for this load on a left span. Ordinate $n$–1 represents the moment about 1 of the reaction at 2, which is $(wL/2)(L) = wL^2/2$. Area $n$–1–2 then equals $(\frac{1}{2})(L)(wL^2/2) = wL^3/4$. Similarly, ordinate $n'$–1 represents the moment about 1 of the load on the beam, which is $(wL)(L/2) = wL^2/2$. Area $n'$–1–2 then equals $(\frac{1}{3})(L)(wL^2/2) = wL^3/6$. Now, $A\bar{a}$ equals the moment about 1 of these two areas, or

$$[M_1] \qquad A\bar{a} = \left(\frac{wL^3}{4}\frac{L}{3}\right) - \left(\frac{wL^3}{6}\frac{L}{4}\right) = \frac{wL^4}{24} \qquad (b)$$

The load term is now obtained by multiplying Eq. ($b$) by $6/L$. That is,

**Case 2, Uniform Load, Any-Span Load Term:**

$$\frac{6A\bar{a}}{L} = \frac{6A\bar{b}}{L} = \frac{wL^3}{4} = \frac{WL^2}{4} \qquad (10\text{-}4)$$

**Table 10-1 LOAD TERMS FOR USE WITH THE THREE-MOMENT EQUATION**

$w$ = kips/ft (or lb/ft) of span; $W$ = total load on span.

| Left span $\dfrac{6A\bar{a}}{L}$ | Load no. | Type of loading | $\dfrac{6Ab}{L}$ right span |
|---|---|---|---|
| $\dfrac{Pa}{L}(L^2 - a^2)$ | 1 | | $\dfrac{Pb}{L}(L^2 - b^2)$ |
| $\dfrac{wL^3}{4} = \dfrac{WL^2}{4}$ | 2 | | $\dfrac{wL^3}{4} = \dfrac{WL^2}{4}$ |
| $\frac{8}{60}wL^3 = \frac{8}{30}WL^2$ | 3 | | $\frac{7}{60}wL^3 = \frac{7}{30}WL^2$ |
| $\frac{7}{60}wL^3 = \frac{7}{30}WL^2$ | 4 | | $\frac{8}{60}wL^3 = \frac{8}{30}WL^2$ |
| $\frac{5}{32}wL^3 = \frac{5}{16}WL^2$ | 5 | | $\frac{5}{32}wL^3 = \frac{5}{16}WL^2$ |
| $\dfrac{w(b^2 - a^2)}{4L}[2L^2 - (b^2 + a^2)]$ | 6 | | $\dfrac{w(d^2 - c^2)}{4L}[2L^2 - (d^2 + c^2)]$ |

The moment diagram (by parts) for a triangular load (Fig. 10-4c) on a *right span* is shown in Fig. 10-4d. When moments are taken about point 3, we obtain

$[M_3]$
$$A\bar{b} = \left(\frac{wL^3}{12}\frac{L}{3}\right) - \left(\frac{wL^3}{24}\frac{L}{5}\right) = \frac{wL^4}{36} - \frac{wL^4}{120} \qquad (c)$$

or
$$A\bar{b} = \frac{10wL^4}{360} - \frac{3wL^4}{360} = \frac{7wL^4}{360} = \frac{7WL^3}{180} \qquad (d)$$

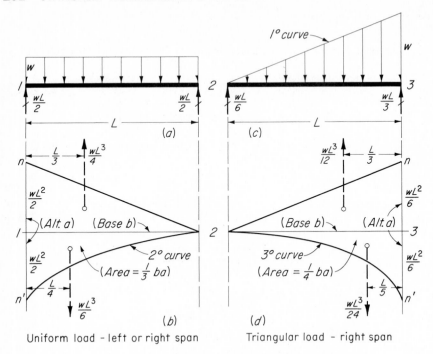

Uniform load - left or right span          Triangular load - right span

Fig. 10-4 Determination of the load terms.

since $W = wL/2$. Multiplying these terms by $6/L$ gives

**Case 3, Triangular Load, Right-Span Load Term:**

$$\frac{6A\overline{b}}{L} = \frac{7wL^3}{60} = \frac{7WL^2}{30} \tag{10-5}$$

The left-span load term for case 3 is $8wL^3/60 = 8WL^2/30$.

### 10-4 Applications of the Three-Moment Equation

When applied to a two-span beam with only one unknown moment at the middle support, a single application and solution of the three-moment equation will give this moment, as is illustrated in Prob. 10-1. For all other multiple-span beams, one three-moment equation must be written for every unknown moment. These equations are then solved simultaneously, as is illustrated in Prob. 10-2.

### ILLUSTRATIVE PROBLEMS

**10-1.** Using the three-moment equation, solve for the unknown moment $M_B$ at support $B$ of the two-span beam shown in Fig. 10-5. Disregard weight of beam.

FIG. 10-5 Application of the three-moment equation.

*Solution:* A restatement here of the three-moment equation is convenient:

$$M_1L_1 + 2M_2(L_1 + L_2) + M_3L_2 = -\frac{6A_1\bar{a}_1}{L_1} - \frac{6A_2\bar{b}_2}{L_2}$$

We note that $M_1 = M_A = -(4)(5) = -20$ kip-ft (negative because of tension in top fibers), $M_2 = M_B$ (unknown), and $M_3 = M_C = 0$. From Table 10-1 the load term (No. 2) for the left span $(AB)$ is

$$-\frac{6A_1\bar{a}_1}{L_1} = -\frac{wL_1{}^3}{4} = -\frac{(2)(16)(16)(16)}{4} = -2,048$$

and for the right span $(BC)$ the two load terms (No. 1) are

$$\frac{6A_2\bar{b}_2}{L_2} = -\frac{Pb}{L}(L^2 - b^2) = -\frac{(7)(10)}{14}[(14)(14) - (10)(10)] = -480$$

and

$$-\frac{6A_2\bar{b}_2}{L_2} = -\frac{Pb}{L}(L^2 - b^2) = -\frac{(14)(4)}{14}[(14)(14) - (4)(4)] = -720$$

The subscripts 1 and 2 indicate that the quantities refer to spans 1 and 2, respectively.

Substituting the appropriate values[1] in the three-moment equation, we obtain, since $M_1 = -20$,

$$(-20)(16) + 2M_B(16 + 14) + 0 = -2,048 - 480 - 720$$

or

$$60M_B = -3,248 + 320 = -2,928 \quad \text{and} \quad M_B = -48.8 \text{ kip-ft} \quad Ans.$$

**10-2.** Calculate the moments at supports $B$ and $C$ of the continuous beam shown in Fig. 10-6. Neglect weight of beam.

*Solution:* We note that $M_A$ and $M_D$ both are zero. Since there are two unknown support moments, $M_B$ and $M_C$, two three-moment equations must be written, one involving points $A$, $B$, and $C$ and one involving points $B$, $C$, and $D$. These equations are then solved simultaneously. They are

$$M_AL_1 + 2M_B(L_1 + L_2) + M_CL_2 = -\frac{6A_1\bar{a}_1}{L_1} - \frac{6A_2\bar{b}_2}{L_2} \tag{a}$$

---

[1] Two types of errors are commonly made where a beam overhangs a support. It is frequently forgotten (1) that the moment due to load on overhanging portion of beam is usually *negative* and (2) that this moment must be multiplied by the adjacent full-span length.

FIG. 10-6 Application of the three-moment equation.

$$M_B L_2 + 2M_C(L_2 + L_3) + M_D L_3 = -\frac{6A_2\bar{a}_2}{L_2} - \frac{6A_3\bar{b}_3}{L_3} \tag{b}$$

Evaluating the load terms, we obtain

Left span 1: $\quad -\dfrac{6A_1\bar{a}_1}{L_1} = -\dfrac{8}{60}wL^3 = -\dfrac{8}{60}(3)(10)(10)(10) = -400$

Right span 2: $\quad -\dfrac{6A_2\bar{b}_2}{L_2} = -\dfrac{Pb}{L}(L^2 - b^2) = -\dfrac{(10)(8)}{20}(400 - 64) = -1{,}344$

Left span 2: $\quad -\dfrac{6A_2\bar{a}_2}{L_2} = -\dfrac{Pa}{L}(L^2 - a^2) = -\dfrac{(10)(12)}{20}(400 - 144) = -1{,}536$

Right span 3: $\quad -\dfrac{6A_3\bar{b}_3}{L_3} = -\dfrac{wL^3}{4} = -\dfrac{(2)(16)(16)(16)}{4} = -2{,}048$

Substituting the appropriate values in Eqs. (a) and (b), respectively, we obtain

$$0 + 2M_B(10 + 20) + 20M_C = -400 - 1{,}344$$

and $\qquad 20M_B + 2M_C(20 + 16) + 0 = -1{,}536 - 2{,}048$

or $\qquad\qquad\qquad 60M_B + 20M_C = -1{,}744 \tag{c}$

and $\qquad\qquad\qquad 20M_B + 72M_C = -3{,}584 \tag{d}$

Multiplying Eq. (c) by 3.6 and subtracting Eq. (d) gives

$$216M_B + 72M_C = -6{,}278 \tag{e}$$
$$\underline{20M_B + 72M_C = -3{,}584} \tag{d}$$
$$196M_B \qquad\quad = -2{,}694 \quad \text{and} \quad M_B = -13.74 \text{ kip-ft} \quad \textit{Ans.}$$

and, multiplying Eq. (d) by 3 and subtracting Eq. (c) gives

$$60M_B + 216M_C = -10{,}752 \tag{f}$$
$$\underline{60M_B + 20M_C = -1{,}744}$$
$$196M_C = -9{,}008 \quad \text{and} \quad M_C = -45.96 \text{ kip-ft} \quad \textit{Ans.}$$

Substituting the values of $M_B$ and $M_C$ in either Eq. (c) or Eq. (d) will check only the correctness of the simultaneous solution. Such a check, however, will not disclose possible errors in Eqs. (c) and (d). Substituting in Eq. (c), we have

$$(60)(-13.74) + (20)(-45.96) = -1{,}744$$

or $\qquad -824.4 - 919.2 = -1{,}744 \qquad \text{and} \qquad -1{,}743.6 = -1{,}744 \qquad \textit{Check.}$

## PROBLEMS

NOTE: In each of the following problems, determine all unknown bending moments at the supports.

**10-3**

FIG. 10-7

**10-4**

FIG. 10-8

**10-5**

FIG. 10-9

**10-6**

FIG. 10-10

**10-7**

FIG. 10-11

**10-8**

FIG. 10-12

**10-9**

FIG. 10-13

**10-10**

FIG. 10-14

**10-11**

FIG. 10-15

## 10-5 Indeterminate Beams with Fixed Ends

The application of the three-moment equation to indeterminate beams with fixed ends is made possible by the addition, at the fixed end, of an "imaginary span." The following example will establish proof of this statement.

Let Fig. 10-16 represent the right-end span of any multiple-span beam, terminating in a fixed-end support at the right end $D$. At $C$, the left end of that span, the beam has been cut barely to the right of support reaction $R_C$, and $V_C$ and $M_C$ are, respectively, the shear and moment at the cut section.

We shall assume always that any fixed end of a beam is horizontal and that a fixed end means that a tangent to the elastic curve at the fixed end must also be horizontal. We shall now remove the fixed-end support and shall add an imaginary "reflected span" at the right, in such manner that the two spans

FIG. 10-16 Indeterminate beams with fixed ends.

with loads are perfectly symmetrical about $D$. When this symmetry exists, a tangent to the elastic curve at $D$ must necessarily be horizontal, and we have, therefore, created a condition that is mathematically and physically equivalent to that of a fixed end, insofar as the original beam to the left of $D$ is concerned.

The original form of the three-moment equation (10-1) is

$$M_1L_1 + 2M_2(L_1 + L_2) + M_3L_2 = -\frac{6A_1\bar{a}_1}{L_1} - \frac{6A_2\bar{b}_2}{L_2} \qquad (a)$$

When this equation is applied to the two spans in Fig. 10-16b, with the proper substitution of subscripts, we obtain

$$M_CL + 2M_D(L + L) + M_CL = -\frac{6A\bar{a}}{L} - \frac{6A\bar{a}}{L} \qquad (b)$$

or

$$2M_CL + 4M_DL = 2\left[-\frac{6A\bar{a}}{L}\right] \qquad (c)$$

Dividing each term by 2 gives

$$M_CL + 2M_DL = -\frac{6A\bar{a}}{L} \qquad (d)$$

But note now that Eq. ($d$) could have been obtained directly from Eq. ($a$), if the load on the imaginary span had been considered to be zero and if the length approached zero. Equation ($a$) would then become

$$M_CL + 2M_D(L + 0) + M_C(0) = -\frac{6A\bar{a}}{L} - 0 \qquad (e)$$

or

$$M_CL + 2M_DL = -\frac{6A\bar{a}}{L} \qquad (f)$$

which is the same as that previously obtained by Eq. ($d$). Consequently correct results will be obtained by the three-moment equation when at any fixed-end support an imaginary span is added whose length is short and load is zero.

## ILLUSTRATIVE PROBLEM

**10-12.** Using the three-moment equation, calculate the moment $M_B$ at the fixed-end support $B$ of the beam shown in Fig. 10-17. Weight of beam is not to be considered.

*Solution:* An imaginary span $L_2$ of extremely short length and with zero load is added at the fixed end as shown. For the resulting two-span beam, the three-moment equation is

$$M_AL_1 + 2M_B(L_1 + L_2) + M_CL_2 = -\frac{6A_1\bar{a}_1}{L_1} - \frac{6A_2\bar{b}_2}{L_2} \qquad (a)$$

FIG. 10-17 Moment at fixed end of indeterminate beam.

The moment $M_A$ at $A$ can readily be evaluated and is $(-5)(6) = -30$ kip-ft. Since $L_1 = 15$ ft, $L_2 \approx 0$, and $A_2 = 0$, Eq. (a) reduces to

$$(-30)(15) + 2M_B(15 + 0) + 0 = -\frac{(18)(5)}{15}(225 - 25) - 0 \qquad (b)$$

and

$$30M_B = 450 - 1,200 = -750 \qquad \text{and} \qquad M_B = -25 \text{ kip-ft} \qquad Ans.$$

## PROBLEMS

In each of the following problems, determine all unknown bending moments at the supports.

**10-13.** Refer to Fig. 10-18. $P = 3$ kips, $w = 4$ kips per ft, $L_1 = 4$ ft, and $L_2 = 12$ ft.          Ans. $M_A = -12$ kip-ft; $M_B = -66$ kip-ft

**10-14.** Refer to Fig. 10-18. $P = 6$ kips, $w = 2$ kips per ft, $L_1 = 5$ ft, and $L_2 = 10$ ft.

FIG. 10-18 Probs. 10-13 and 10-14.

FIG. 10-19 Probs. 10-15 and 10-16.

**10-15.** Refer to Fig. 10-19. $P_1 = P_2 = 0$, $w = 2$ kips per ft, $L_1 = 4$ ft, and $L_2 = 10$ ft.          Ans. $M_A = -16$ kip-ft; $M_B = -17$ kip-ft

**10-16.** Refer to Fig. 10-19. $P_1 = 6$ kips, $P_2 = 10$ kips, $w = 4$ kips per ft, $a = 10$ ft, $L_1 = 5$ ft, and $L_2 = 15$ ft.

**10-17.** Refer to Fig. 10-20. $L_1 = 6$ ft, $L_2 = 12$ ft, and $w = 6$ kips per ft.          Ans. $M_A = -12$ kip-ft; $M_B = -68.4$ kip-ft

**10-18.** Refer to Fig. 10-20. $L_1 = 3$ ft, $L_2 = 9$ ft, and $w = 12$ kips per ft.

**10-19.** Refer to Fig. 10-21. $L = 20$ ft, $w_1 = 2$ kips per ft, and $w_2 = 5$ kips per ft.          Ans. $M_B = -180$ kip-ft

FIG. 10-20 Probs. 10-17 and 10-18.    FIG. 10-21 Probs. 10-19 and 10-20.

**10-20.** Refer to Fig. 10-21. $L = 20$ ft, $w_1 = 3$ kips per ft, and $w_2 = 5$ kips per ft.

**10-21.** Refer to Fig. 10-22. Use symbols.

**10-22.** Refer to Fig. 10-22. $L = 16$ ft and $w = 4$ kips per ft.

FIG. 10-22 Probs. 10-21 and 10-22.    FIG. 10-23 Probs. 10-23 and 10-24.

**10-23.** Refer to Fig. 10-23. Use symbols.

**10-24.** Refer to Fig. 10-23. $P = 10$ kips, $a = 5$ ft, $b = 10$ ft, and $L = 15$ ft.

**10-25.** Refer to Fig. 10-24. $L_1 = 10$ ft, $L_2 = 12$ ft, and $w = 3$ kips per ft.

*Ans.* $M_B = -27.6$ kip-ft; $M_C = -40.2$ kip-ft

FIG. 10-24 Prob. 10-25.    FIG. 10-25 Probs. 10-26 and 10-27.

**10-26.** Refer to Fig. 10-25. Use symbols.

**10-27.** Refer to Fig. 10-25. $L = 18$ ft and $w = 5$ kips per ft.

**10-28.** Refer to Fig. 10-26. $P_1 = 5$ kips, $P_2 = 9$ kips, $w = 2$ kips per ft, $L_1 = 6$ ft, $L_2 = 18$ ft, and $L_3 = 12$ ft.     *Ans.* $M_B = -18.7$ kip-ft; $M_C = -26.6$ kip-ft

**10-29.** Refer to Fig. 10-27. Use symbols.

**10-30.** Refer to Fig. 10-27. $L = 24$ ft and $w = 3$ kips per ft.

**10-31.** Refer to Fig. 10-28. $P = 12$ kips, $w = 2$ kips per ft, $L_1 = 6$ ft, $L_2 = 20$ ft, and $L_3 = 15$ ft.

FIG. 10-26 Prob. 10-28.

FIG. 10-27 Probs. 10-20 and 10-30.

FIG. 10-28 Prob. 10-31.

FIG. 10-29 Probs. 10-32 and 10-33.

**10-32.** Refer to Fig. 10-29. $P = 18$ kips, $a = 10$ ft, and $b = 5$ ft.

$\qquad$ Ans. $M_A = -20$ kip-ft; $M_B = -40$ kip-ft

**10-33.** Refer to Fig. 10-29. $P = 10$ kips, $a = 9$ ft, and $b = 16$ ft.

**10-34.** Refer to Fig. 10-30. $P = 10$ kips, $w = 3$ kips per ft, $L_1 = 6$ ft, $L_2 = 16$ ft, and $L_3 = 12$ ft.

FIG. 10-30 Prob. 10-34.

FIG. 10-31 Probs. 10-35 and 10-36.

**10-35.** Refer to Fig. 10-31. $L = 12$ ft and $w = 6$ kips per ft.

$\qquad$ Ans. $M_A = M_B = -45$ kip-ft

**10-36.** Refer to Fig. 10-31. $L = 15$ ft and $w = 5$ kips per ft.

**10-37.** Refer to Fig. 10-32. $L_1 = 10$ ft, $L_2 = 12$ ft, and $w = 6$ kips per ft.

$\qquad$ Ans. $M_A = -44$ kip-ft; $M_B = -62$ kip-ft; $M_C = -77$ kip-ft

**10-38.** Refer to Fig. 10-33. $L_1 = 20$ ft, $L_2 = 15$ ft, and $w = 3$ kips per ft.

$\qquad$ Ans. $M_A = -121.5$ kip-ft; $M_B = -57.1$ kip-ft; $M_C = +28.5$ kip-ft

FIG. 10-32 Prob. 10-37.

FIG. 10-33 Prob. 10-38.

## 10-6 Determination of Support Reactions

After the unknown moments at the supports of an indeterminate beam have been determined, the values of the support reactions are readily calculated, as is illustrated in the following problems.

### ILLUSTRATIVE PROBLEMS

**10-39.** Calculate the reactions at supports $A$, $B$, and $C$ of the beam shown in Fig. 10-34 (see Prob. 10-1); $M_B = -48.8$ kip-ft.

FIG. 10-34 Calculation of support reactions of continuous beams.

*Solution:* In Fig. 10-35 the beam is shown cut at $B$ and the *negative* moment there ($M_B = -48.8$ kip-ft) is shown so as to put tension in the top fibers. The forces $V_{B_L}$ and $V_{B_R}$ are the shears barely at the left and right, respectively, of reaction $R_B$. Their algebraic sum equals $R_B$.

FIG. 10-35 Calculation of support reactions of continuous beams.

For the left part, by moments about $B$, we obtain

$[\Sigma M_B = 0]$
$$16R_A = (32)(8) + (4)(21) - 48.8 = 291.2 \quad \text{and} \quad R_A = 18.2 \text{ kips} \quad \textit{Ans.}$$

For the right part, also by moments about $B$, we obtain

$[\Sigma M_B = 0]$
$$14R_C = (7)(4) + (14)(10) - 48.8 = 119.2 \quad \text{and} \quad R_C = 8.51 \text{ kips} \quad \textit{Ans.}$$

Using the value of $R_A$, $R_B$ may now be found by moments about $C$, in Fig. 10-34, after which a check by $\Sigma V$ should be made. That is,

$[\Sigma M_C = 0] \qquad 14R_B + (30)(18.2) = (14)(4) + (7)(10) + (32)(22) + (4)(35)$

or

$$14R_B = 56 + 70 + 704 + 140 - 546 = 424 \quad \text{and} \quad R_B = 30.29 \text{ kips} \quad \textit{Ans.}$$

$[\Sigma V = 0] \qquad 18.2 + 30.29 + 8.51 = 4 + 32 + 7 + 14 \quad \text{or} \quad 57 = 57 \qquad \textit{Check.}$

**10-40.** Calculate the reactions $R_A$ and $R_B$ at supports $A$ and $B$ of the fixed-end beam shown in Fig. 10-36 ($M_A = -12$ kip-ft and $M_B = -24$ kip-ft).

*Discussion:* The actual distribution of reacting forces around the imbedded end of a fixed-end beam, the resultant of which forces would be the single-force reaction, is rather complex. It is customary, therefore, to show a reaction at a fixed end as a force acting at the surface where the beam enters the supporting wall or abutment, as illustrated in Fig. 10-36. If the beam were cut at the wall, the shear $V$ at the cut section would equal that reaction. Hence, when a reaction at a fixed end is called for, it may be thought of as actually being the shear $V$, or vice versa.

FIG. 10-36 Reactions at fixed ends.

FIG. 10-37 Calculation of reactions.

*Solution:* The beam is cut at the supporting walls and is isolated as a free body, as shown in Fig. 10-37. The known negative moments at $A$ and $B$ are applied so as to produce the required tension on the top fibers. Solving for $V_A$ and $V_B$, which respectively equal $R_A$ and $R_B$, we obtain

$[\Sigma M_B = 0] \qquad 6V_A + 24 = (27)(2) + 12$
$$6V_A = 54 + 12 - 24 = 42 \quad \text{and} \quad V_A = 7 \text{ kips} \quad \textit{Ans.}$$

$[\Sigma M_A = 0] \qquad 6V_B + 12 = (27)(4) + 24$
$$6V_B = 108 + 24 - 12 = 120 \quad \text{and} \quad V_B = 20 \text{ kips} \quad \textit{Ans.}$$

$[\Sigma V = 0] \qquad\qquad 7 + 20 = 27 \qquad \textit{Check.}$

## PROBLEMS

**10-41.** Calculate the support reactions of the beam referred to in Prob. 10-3 $(M_B = -47.0 \text{ kip-ft})$.      *Ans.* $R_A = 12.9 \text{ kips}; R_B = 30.8 \text{ kips}; R_C = 10.3 \text{ kips}$

**10-42.** Determine all support reactions of the beam referred to in Prob. 10-4 $(M_B = -44 \text{ kip-ft})$.

**10-43.** Evaluate the support reactions of the beam referred to in Prob. 10-5 $(M_B = -81.9 \text{ kip-ft})$.      *Ans.* $R_A = 18.9 \text{ kips}; R_B = 37.2 \text{ kips}; R_C = 6.9 \text{ kips}$

**10-44.** Calculate the four reactions at the supports of the beam referred to in Prob. 10-7.

**10-45.** Evaluate the support reactions of the beam referred to in Prob. 10-8.

**10-46.** Determine the reactions at all supports of the beam referred to in Prob. 10-9 $(M_B = M_C = -54 \text{ kip-ft})$.

*Ans.* $R_A = R_D = 31.1 \text{ kips}; R_B = R_C = 30.9 \text{ kips}$

**10-47.** Calculate the reactions at all supports of the beam referred to in Prob. 10-11 $(M_B = -69.03 \text{ kip-ft}; M_C = -67.19 \text{ kip-ft})$.

**10-48.** Evaluate reactions $R_A$ and $R_B$ of the beam in Prob. 10-12 $(M_B = -25 \text{ kip-ft})$

*Ans.* $R_A = 17.33 \text{ kips}; R_B = 5.67 \text{ kips}$

**10-49.** Determine the reactions at supports $A$ and $B$ of the beam referred to in Prob. 10-27 $(M_B = -108 \text{ kip-ft})$.

**10-50.** Calculate the support reactions at $A$ and $B$ of the beam referred to in Prob. 10-19.

**10-51.** Show that in Prob. 10-21 $R_A = 3wL/8$ and $R_B = 5wL/8$ $(M_B = -wL^2/8)$.

**10-52.** Evaluate the reactions at all supports of the beam referred to in Prob. 10-25.

**10-53.** Calculate support reactions $R_A$, $R_B$, and $R_C$ of the beam referred to in Prob. 10-28.      *Ans.* $R_A = 9.63 \text{ kips}; R_B = 15.71 \text{ kips}; R_C = 12.66 \text{ kips}$

**10-54.** Determine the support reactions of the beam of Prob. 10-32.

**10-55.** In Prob. 10-37, calculate the reactions $R_A$, $R_B$, and $R_C$.

**10-56.** Calculate the support reactions $R_A$, $R_B$, and $R_C$ of the beam referred to in Prob. 10-38.

**10-57.** Determine the reactions $R_A$ and $R_B$ of the beam in Prob. 10-17.

## 10-7 Shear and Moment Diagrams

While shear and moment diagrams are very useful for the determination of shears and moments in statically *determinate* beams, they cannot be used to determine shears and moments in statically *indeterminate* beams. After the support moments and reactions have been determined, however, as outlined in the preceding articles, the load, shear, and moment diagrams may readily be drawn; they are very helpful to the designer, since they show so clearly the variation in shear and moment over the full length of the beam and also indicate clearly the value and location of maximum shear and maximum moment.

The maximum moment in an indeterminate beam generally occurs at a support and is negative. Under certain load conditions, however, the maximum moment could occur at some point between two supports and might then be positive. A complete set of diagrams is the best safeguard against overlooking the *actual* maximum moment.

## ILLUSTRATIVE PROBLEM

**10-58.** Draw a complete set of load, shear, and moment diagrams for the beam shown in Fig. 10-34. Calculate and show the maximum positive moments between supports; $M_B = -48.8$ kip-ft, $R_A = 18.20$ kips, $R_B = 30.29$ kips, $R_C = 8.51$ kips (see Probs. 10-1 and 10-39).

FIG. 10-38 Load, shear, and moment diagrams for beams in Probs. 10-1 and 10-39.

*Solution:* The shear diagram clearly reveals a section of zero shear at $D$ and a change in sign of shear at $E$, with consequent maximum positive moments at these points. The moment at $A$ is $-20$ kip-ft. The *difference* between this moment and the moment at $D$ is $(\frac{1}{2})(7.1)(14.2) = 50.41$, and $M_D = -20 + 50.41 = +30.41$. Since $M_C = 0$ and the difference between $M_E$ and $M_C = (-8.51)(4) = -34.04$, $M_E$ must equal $+34.04$.

Clearly, now, the maximum moment in the beam is $-48.8$ kip-ft at support $B$. If these diagrams are completed using the given loads and the calculated reactions, and the moments $M_A = -20$, $M_B = -48.8$, and $M_C = 0$ are obtained, it will be a check on the correctness of the support reactions only. These diagrams cannot check the correctness of $M_B$.

## PROBLEMS

NOTE: Instructors will find it convenient to have students draw the load, shear, and moment diagrams for Probs. 10-40, 10-41, 10-44, 10-46, and 10-53, for which the support moments and reactions have already been calculated.

## SUMMARY
### (*By article number*)

**10-1. Statically indeterminate beams** comprise that group of beams having more than two unknown reacting forces or moments at the supports, thus making the simple laws of statics, $\Sigma F = 0$ and $\Sigma M = 0$, insufficient for the solution. Beams having three or more supports, called **continuous** beams, fall in this category.

**10-2.** The **three-moment equation** establishes a mathematical relationship among the moments at any three consecutive supports of a "continuous" beam. Its application makes possible the determination of the moments and reactions at all supports of any continuous beam having more than two supports. Its simplest form is

$$* \quad M_1L_1 + 2M_2(L_1 + L_2) + M_3L_2 = -\frac{6A_1\bar{a}_1}{L_1} - \frac{6A_2\bar{b}_2}{L_2} \tag{10-1}$$

in which $L_1$ and $L_2$ are any two consecutive spans of a continuous beam and $M_1$, $M_2$, and $M_3$ are the bending moments at the three consecutive supports of those two spans. The numerical values of the two *load terms* in the right side of Eq. (10-1) are determined by the type and magnitude of the load and also upon whether the load is on the *left* or the *right* span of a given two-span section of a beam. The six types of loading, for which load terms are given in Table 10-1, will cover satisfactorily all situations commonly met in practice.

Other methods are available for "solving" statically indeterminate beams. Of these, the most popular, and perhaps the most useful, is the so-called moment-distribution method. Further study along this line, however, is usually reserved for books on structural analysis and design.

**10-5. Indeterminate beams with fixed ends** are solved simply by the three-moment equation by adding, at the fixed end, an additional "imaginary" span whose length and load both are zero. Consequently, the terms containing the length of and load on this imaginary span become zero and therefore drop out of the three-moment equation.

**10-6. Determination of support reactions** is a simple matter after the bending moments at the supports have been calculated. The beam may be cut successively at the different supports in such manner that only one unknown, a reaction, needs to be calculated.

**10-7. Shear and moment diagrams** cannot in themselves solve for the bending moments and the reactions at the supports of indeterminate beams. They are extremely useful, however, as a check upon calculated reactions and to determine the maximum positive moments *between* the supports that are not disclosed by the application of the three-moment equation.

### REVIEW PROBLEMS

**10-59.** Using the three-moment equation, calculate the bending moment at support $B$ of the beam shown in Fig. 10-39.                    *Ans.* $M_B = -27.55$ kip-ft

FIG. 10-39 Probs. 10-59 and 10-69.

FIG. 10-40 Probs. 10-60 and 10-70.

**10-60.** Determine the moment at support $B$ of the beam in Fig. 10-40.
**10-61.** Calculate the moment at support $B$ of the beam shown in Fig. 10-41.
*Ans.* $M_B = -60$ kip-ft

FIG. 10-41 Probs. 10-61 and 10-71.

FIG. 10-42 Probs. 10-62 and 10-72.

**10-62.** Evaluate the moment at support $B$ of the beam shown in Fig. 10-42.
**10-63.** Determine the moment at support $B$ of the beam shown in Fig. 10-43.
**10-64.** Calculate the moment at supports $A$ and $B$ of the beam shown in Fig. 10-44.
*Ans.* $M_A = -120$ kip-ft; $M_B = -80$ kip-ft
**10-65.** In Fig. 10-45, show that $M_A = -Pab^2/L^2$.
**10-66.** Show that $M_B$ in Fig. 10-45 equals $-Pa^2b/L^2$.

FIG. 10-43 Probs. 10-63 and 10-73.

FIG. 10-44 Probs. 10-64 and 10-74.

FIG. 10-45 Probs. 10-65 and 10-66.

FIG. 10-46 Probs. 10-67 and 10-75.

FIG. 10-47 Prob. 10-68.

**10-67.** Determine the moments at supports $A$, $B$, and $C$ of the beam shown in Fig. 10-46.

**10-68.** Calculate the moments at supports $A$, $B$, and $C$ of the beam shown in Fig. 10-47.    *Ans.* $M_A = -12.38$ kip-ft; $M_B = -55.24$ kip-ft; $M_C = -72.38$ kip-ft
NOTE: In the following problems, instructors may find it convenient to have students draw the load, shear, and moment diagrams and calculate the maximum positive moment in each span, in addition to the reactions called for.

**10-69.** Calculate the reactions $R_A$, $R_B$, and $R_C$ of the beam shown in Fig. 10-39 ($M_B = -27.55$ kip-ft).    *Ans.* $R_A = 7.25$ kips; $R_B = 21.05$ kips; $R_C = 11.70$ kips
**10-70.** Determine the reactions at supports $A$, $B$, and $C$ of the beam shown in Fig. 10-40 ($M_B = -71.4$ kip-ft).
**10-71.** Evaluate the reactions at supports $A$ and $B$ of the beam shown in Fig. 10-41 ($M_B = -60$ kip-ft).    *Ans.* $R_A = 49$ kips; $R_B = 17$ kips
**10-72.** Calculate the support reactions $R_A$ and $R_B$ of the beam shown in Fig. 10-42 ($M_B = -69$ kip-ft).
**10-73.** Compute the reactions at supports $A$ and $B$ of the beam shown in Fig. 10-43 ($M_B = +14.4$ kip-ft).
**10-74.** Determine the support reactions $R_A$ and $R_B$ of the beam shown in Fig. 10-44 ($M_A = -120$ kip-ft; $M_B = -80$ kip-ft).    *Ans.* $R_A = 43$ kips; $R_B = 17$ kips
**10-75.** Calculate the reactions at supports $A$, $B$, and $C$ of the beam shown in Fig. 10-46 ($M_B = -37.33$ kip-ft; $M_C = +18.67$ kip-ft).
**10-76.** Evaluate the reactions at $A$, $B$, and $C$ of the beam referred to in Prob. 10-68.
**10-77.** Compute the four reactions of the beam referred to in Prob. 10-2.
    *Ans.* $R_A = 3.63$ kips; $R_B = 13.76$ kips; $R_C = 26.49$ kips; $R_D = 13.12$ kips

## REVIEW QUESTIONS

**10-1.** When is a beam statically indeterminate?
**10-2.** What is meant by a "continuous" beam?
**10-3.** Discuss briefly the essence of the second moment-area principle.
**10-4.** What relationship is expressed by the three-moment equation?
**10-5.** Is the three-moment equation applicable to a continuous beam with different $I$ values in each span?
**10-6.** What manner of loading will give the same load terms for both left and right spans?
**10-7.** What errors are commonly made in the application of the three-moment equation?

**10-8.** How many applications of the three-moment equation are required to solve for four unknown support moments in a beam?

**10-9.** What is done to make the three-moment equation applicable to a fixed-end support?

**10-10.** Can support moments and reactions of indeterminate beams be found by use of beam diagrams?

**10-11.** In what manner are beam diagrams useful in connection with indeterminate beams?

**10-12.** How may positive moments between supports readily be evaluated?

# CHAPTER 11

# Combined Stresses

## 11-1 Combined Axial and Bending Stresses

The problems of the determination of direct (axial) tensile and compressive stresses in structural members and of the design of simple members to withstand direct (axial) tensile and compressive loads were discussed in Chap. 1. Bending stresses and the design of members to withstand bending loads were discussed in Chaps. 7 and 8.

In many instances, however, structural members are subjected both to axial and bending loads, as, for example, the cantilevered bar shown in Fig. 11-1a. The load $P$ clearly is a direct, axial load producing a unit axial stress of $P/A$ at any cross section $AA$ of the bar and $Q$ is a bending load producing a unit bending stress of $Mc/I$, which is tensile at the top of the bar and compressive at the bottom. The maximum combined stress, axial and bending, then is $s = s_d + s_b$, or

$$s = \frac{P}{A} \pm \frac{Mc}{I} \tag{11-1}$$

If the axial stress is tensile, it will add to the tensile bending stress; if it is compressive, it will add to the compressive bending stress. In either case the total, or maximum, stress is always the sum of the axial and the bending stresses. The bending-stress term is here kept in the form $s = Mc/I$ in order to make it possible to calculate either the tensile or compressive bending stress, especially in sections that are unsymmetrical about the neutral axis, where the distances $c$ to the most highly stressed tensile and compressive fibers are not equal.

269

FIG. 11-1 Combined axial and bending stress. (a) Deflection disregarded. (b) Deflection exaggerated.

The bending load $Q$ will cause a deflection $\delta$ of the bar, as indicated, and greatly exaggerated, in Fig. 11-1b. The axial load $P$ will, however, reduce this deflection, since the moment $P\delta$ is opposite in sense to the bending moment $QL$ and hence will actually reduce the bending stress. Had $P$ been a compressive force, however, its moment $P\delta$ would have added to the moment $QL$ and hence would have increased the bending stress and the lateral deflection.

When the deflection $\delta$ is relatively small, it may be, and customarily is, disregarded without appreciable error.

In textbook problems, ideal conditions and absolute precision are necessarily presumed. In practice, however, such presumed precision rarely exists. In most ordinary problems, the effect of unknown errors due to ordinary lack of precision is often far greater than the effect of some known secondary condition which may purposely be disregarded. Secondary effects, such as the change in stress due to the deflection $\delta$, may be disregarded in "relatively short" members but not in "relatively long" members, which, of course, is somewhat ambiguous. Only study and experience will enable a designer to detect the line of demarcation. As a guide, however, the structural members mentioned in all problems in this chapter are of the "relatively short" type.

Since three basic stresses exist—axial, bending, and torsional (shear, as in shafts)—four combinations of stress are possible: (1) axial and bending; (2) axial and torsional; (3) axial, bending, and torsional; and (4) bending and torsional. Only cases 1 and 4, the most commonly occurring ones, are discussed here. The others are usually considered in books on machine design.

### ILLUSTRATIVE PROBLEM

**11-1.** Assume that the bar in Fig. 11-1 is of structural steel 4 in. deep (vertical) and $\frac{3}{4}$ in. thick, that $P = 6,000$ lb, $Q = 1,000$ lb, and $L = 30$ in. Then calculate (a) the maximum combined bending and direct stress at section $A$ and (b) the deflection $\delta$ and percentage of error in stress involved in disregarding $\delta$.

*Solution:* The cross-sectional area of the bar $A = (0.75)(4) = 3$ sq in., the moment of inertia $I = bd^3/12 = (0.75)(4)^3/12 = 4$ in.$^4$, and $c = \frac{4}{2} = 2$ in. Hence

$$\left[s = \frac{P}{A} + \frac{Mc}{I}\right]$$

$$s = \frac{6,000}{3} + \frac{(1,000)(30)(2)}{4} = 2,000 + 15,000 = 17,000 \text{ psi} \qquad Ans.$$

The maximum deflection (case 1, Table 9-1) is

$$\left[\delta = \frac{QL^3}{3EI}\right] \qquad \delta = \frac{(1,000)(27,000)}{(3)(29,000,000)(4)} = 0.0775 \text{ in.} \qquad Ans.$$

The decrease in moment due to this deflection is $P\delta = (6,000)(0.0775) = 465$ lb-in., and the decrease in bending stress is $Mc/I = (465)(2)/4 = 233$ psi. The percentage of error then is

$$100 - \left(\frac{16,767}{17,000}\right)(100) = 100 - 98.7 = 1.3 \text{ percent} \qquad Ans.$$

So small an error may well be disregarded in this case, especially so because bending accounts for about 88 percent of the total stress.

### PROBLEMS

**11-2.** In Fig. 11-1, let $P = 18,000$ lb, $Q = 2,250$ lb, and $L = 40$ in. and let the bar be of structural steel 1 in. by 6 in., which latter dimension is measured parallel to force $Q$. Then calculate (a) the maximum combined bending and direct stress at section $A$, and (b) the deflection $\delta$ and percentage of error involved in disregarding $\delta$.

**11-3.** The steel plate $A$ in Fig. 11-2 is 4 in. wide and $\frac{5}{8}$ in. thick. Calculate the weight $W$ which will produce a maximum combined stress of 10,000 psi in the plate.

Ans. $W = 2,940$ lb

**11-4.** A $1\frac{1}{2}$-in.-diameter standard steel pipe extends vertically out of a solid support as shown in Fig. 11-3. Calculate the value of $P$ which will produce a combined stress of 12,000 psi when $h = 20$ in. and $\theta = 30°$.

FIG. 11-2 Prob. 11-3.

FIG. 11-3 Prob. 11-4.

FIG. 11-4 Probs. 11-5 and 11-35.          FIG. 11-5   Probs. 11-6 and 11-7.

**11-5.** In Fig. 11-4 is illustrated the cross section of an earth-retaining wall which is supported by round timber piles. The portion $h$ of each pile extending above the ground level is assumed to act as a cantilever beam with complete fixity at ground level. Each pile is subjected to a lateral earth load whose resultant is $P$ and a vertical axial load $R$ from a beam resting on top of the pile. Calculate the maximum combined stress in one 8-in.-diameter pile when $P = 450$ lb, $R = 10,000$ lb, and $h = 6$ ft.

**11-6.** In the type of timber truss shown in Fig. 11-5, closely spaced joists are often laid directly on the upper chord, thus subjecting the chord to bending as well as to direct stress, as illustrated by the chord section $AB$, shown isolated above the truss. If the chord measures $9\frac{1}{2}$ by $11\frac{1}{2}$ in. (vertical) in cross section and if $P = 25,000$ lb, $w = 1,000$ lb per ft, and $L = 14$ ft, what is the maximum stress $(a)$ in the top fibers and $(b)$ in the bottom fibers of the chord?

Ans. $(a)$ 1,634 psi; $(b)$ 1,176 psi

**11-7.** Solve Prob. 11-6 when the chord is a $7\frac{1}{2}$- by $9\frac{1}{2}$-in. (vertical) timber, $P = 22,400$ lb, $w = 800$ lb per ft, and $L = 10$ ft.

**11-8.** Member $AB$ of the crane shown in Fig. 11-6 is a steel wide-flange section, 6 by 6 in. by 15.5 lb per ft, and its length $L$ is 8 ft. Calculate $(a)$ the maximum combined stress in $AB$ when a load of 5,000 lb is applied at the mid-point and weight of beam and its deflection are neglected and $(b)$ the maximum deflection $\delta$ and the increase in bending stress caused by the moment $P\delta$.

Ans. $(a)$ $s = 13,170$ psi; $(b)$ $\delta = 0.105$ in.; $s_b = 62.3$ psi

FIG. 11-6 Probs. 11-8 and 11-9.

FIG. 11-7 Probs. 11-10 and 11-36.   FIG. 11-8 Probs. 11-11 and 11-12.   FIG. 11-9 Probs. 11-13 and 11-37.

**11-9.** The horizontal member $AB$ of the crane shown in Fig. 11-6 is 12 ft long and must be capable of supporting a load $W$ of 15,000 lb at its mid-point. What is the most economical 10-in. wide-flange section to use if the allowable combined stress is 18,000 psi? (NOTE: A trial selection may be based on bending only.)

**11-10.** The bracket $BEC$, attached to the vertical member $AD$ in Fig. 11-7, must be capable of supporting a load $P$ of 600 lb. $AD$ is a 2- by 4-in. timber S4S whose greatest dimension is in the plane of bending. Calculate the maximum combined stress in $AD$ when $a = 4$ ft, $b = 4$ ft, $c = 4$ ft, and $d = 2$ ft (a) barely above $C$ and (b) barely below $C$.

**11-11.** The ladder shown in Fig. 11-8 leans against what may be considered to be a frictionless edge support at $C$. Hence $R_C$ is perpendicular to $AC$. Each of the two identical sidepieces measures 1 by 3 in. Holes 1 in. in diameter for the rungs are bored part way through both sidepieces, as indicated in the cross section, but should be considered to extend all through for the purpose of calculating the moment of inertia of the sidepieces. What maximum combined stress is produced in one sidepiece when $P = 260$ lb and $L = 10$ ft and load is centered on ladder rung?
*Ans. $s_c = -60 - 1,038 = -1,098$ psi*

**11-12.** What vertical load $P$ can safely be supported at the mid-point $B$ of the ladder described in Prob. 11-11 if $L = 13$ ft and safe allowable combined stress is 1,200 psi? See Fig. 11-8.

**11-13.** The loading-yard boom shown in Fig. 11-9 is a 10- by 10-in. S4S timber. Assume that $a = 15$ ft, $b = 2\frac{1}{2}$ ft, and $W = 10,000$ lb. Calculate the maximum combined stress in this boom (a) barely above $B$ and (b) barely below $B$.
*Ans. (a) $s = 88 + 1,260 = 1,348$ psi; (b) $s = 147 + 1,260 = 1,407$ psi*

## 11-2 Combined Stresses Caused by Eccentric Axial Load
In short tension and compression members subjected to axial (concentric) loads, the resulting stress is assumed to be uniformly distributed over the entire

FIG. 11-10 Combined stresses caused by eccentric load.

cross-sectional area. When the action line of the load is parallel to, but does not coincide with, the centroidal axis of the supporting member, as, for example, in Fig. 11-10a and b, P is referred to as an **eccentric load** and the distance e is called the **eccentricity** of the load or the **eccentric distance.** In such instances, then, the product Pe becomes an eccentric moment which will bend the post about the axis XX, thus adding tensile and compressive bending stresses to the already existing compressive direct stresses.

If dimension d of the post is greater than b, the moment of inertia of the post will be greatest with respect to axis XX, which, therefore, is called the **major axis,** while YY is called the **minor axis.** We might, then, locate load P by stating that it is applied on the minor axis at a point a distance e from the major axis.

A clearer understanding of these combined stresses may be obtained by a study of diagrams c, d, e, and f of Fig. 11-10. The upper portion of the post is shown in diagram c. As an aid in this development, we shall now apply two equal and opposite forces, P' and P'', along the axis of the post. These forces have the magnitude of P. This may be done without in any way changing the resultant force system or its effect. Clearly, now, the post is subjected to a direct or axial force P' and to a bending moment Pe (P and P'' form a couple whose moment M = Pe). This bending moment M acting at the top of the post is counteracted by an equal and opposite resisting moment M at any horizontal plane between the top and bottom of the post, thus indicating that the changes in unit stresses caused by the eccentric moment are constant at all transverse sections in the post.

If − indicates compression and + tension, the uniformly distributed direct stress then is $-P/A$, as shown in diagram $d$. The bending stress caused by the moment $Pe$ is $\pm Mc/I$, or $Pec/I$, and is clearly compressive $(-)$ on the right side of the post and tensile $(+)$ on the left side, as indicated in diagram $e$. The combined axial and bending stress on the right side then is $-P/A - Pec/I$ and on the left side is $-P/A + Pec/I$, as indicated in diagram $f$. When, as in this illustration, the unit tensile stress caused by bending exceeds the unit compressive stress caused by the axial load, the resultant will actually be a tensile stress in the post, which in general appearance is subjected only to compressive stresses. Consequently such eccentricity of load should be avoided wherever possible, but must be taken into account when it is known to exist.

### ILLUSTRATIVE PROBLEM

**11-14.** In Fig. 11-10, let the rectangular post be of timber, let $b = 8$ in., $d = 12$ in., and $e = 3$ in., and let the load $P$ transmitted from the beam to the post be 24,000 lb. Then calculate the combined stress in the extreme fibers on the right and left sides of the post, respectively.

*Solution:* The moment of inertia $I$ with respect to the axis of bending $XX$ is $bd^3/12 = (8)(12)^3/12 = 1,152$ in.$^4$, and distance $c$ to extreme fiber is 6 in. Hence

$$\frac{P}{A} = -\frac{24,000}{(8)(12)} = -250 \text{ psi} \quad \text{and} \quad \frac{Mc}{I} = \pm\frac{(24,000)(3)(6)}{1,152} = \pm375 \text{ psi}$$

On the right side

$$s = -250 - 375 = -625 \text{ psi compression} \quad \textit{Ans.}$$

and on the left side

$$s = -250 + 375 = +125 \text{ psi tension} \quad \textit{Ans.}$$

### PROBLEMS

**11-15.** The timber post shown in Fig. 11-10 measures 8 in. ($b$) by 10 in. ($d$) and is 4 ft long. A load $P$ of 40,000 lb is applied at a point located by measuring 2 in. along the minor axis from the centroid and then 1 in. at right angles to that axis. Calculate the maximum unit stress caused by this doubly eccentric load. (NOTE: Add one additional bending-stress diagram to the three shown in Fig. 11-10.)

**11-16.** A section of $1\frac{1}{2}$-in. standard steel pipe is solidly imbedded in a concrete footing, as shown in Fig. 11-11. Calculate the unit stresses on the left and right side of the vertical portion of the pipe when $P = 60$ lb and $d = 20$ in.

**11-17.** A 6- by 6-in. timber post 5 ft long is subjected to an axial load $P$, as illustrated in Fig. 11-12. It is found that the axis is 0.1 in. out of straight line at the mid-point under a load of 36,000 lb. Calculate the maximum combined stress and the percentage increase in stress due to deviation of axis.

$\quad$ *Ans.* 1,000 psi + 100 psi; 10 percent

FIG. 11-11 Probs. 11-16 and 11-39.    FIG. 11-12 Probs. 11-17 and 11-40.

**11-18.** The bearing bracket shown in Fig. 11-13 has a symmetrical cross section and is subjected to a maximum load $P$ of 1,000 lb. Let $b = 10$ in. and, at section $AA$, $c = 1$ in., $d = 4$ in., $I_A = 10$ in.$^4$, and $A = 8$ sq in. Calculate the combined unit stresses at the top and bottom of the bracket at section $AA$.

**11-19.** The cast-iron bracket bearing in Fig. 11-13 is subjected to a maximum load of 500 lb. Assume that $b = 8$ in. and that, at section $AA$, $c = 2$ in., $d = 4$ in., $I_A = 8$ in.$^4$, and $A = 6$ sq in. Calculate the unit combined stresses at the top and bottom of the bracket at section $AA$.        *Ans. $s = +1,050$ psi; $s = -1,150$ psi*

**11-20.** The steel link shown in Fig. 11-14 is subjected to the two tensile pulls $P$ of 130 lb each. The vertical cross section at $AA$ measures $\frac{1}{2}$ by $\frac{1}{2}$ in. Calculate the maximum combined unit stress at $AA$.        *Ans. $s_t = 480$ psi*

**11-21.** Calculate the maximum tensile and compressive unit combined stresses at section $AA$ of the clamp shown in Fig. 11-15 when $e = 2$ in. and $P = 100$ lb (area = 0.375 sq in.; $I = 0.0332$ in.$^4$).

**11-22.** In the cast-iron press frame shown in Fig. 11-16, $a = 6$ in., $b = 6$ in., $t = 2$ in., and $d = 17$ in. Calculate the maximum tensile and compressive unit combined stresses in the vertical member of this frame when $P = 10,000$ lb.

*Ans. $+4,837$ psi, $-6,933$ psi*

**11-23.** If in Fig. 11-16 $a = 8$ in., $b = 6$ in., $t = 3$ in., and $d = 20$ in., what maximum value of $P$ may the press be subjected to without exceeding the following unit stresses: 3,000 psi in tension and 10,000 psi in compression?

FIG. 11-13 Probs. 11-18 and 11-19.    FIG. 11-14 Prob. 11-20.

FIG. 11-15 Probs. 11-21 and 11-41.

FIG. 11-16   Probs. 11-22 and 11-23.

**11-24.** The concrete footing shown in Fig. 11-17 supports a timber post carrying an axial compressive load $P$ of 16,650 lb. The concrete weighs 150 pcf; weight of the post may be disregarded. The load is centered on one of the two centroidal axes at a point $\frac{1}{2}$ ft away from the other. (a) Calculate the maximum unit soil pressure under footing in psf. (NOTE: Calculate $I$ in feet[4].) (b) Considering the solution to part (a), what will be the maximum soil pressure if the post is moved $\frac{1}{2}$ ft toward point $a$, thereby giving the load double eccentricity? A simple addition will give the answer.

**11-25.** A small concrete diversion dam has the cross-sectional dimensions shown in Fig. 11-18. Assume that concrete weighs 150 pcf. Consider a 1-ft length of dam and calculate the unit combined stresses at $a$ and $b$ due to (a) weight of dam only, if $h = 12$ ft, and (b) weight of dam plus the effect of 9 ft of water against right side.

*Ans.* (a) $-0.79$ psi at $a$, $-14.85$ psi at $b$; (b) $-20.55$ psi at $a$, $+4.91$ psi at $b$

## 11-3 Stresses on Inclined Cutting Planes

While the principal stress in a bar subjected to an axial pull $P$ is tension, one should be aware of the fact that shearing stresses exist on diagonal planes.

Section at base of dam for 1-ft length

FIG. 11-17 Prob. 11-24.

FIG. 11-18 Prob. 11-25.

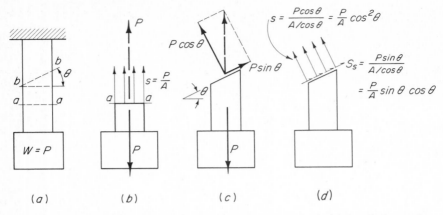

FIG. 11-19 Diagonal tension and shear due to axial load. ($a$) Loaded member. ($b$) Axial forces. ($c$) Forces on inclined plane. ($d$) Stresses on inclined plane.

As a matter of fact, many materials weak in shear fail along surfaces inclined at 45° to the geometric axis when tested in axial compression or tension.

The forces acting on an inclined plane through an axially loaded member are shown in Fig. 11-19$c$. The area over which these forces are distributed is $A/\cos\theta$. Therefore the diagonal shearing stress is $P/A \sin\theta \cos\theta$, and the diagonal tension is $P/A \cos^2\theta$, as shown in Fig. 11-19$d$. It may be of interest to note that the diagonal shearing stress is maximum when $\theta$ equals 45°.

Now the diagonal shearing stresses can arise from bending stresses as well as from stresses due to axial loads. Furthermore, diagonal tensile stresses are caused by shear. This explains the diagonal tension cracks sometimes observed in regions of high shear in concrete beams. Steel stirrups are usually placed in concrete beams to control the diagonal tension cracking.

Consider the body shown in Fig. 11-20$a$. The stresses at point $O$ in the body depend upon the orientation of the cutting planes used to expose them. If cuts are made in the $X$- and $Y$-directions, the stresses are as shown in Fig. 11-20$b$. Suppose we know these stresses but wish to compute the stresses found on a cutting plane making an angle $\theta$ to the $XZ$ plane. We may cut out a wedge-shaped element (see Fig. 11-20$c$) from the cube shown in Fig. 11-20$b$. Assume that the sloping face on the wedge has an area $\Delta A$. The areas of the right face and the bottom face will be $\Delta A \sin\theta$ and $\Delta A \cos\theta$ respectively. The **forces** acting on each face of the wedge are shown in Fig. 11-20$c$.

To find the force $s_N \Delta A$, let us sum forces in the $N$-direction.

$$s_N \Delta A = (s_X \Delta A \sin\theta)\sin\theta + (s_S \Delta A \sin\theta)\cos\theta$$
$$+ (s_Y \Delta A \cos\theta)\cos\theta + (s_S \Delta A \cos\theta)\sin\theta$$

Dividing all terms by $\Delta A$ gives

$$s_N = s_X \sin^2 \theta + s_Y \cos^2 \theta + 2s_S \sin \theta \cos \theta \qquad (11\text{-}2)$$

Equation (11-2) gives the normal stress on the inclined cutting plane in terms of the stresses on the $X$- and $Y$-planes. A summation of forces in the $\theta$-direction will yield

$$(s_S)_\theta = s_S (\cos^2 \theta - \sin^2 \theta) + (s_X - s_Y) \sin \theta \cos \theta \qquad (11\text{-}3)$$

Equation (11-3) gives the shearing stress on the inclined cutting plane.

By means of calculus, it can be shown that the maximum normal stress at point $O$ is

$$\text{Max } s = \frac{s_Y + s_X}{2} + \sqrt{s_S^2 + \left(\frac{s_X - s_Y}{2}\right)^2} \qquad (11\text{-}4)$$

and the maximum shearing stress is

$$\text{Max } s_S = \sqrt{s_S^2 + \left(\frac{s_X - s_Y}{2}\right)^2} \qquad (11\text{-}5)$$

The principal (maximum) tensile and compressive stresses in a shaft, beam, or other structural element are given by Eq. (11-4) and the principal shearing stresses are given by Eq. (11-5). The cutting planes that expose these stresses are called the principal planes.

One must be careful about signs when using the above equations. Positive stresses have been indicated in Fig. 11-20b. If any stress is opposite in direction to that shown in the sketch, it should be assigned a negative sign. Also recall that the square root of a number can be either plus or minus, as required to fit the physical problem.

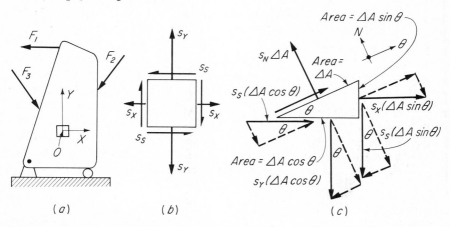

FIG. 11-20 Stresses on inclined cutting planes. ($a$) Body. ($b$) Stresses on $X$-$Y$ cutting planes. ($c$) Forces on wedge-shaped element cut from cube shown in ($b$).

The principal tension in a beam may be greater than the maximum tensile bending stress (see, for example, Prob. 11-29). In beams of the usual cross section and span, however, bending stresses and shearing stresses on horizontal and vertical planes are usually maximum stresses.

FIG. 11-21 (*a*) Stresses on *X-Y* planes. (*b*) Stresses on $\theta$ plane.

## ILLUSTRATIVE PROBLEMS

**11-26.** The bending and shearing stresses at a certain point in a beam are as shown in Fig. 11-21*a*. (*a*) Calculate the normal and shearing stresses at this same point on a cutting plane making an angle of 30° with the *XZ* plane. (*b*) Calculate the maximum tensile, compressive, and shearing stresses at this point in the beam.

*Solution:* The normal and shearing stresses on the 30° plane are

$$[s_N = s_X \sin^2\theta + s_Y \cos^2\theta + 2 s_S \sin\theta \cos\theta]$$
$$s_N = (-1,000)(0.5)^2 + 0(0.866)^2 + 2(8,000)(0.5)(0.866)$$
$$s_N = -250 + 6,930 = +6,680 \text{ psi (tension)} \qquad Ans.$$

$$[(s_S)_\theta = s_S(\cos^2\theta \sin^2\theta) + (s_X - s_Y)\sin\theta \cos\theta]$$
$$(s_S)_\theta = 8,000(0.866^2 - 0.5^2) + (-1,000)(0.5)(0.866)$$
$$(s_S)_\theta = +4,000 - 433 = +3,567 \text{ psi} \qquad Ans.$$

The plus sign on the shearing stress indicates that the direction of the computed stress on the inclined plane is that shown on the derivation sketch, that is, downward to the left.

The maximum tensile and compressive stresses are

$$\left[\text{Max } s = \frac{s_X + s_Y}{2} + \sqrt{s_S^2 + \left(\frac{s_Y - s_X}{2}\right)^2}\right]$$

$$\text{Max } s = \frac{-1,000 + 0}{2} + \sqrt{(8,000)^2 + \left(\frac{0 - (-1,000)}{2}\right)^2}$$

$$\text{Max } s = -500 + \sqrt{64,250,000} = -500 + 8,012$$
$$= 7,512 \text{ psi tension} \qquad Ans.$$

$$\text{Min } s = \frac{-1,000 + 0}{2} - \sqrt{(8,000)^2 + \left(\frac{0 - (-1,000)}{2}\right)^2}$$

$$= -500 - \sqrt{64,250,000} = -8,512 \text{ psi compression} \qquad Ans.$$

The maximum shear stress is

$$\left[\text{Max } s_S = \sqrt{s_S{}^2 + \left(\frac{s_Y - s_X}{2}\right)^2}\right]$$

$$\text{Max } s_S = \sqrt{(8,000)^2 + \left(\frac{0 - (-1,000)}{2}\right)^2} = \sqrt{64,250,000} = 8,012 \text{ psi} \qquad Ans.$$

Note that all signs must be handled with care.

FIG. 11-22 (*a*) Bracket. (*b*) A cross section *AA*. (*c*) Vertical and horizontal stresses at *O*.

**11-27.** The bracket in Fig. 11-22*a* is subjected to the pull $P = 5,000$ lb at an angle $\theta = 45°$. Calculate the principal tensile stress at *O*.

*Solution:* The bending moment on section *AA* is the horizontal component of the force *P* multiplied by 2 in. That is, $M = (0.707)(5,000)(2) = 7,070$ lb-in. The vertical tensile stress at point *O* is (see Fig. 11-22*c*)

$$\left[s_Y = \frac{P}{A} + \frac{My}{I}\right] \qquad s_Y = \frac{0.707(5,000)}{2} + \frac{7,070(\frac{1}{2})}{\dfrac{(1)(2)^3}{12}} = 1,768 + 5,302 = 7,070 \text{ psi}$$

The horizontal and vertical shearing stresses at *O* are

$$\left[s_S = \frac{VQ}{Ib}\right] \qquad s_S = \frac{0.707(5,000)\frac{1}{2}(1)(\frac{3}{4})}{\dfrac{(1)(2)^3}{12}(1)} = 1,988 \text{ psi}$$

The maximum tensile stress at *O* is

$$\left[\text{Max } s = \frac{s_Y + s_X}{2} + \sqrt{s_S{}^2 + \left(\frac{s_Y - s_X}{2}\right)^2}\right]$$

$$\text{Max } s = \frac{7,070}{2} + \sqrt{1,988^2 + \left(\frac{7,070 - 0}{2}\right)^2}$$

$$= +3,535 + \sqrt{16,400,000} = 3,535 + 4,055 = 7,590 \text{ psi} \qquad Ans.$$

FIG. 11-23 (*a*) Loaded beam. (*b*) Cross section.

## PROBLEMS

**11-28.** Solve Prob. 11-27 if $P = 8,000$ lb and $\theta = 30°$.

*Ans.* Max $s = 14,100$ psi

**11-29.** Calculate the magnitude of the uniform load $w$ lb per ft that will cause a maximum bending stress of 10,000 psi in the beam shown in Fig. 11-23. Then calculate the principal tension at point $A$ (in the web just below the flange).

*Ans.* $w = 23,200$ lb per ft; $s = 28,990$ psi

**11-30.** Solve Prob. 11-27 for the principal tension at a point $\frac{1}{2}$ in. to the left of $O$, that is, on the centerline of the bracket.

### 11-4 Combined Shearing and Bending Stresses in Circular Shafts

Shearing stresses in circular shafts caused by torque were discussed in Chap. 4. However, horizontal shafts as well as other nonvertical shafts are also subjected to bending due to their own weight and the weight of pulleys. Belt tensions, which most often act in directions other than the vertical, are the greatest single cause of bending in vertical shafts. Shafts supported by more than two bearings are continuous beams. Only shafts supported by two bearings will be dealt with here.

The maximum shearing stress in a circular shaft due to a combined torque $T$ and a bending moment $M$ will be found on an inclined plane within the shaft. It can be shown that this maximum shearing stress is equal in magnitude to that on planes normal to the axis caused by a single hypothetical resultant torque $T_R$ equal to

$$T_R = \sqrt{T^2 + M^2} \tag{11-6}$$

This simple relationship enables us to use Eqs. (4-2) and (4-3) in the following forms:

**Maximum Resultant Shearing Stress:**

$$s_R = \frac{T_R D}{2J} = \frac{5.09 T_R}{D^3} \tag{11-7}$$

**Required Diameter of Solid Shaft:**

$$D = \sqrt[3]{\frac{16T_R}{\pi s_R}} = \sqrt[3]{\frac{5.09T_R}{s_R}} \tag{11-8}$$

In any given situation the point of maximum stress remains fixed in space, but moves around the surface of the shaft as different points on the surface pass through that fixed point. This continual reversal of stress (due to bending) calls for lower allowable design stresses.

### ILLUSTRATIVE PROBLEM

**11-31.** The shaft shown in Fig. 11-24a is subjected to the following loads: $W_1 = 60$ lb, $W_2 = 40$ lb, $T_1 = 50$ lb, $T_2 = 10$ lb, $T_3 = 50$ lb, and $T_4 = 110$ lb. The principal dimensions are $a = 6$ in., $b = 12$ in., $c = 18$ in., $r_1 = 9$ in., and $r_2 = 6$ in. Neglect the weight of the shaft itself. Calculate the required diameter $D$ of shaft if the combined shearing stress $s_R$ must not exceed 10,000 psi. Small bearing friction may be disregarded.

*Solution:* The shear and bending moment diagrams are shown in Fig. 11-24b. Note that the torque $T$ between the pulleys is

$$T = (50 - 10)(9) = (110 - 50)(6) = 360 \text{ lb-in.}$$

The resultant torque $T_R$ that will give a stress by Eq. (4-2) equivalent to that caused by the combined torque $T$ and the bending moment is

$$[T_R = \sqrt{T^2 + M^2}] \qquad T_R = \sqrt{(360)^2 + (2,160)^2} = 2,190 \text{ lb-in.}$$

(a)                                                     (b)

FIG. 11-24 Shaft subjected to combined bending and torsion (Prob. 11-31). (a) Shaft and pulleys. (b) Shear and moment diagrams.

The required diameter $D$ of shaft is, then,

$$\left[ D = \sqrt[3]{\frac{5.09 T_R}{s_R}} \right] \qquad D = \sqrt[3]{\frac{(5.09)(2,190)}{10,000}} = 1.04 \text{ in.} \qquad Ans.$$

## PROBLEMS

**11-32.** In Prob. 11-31 let $a = 6$ in., $b = 12$ in., $c = 15$ in., $r_1 = 8$ in., $r_2 = 4$ in., $W_1 = 60$ lb, $W_2 = 50$ lb, $T_1 = 30$ lb, $T_2 = 10$ lb, $T_3 = 30$ lb, and $T_4 = 70$ lb. Calculate the required shaft diameter $D$ if the allowable stress $s_R$ is 8,000 psi.

*Ans. $D = 0.987$ in.*

**11-33.** In Fig. 11-24$a$ let $a = 4$ in., $b = 8$ in., $c = 5$ in., $r_1 = 10$ in., $r_2 = 20$ in., $W_1 = 30$ lb, $W_2 = 20$ lb, $T_1 = 55$ lb, $T_2 = 5$ lb, $T_3 = 10$ lb, $T_4 = 110$ lb, and $D = 1$ in. Calculate the absolute maximum shearing stress $s_R$ due to the combined bending and torsion.

*Ans. $s_R = 5,940$ psi*

**11-34.** A 2-in.-diameter shaft is subjected to a bending moment $M$ of 7,500 in.-lb. If the allowable absolute maximum shearing stress $s_R$ cannot exceed 10,000 psi, what torque $T$ may be applied to the shaft in addition to the above bending moment $M$?

## SUMMARY
### (By article number)

**11-1.** When relatively short members are subjected to loads which produce both direct (axial) and bending unit stresses, the combined unit stress at any point is the algebraic sum of these stresses. That is,

$$\textbf{Combined unit stress} = \frac{P}{A} \pm \frac{Mc}{I} \tag{11-1}$$

In such members, deflections due to loads give rise to secondary bending stresses which are generally small and may usually be disregarded.

**11-2.** An eccentric load $P$ applied along an axis parallel to the centroidal axis of a member will produce bending stresses equal to $\pm Mc/I$ in addition to the direct stress $P/A$, where the moment $M = Pe$ and $e$ is the distance between the two axes.

**11-3.** The stresses at a point in a loaded structure depend upon the orientation of the cutting planes that expose the stresses. Such cuts are usually made in the vertical or horizontal direction. The maximum stresses (principal stresses) may occur on inclined cutting planes (the principal planes).

**11-4.** A shaft subjected to both bending and torque may be treated as though it were subjected to a single torque of

$$T_R = \sqrt{T^2 + M^2} \tag{11-6}$$

## REVIEW PROBLEMS

**11-35.** If, in Prob. 11-5, the pile is 10 in. in diameter and $P = 900$ lb, $R = 12,000$ lb, and $h = 9$ ft, (a) what is the maximum combined unit stress? (b) If the allowable combined stress is 900 psi, can an 8-in. pile be used? Show all work. See Fig. 11-4.

*Ans. (a) $-483$ psi; (b) yes, $s_c = 884$ psi*

**11-36.** Assume in Fig. 11-7 that $a = 4$ ft, $b = 3$ ft, $c = 5$ ft, and $d = 3$ ft. What size of square timber S4S will be required to support a load of 800 lb? Use Douglas fir select. Assume maximum allowable combined stress = 1,600 psi.

**11-37.** In Prob. 11-13, assume that $W = 10,000$ lb and that the boom is a round timber of 8-in. diameter. Then solve for the maximum combined unit stress (a) barely above $B$ and (b) barely below $B$ when boom is in position shown. See Fig. 11-9.

**11-38.** A vertical 6- by 10-in. rough timber post 4 ft long supports a vertical load $P$ of 18,000 lb similar to that shown in Fig. 11-10. $P$ is applied on the minor axis at a point 2 in. from the major axis. Calculate the unit combined stresses on the left and right sides of the post. *Ans.* Left $s_t = 60$ psi; right $s_c = 660$ psi

**11-39.** If, in Fig. 11-11, $P = 140$ lb and $d = 24$ in., what size standard steel pipe should be used to keep the maximum combined unit stress less than 5,000 psi?

**11-40.** The longitudinal axis of a short structural-steel post, W 8 × 31, is found to deviate 0.1 in. from straight in the plane of the major axis under a load of 150,000 lb. The longitudinal axis is straight in the plane of the minor axis. Calculate the maximum combined stress in post. See Fig. 11-12.

*Ans.* $s_c = 18,100$ psi

**11-41.** If, in Fig. 11-15, $e = 3$ in., what maximum force $P$ can be exerted by the clamp without exceeding a combined stress of 10,000 psi at section $AA$? The cross-sectional area at $AA$ is 0.375 sq in., and its moment of inertia about the neutral axis is 0.0332 in.[4]. *Ans.* $P = 169$ lb

**11-42.** The three-hinged arch shown in Fig. 11-25 is made of two identical sections of laminated wood with constant cross-sectional areas measuring 5 by 12 in., the latter dimension lying in the vertical plane. The maximum lateral deviation $d$ of the longitudinal axis is 2 ft at the mid-point under a load $P$ of 6,000 lb. Calculate the maximum combined stress in one section when $s = 20$ ft and $h = 8$ ft.

Fig. 11-25 Prob. 11-42.          Fig. 11-26 Prob. 11-43.

**11-43.** The cast-iron frame of a jigsaw is shown in Fig. 11-26. The exterior dimensions of the hollow section at $AA$ are 2 by 4 in., the walls are $\frac{1}{4}$ in. thick, and $d = 18$ in. What maximum load $P$ may be exerted if the tensile stress at section $AA$ is not to exceed 3,000 psi? *Ans.* $P = 418$ lb

**11-44.** A $1\frac{1}{2}$-in.-diameter vertical shaft, 6 ft long, is supported at its ends by bearings. Belt tensions cause a lateral force of 200 lb at mid-height. The shaft transmits 60 hp to the belt when turning at 1,800 rpm. The shaft is driven from the top. Compute the absolute maximum shearing stress $s_R$ due to the combined torque and bending.

**11-45.** If the load $w$ in Fig. 11-23 is 10,000 lb per ft, calculate the principal tension at a point 6 in. above the bottom of the beam at a section adjacent to the wall.

*Ans.* $s = 11,080$ psi.

## REVIEW QUESTIONS

**11-1.** In general, what is meant by the expression "combined stress"?

**11-2.** Name the four possible combinations of stress and cite examples of each.

**11-3.** Give the formula for the combination of direct and bending stresses.

**11-4.** Explain the combination of stress due to an eccentric load.

**11-5.** In a section symmetrical about each of two rectangular axes, which is referred to as the "major axis" and which the "minor"?

**11-6.** If an 8- by 12-in. timber post is to support an eccentric load $P$ applied on one of the axes at a distance 2 in. away from the other, on which axis should the load be applied to give the least combined stress?

**11-7.** In Fig. 11-16, should the direct and bending stresses be added to give the maximum compressive combined stress?

**11-8.** In Fig. 11-24, should the direct and bending stresses be added to give the maximum tensile combined stress?

**11-9.** Consider the cast-iron frame in Fig. 11-16, and discuss some reasons why the flange of the T-section is on the inside instead of the outside.

**11-10.** If the clamp in Fig. 11-15 is of malleable iron, would the reasons advanced for the cast-iron frame mentioned in question 11-9 hold also for the clamp?

**11-11.** Show by means of an appropriate free-body diagram that shearing stresses exist on inclined planes through short columns under axial compressive loads.

**11-12.** How can we take into account the combined effects of torque and bending on solid circular shafts?

# CHAPTER 12

# Columns

## 12-1 Introduction

A column is a structural member subjected to an end load whose action line parallels that of the member and whose length is generally ten or more times its least lateral dimension. When the action line of the end load coincides with the axis of the member, it is said to be **axially** or **concentrically loaded;** when this action line and the axis do not coincide, the member is said to be **eccentrically loaded.** A column is presumed to be axially loaded unless it is known not to be.

The **ideal column** is perfectly straight, its material is perfectly homogeneous, it is entirely free from any flaws of manufacture, and it is perfectly axially loaded. This ideal column would fail by direct crushing of the material in a manner similar to that of a short block. It would never fail by buckling sidewise. But the ideal column does not exist. The columns that we actually use are never perfectly straight, their material is not perfectly homogeneous, they contain unavoidable flaws of manufacture, however small, and they are rarely, if ever, perfectly axially loaded. Consequently columns used in practice fail by buckling sidewise in the direction of their least strength and stiffness. **Buckling** of a column is here defined as elastic instability resulting in lateral buckling due to an axial load.

## 12-2 Types of Column Cross Sections

The type of section most suitable for a given column depends largely upon its particular function. The piston rod of an engine, the leg of a chair, and the boom of a derrick are all compression members coming under the classifi-

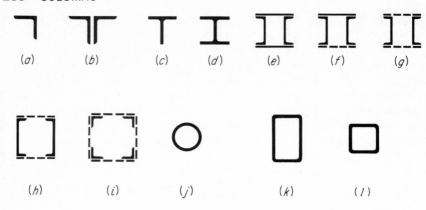

*(a)*        *(b)*        *(c)*        *(d)*        *(e)*        *(f)*        *(g)*

*(h)*        *(i)*        *(j)*        *(k)*        *(l)*

FIG. 12-1 Typical column cross sections.

cation of columns. The columns with which we are perhaps most familiar are those commonly used in buildings of timber, steel, or reinforced concrete. A group of commonly used structural-steel sections is shown in Fig. 12-1. The solid lines indicate sections that extend the full length of the member. The dash lines indicate lacing bars or tie plates whose sole function is to hold rigid the load-carrying elements of the column, as indicated in Fig. 12-2.

The single angle shown in *a* is often used for web members in light trusses and for bracing in general, the double-angle section in *b* is generally used as a chord member in riveted trusses, and the section in *c* as a chord member in welded trusses. The wide-flange section in *d* is the standard structural column in steel-frame buildings, but is also used in heavier types of trusses. The types in *e*, *f*, *g*, and *h* may be used in buildings or bridges. Type *g* is frequently used for long columns supporting moderate loads. Type *i* is very suitable for derrick booms and for light towers. The hollow tube in *j* is extensively used in aircraft work and is the familiar pipe column so widely used in light buildings. Structural tubing, both rectangular *(k)* and square *(l)*, is available in many sizes and thicknesses. The dimensions and properties of these tubing sections are now listed in the steel handbooks. Many other sections have been developed to meet requirements in specialized work.

Channels are very suitable for built-up columns of the latticed type. When they are so used, the spacing back to back is generally such that the column

Tie plate        Double lacing    Tie plate        Single lacing

FIG. 12-2 Latticed columns.

has equal moment of inertia with respect to the two major axes. The use of lattice bars saves material, gives columns of relatively great stiffness, and also provides access for painting inside surfaces.

The design of a column cannot be accomplished in as simple a manner as that of an axially loaded tension member or short compression member, where $A = P/s$, or of a beam, where $Z = M/s$, in both of which cases intensity of stress is directly proportional to the magnitude of the load. Long columns exhibit no such linearity. Short columns fail by yielding while long columns fail by elastic instability, that is, by lateral buckling. The load-carrying ability of intermediate length columns is influenced by both the strength and elastic properties of the column material. Empirical design formulas, based on much authoritative study and numerous actual tests, have been devised which satisfactorily cover the design of columns within the limits of each group. These formulas differ also with the material used, and thus we find separate sets of formulas for wood, steel, and aluminum. To complicate matters still further, various authorities have devised special formulas for use within their own particular fields of work. While such a multiplicity of column formulas may temporarily confuse students, designers experience little difficulty, since they soon limit themselves to those formulas that are most applicable to their particular field of work.

## 12-3 Slenderness Ratio and Column Groups
While the three groups of columns are referred to above as (1) short, (2) intermediate, and (3) long, the length of a column alone cannot determine the group in which it belongs; rather, its slenderness is the determining factor.

The slenderness ratios of metal columns, such as pipe sections, rolled structural-steel or aluminum sections, and others, is measured by $L/k$, where $k$ is the radius of gyration (see Art. 5-8), often given in handbooks under the symbol $r$. The slenderness ratio of timber columns is measured by $L/d$ where $d$ is the least dimension of a square or rectangular cross section.

## 12-4 Critical or Buckling Loads on Columns
The critical or buckling load is here defined as the maximum load a column can support. The buckling load for any given column depends upon many items: (1) whether the column is short, intermediate, or long, and (2) whether the ends are free to rotate or are completely or partially restrained against rotation. For the purpose of this discussion, we shall select an "ideal" column, perfect in every detail, with ends free to rotate and falling within the "long" range of columns.

We shall now place this slender column in a vertical position and shall apply a horizontal force $H$ at its mid-point until a lateral deflection $\Delta$ has been produced, as shown in Fig. 12-3a. In Chap. 6 we learned that the resisting moment $M$ was proportional to the load $H$ and in Chap. 9 that the deflection

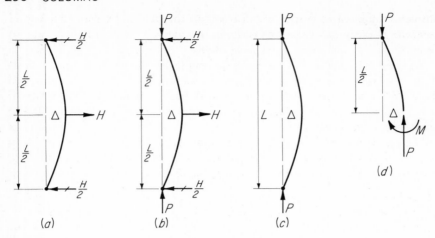

FIG. 12-3 Critical, or buckling, load on columns.

$\Delta$ (case 6, Table 9-1) is also proportional to the load $H$. Hence in Fig. 12-3 the deflection $\Delta$ at the mid-point is proportional to the moment $M$ at that point.

Let us now gradually apply two equal and opposite forces $P$ at the ends of the column, as in Fig. 12-3b, while simultaneously we decrease $H$ so as to maintain a constant deflection $\Delta$. When $H$ finally becomes zero, the forces $P$ alone maintain the column in its bent position, as in Fig. 12-3c, and the resisting moment $M$ at the mid-point (Fig. 12-3d) then is

$$M = P\Delta \tag{12-1}$$

But we concluded above that $M$ was proportional to $\Delta$. Hence, for *all* values of $M$ and $\Delta$, $P$ will remain constant. If $P$ is now increased, $M$ and $\Delta$ will both increase fairly rapidly. It is evident, therefore, that $P$ is the critical or buckling load, that is, the maximum load this slender column can support without buckling. Interpreted further, it means that any load $P$ sufficient to cause the axis of a column (with ends that are not restrained against rotation) to bend or buckle even slightly is the maximum load the column can support. In practice, columns are usually restrained to some degree against rotation of the ends. The effect of such restraint diminishes as the length of the column increases and is usually quite small in columns falling within the "long-column" range.

### 12-5 Euler's Formula for Analysis and Design of Slender Columns

This formula, derived in 1757 by Leonard Euler, a Swiss mathematician, will give the critical or buckling load on columns with ends free to rotate that fall within the "long-column" range. The mathematics involved is somewhat complex and is beyond the intended scope of this book. The formula is therefore given without derivation. It is

$$\frac{P}{A} = \frac{\pi^2 E}{(L/k)^2}$$                    (12-2)

where $P/A$ = average unit stress on cross-sectional area $A$ of column when supporting buckling load $P$, also referred to as average unit load, psi

$E$ = modulus of elasticity of column material, psi

$L/k$ = slenderness ratio in which $L$ is *effective* column length, in., and $k$ is *least* radius of gyration of cross section, in.

It is interesting to note here that in Euler's formula the buckling load $P$ is dependent solely upon the stiffness $E$ (modulus of elasticity) of the material and not upon its strength. This is evident because the formula contains no factor measuring that strength. The explanation is that, in the long, slender columns to which Euler's formula applies, buckling failure occurs *before* the actual maximum unit stress reaches the yield strength of the material. It should be noted here also that Euler's formula contains no design factor of safety, since $P$ is the greatest load a column can support without failure.

## 12-6 Effect of End Restraint on Column Strength. Effective Length

In Fig. 12-4 are shown examples of typical column-end restraints. When both ends are free to rotate (ends rounded, hinged, pinned), as in case *a*, the column

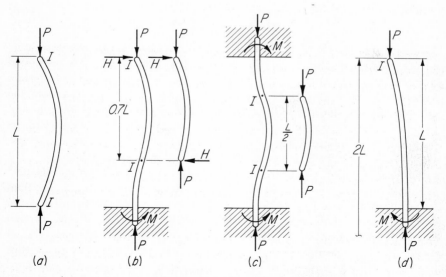

FIG. 12-4 Effect of end restraint on column strength.

may be referred to as a **simple column.** The *actual length* of each of the four columns is $L$, but the **effective length,** or *simple-column length,* of each is the distance between adjacent points of inflection $I$. A point of inflection is found at every column end that is free to rotate and at every inflection point $I$ *between* the ends where there is a change of curvature of the axis.

Hence in case $a$, both ends hinged, the effective length equals the actual length $L$; in $b$, one end hinged and other end fixed, the effective length, when calculated, is found to be $0.7L$ between the top of the column and inflection point $I$; in $c$, both ends fixed, the effective length is $L/2$, the distance between the two inflection points; and in $d$, one end free and laterally unsupported and the other end fixed (flagpole), the free end will sway sideways, and the curvature in the length $L$ will be similar to that of the upper half of the simple column in $a$, which gives type $d$ an effective length of $2L$.

When the effective length for each of the four cases is substituted in Euler's formula, we obtain

Case $a$:
$$\frac{P}{A} = \frac{\pi^2 E}{(L/k)^2} = \frac{(1)\pi^2 E}{(L/k)^2} \tag{12-3}$$

Case $b$:
$$\frac{P}{A} = \frac{\pi^2 E}{(0.7L/k)^2} \approx \frac{2.05\pi^2 E}{(L/k)^2} \tag{12-4}$$

Case $c$:
$$\frac{P}{A} = \frac{\pi^2 E}{(0.5L/k)^2} = \frac{4\pi^2 E}{(L/k)^2} \tag{12-5}$$

Case $d$:
$$\frac{P}{A} = \frac{\pi^2 E}{(2L/k)^2} = \frac{\frac{1}{4}\pi^2 E}{(L/k)^2} \tag{12-6}$$

These equations show that if a column restrained as in case $a$ will support a load $P$, the same column will support twice that load, or $2P$, when restrained as in $b$, four times $P$, or $4P$, when restrained as in $c$, but only $\frac{1}{4}$ of $P$ when restrained as in $d$; *provided, however, that in no case must P/A exceed the value permissible on the "simple-column length" of the column in question.* For various conditions of partial end restraint, the effective length may be satisfactorily estimated to have values between $L$ and $0.5L$. All such estimates should be conservative. Table 12-1 provides a simple summary.

While these values of effective length for the four cases shown in Fig. 12-4 are mathematically correct for the end conditions shown, they are seldom found in practice. In buildings, bridges, roof trusses, and other similar structures, columns and all other compression members are nearly always fastened to other members of the structure with rivets or bolts, or, in the case of steel members, they may be welded to a connection with other members, thus causing some degree of end restraint. An experienced designer will disregard this restraint if it appears to be small, in which case the factor of safety will be increased, or he will estimate its value.

**Table 12-1 EFFECTIVE LENGTH OF COLUMN**   *Relate with Fig. 12-4*

| End condition | Effective length | Number of times stronger than hinged-end column |
|---|---|---|
| Both ends hinged | $L$ | 1 |
| One hinged, other fixed | $0.7L$ | 2 |
| Both ends fixed | $0.5L$ | 4 |
| One free, other fixed | $2L$ | ¼ |
| Either or both ends partially restrained | Between $L$ and $0.5L$ | Between 1 and 4 |

In case $a$, for the end of a compression member to be completely free to rotate, the pin must be lubricated so as to offer no resistance to rotation. This condition is found only in machines, for example, heavy earth-moving and road-grading machines, where extensive use is made of hydraulic jacks to manipulate certain moving parts. Connecting rods in automobile engines and the side bars connecting the driving wheels of a locomotive are other examples of lubricated pinned-end compression members. Such cases are not dealt with in this text.

Case $d$ is sometimes referred to as the "flagpole type," but it occurs only rarely. The column can, relatively speaking, be completely fixed at the bottom. The axial load $P$ at its top would be an unusual situation, such as might be created by the "flagpole sitters" of a few decades ago. However, should such a load exist, the effective length would equal $2L$.

Obviously, therefore, when the degree of end restraint is indeterminate, the designer must rely on accumulated, available experience.

## 12-7 Limitations of Euler's Formula

A column will always buckle in the direction of its *least strength,* which means its greatest $L/k$ ratio. As previously stated, the magnitude of the critical load depends solely upon the slenderness ratio of the column and the stiffness $E$ of the material. But, since $E$ is a constant only up to the proportional limit, Euler's formula holds only for values of $P/A$ up to that limit. For example, for structural steel, with $E$ equal to 29,000,000 psi and a proportional limit of 31,000 psi, the lowest value of $L/k$ for which Euler's formula applies is

$$\frac{L}{k} = \sqrt{\frac{\pi^2 E}{P/A}} = \sqrt{\frac{\pi^2 (29,000,000)}{31,000}} = 96$$

Other limits are given in Table 12-2.

The upper limits of $L/k$ depend entirely upon good judgment and safe design and are usually fixed by code or sound professional practice. For timber columns the upper limit of length divided by least-side dimension $(L/d)$ is generally 50, which corresponds to a slenderness ratio $(L/k)$ of approximately 170. For structural-steel columns the upper limit of $L/k$ is usually 200.

**Table 12-2 LIMITATIONS OF EULER'S FORMULA**

| Material | Modulus of elasticity | Proportional limit | Lowest value of $L/K$ for which Euler's formula applies |
|---|---|---|---|
| Structural steel, low alloy | 29,000,000 | 50,000 | 76.0 |
| Structural steel, carbon | 29,000,000 | 31,000 | 96.0 |
| Douglas fir, southern pine | 1,760,000 | 5,000 | 58.9 |

## 12-8 Analysis of Long Columns by Euler's Formula

A column of given size and length may be analyzed to determine its load-carrying capacity $P$, the average unit stress $P/A$ due to a given load, and other factors. For this purpose, the following form, to which a factor of safety $(FS)$ must be added, seems most suitable:

$$\frac{P}{A} = \frac{\pi^2 E}{(L/k)^2} \quad \text{or} \quad P = \frac{\pi^2 EAk^2}{L^2} = \frac{\pi^2 EI}{kL^2} \tag{12-7}$$

since $I = Ak^2$ (Art. 5-8).

### ILLUSTRATIVE PROBLEM

**12-1.** A 3- by 5-in. wood column of Douglas fir is 12 ft long and is supported laterally at mid-length against buckling in its weakest direction. Assume ends to be hinged. Calculate the maximum safe load $P$ if a factor of safety of 3 is used. Check to make sure that Euler's formula holds for this case.

*Solution:* Support of this column laterally at mid-length has the effect of replacing the 12-ft simple column with two 6-ft simple columns in the weaker direction, while in the stronger direction it remains a 12-ft column. It is possible, therefore, that the 12-ft column may buckle in its strongest direction (parallel to the 5-in. dimension) before the 6-ft column buckles in its weakest direction (parallel to the 3-in. dimension). This may be determined by calculating the safe load on each of the two columns. For Douglas fir, $E = 1,760,000$ psi (Table 8-1). Factor of safety is 3. In the stronger direction, $L = 144$ in. Hence

$$\left[ I = \frac{bd^3}{12} \right] \qquad I = \frac{(3)(5)^3}{12} = 31.2 \text{ in.}^4$$

$$k = \sqrt{\frac{I}{A}} = \sqrt{\frac{31.2}{15}} = 1.44 \quad \text{and} \quad \frac{L}{k} = \frac{144}{1.44} = 100 \quad \text{OK}$$

$$\left[ P = \frac{\pi^2 EI}{L^2(FS)} \right] \qquad P = \frac{(3.1416)^2(1,760,000)(31.2)}{(144)(144)(3)} = 8,710 \text{ lb} \qquad Ans.$$

In the weaker direction, $L = 72$ in. and

$$\left[ I = \frac{bd^3}{12} \right] \qquad I = \frac{(5)(3)^3}{12} = 11.25 \text{ in.}^4$$

$$k = \sqrt{\frac{I}{A}} = \sqrt{\frac{11.25}{15}} = 0.866 \quad \text{and} \quad \frac{L}{k} = \frac{72}{0.866} = 83.1 \quad \text{OK}$$

$$\left[ P = \frac{\pi^2 EI}{L^2(FS)} \right] \qquad P = \frac{(3.1416)^2(1,760,000)(11.25)}{(72)(72)(3)} = 12,540 \text{ lb}$$

showing the column is now actually weakest in the direction parallel to its 5-in. dimension and that the safe load is 8,710 lb. The values of $L/k$ also show that Euler's formula applies.

## PROBLEMS

NOTE: Column ends shall be presumed to be "hinged" or "pinned" unless otherwise stated. Steel columns are formed from carbon structural steel.

**12-2.** In Prob. 12-1, assume the column to be 2 by 6 in. and solve for the safe load $P$.

**12-3.** Standard steel pipe is used making sections of scaffolding commonly used by the building industry. Calculate the safe axial load on a $1\frac{1}{2}$-in. pipe column 78 in. long if a factor of safety of 4 is used.                    *Ans. P* = 3,640 lb

**12-4.** In the preceding problem, what will be the answer if 2- by 4-in. wood uprights (S4S) are used instead of steel pipe? Assume $E = 1,760,000$ psi.

**12-5.** A yardstick is 1 in. wide and $\frac{1}{4}$ in. thick, of southern pine, and exactly 36 in. long. Assuming it to be straight, what axial load $P$ will cause it to buckle if the end restraint is similar (*a*) to case *a* of Fig. 12-4; (*b*) to case *b*; (*c*) to case *c*?

**12-6.** Calculate the safe axial load $P$ on a 3-in.-diameter standard steel pipe column 18 ft long if a factor of safety of 2 is used.

**12-7.** What is the safe axial load $P$ on a rough 6- by 6-in. Douglas fir column 18 ft long if a factor of safety of 3 is used?                    *Ans. P* = 13,390 lb

**12-8.** Calculate the safe axial load $P$ on a steel column, W 8 × 31, 25 ft long if a factor of safety of 3 is used.                    *Ans. P* = 39,200 lb

## 12-9 Design of Long Columns by Euler's Formula

To design a column means to determine its required size. With Euler's formula this is accomplished by solving for the required moment of inertia $I$ in Eq. (12-7). For design, then, the following form is most suitable:

$$I = \frac{PL^2}{\pi^2 E} \tag{12-8}$$

in which $L$ is the *effective* length of the column. $I$ must be multiplied by a suitable factor of safety before a corresponding size is selected, or $P$ may be multiplied by the factor of safety, as in Prob. 12-9.

To make sure that Euler's formula applies to the problem, the slenderness ratio $L/k$ for the member selected should be calculated and checked against the lowest value (Art. 12-7) for which the formula applies.

## ILLUSTRATIVE PROBLEM

**12-9.** A structural-steel column 25 ft long must support an axial load of 75,000 lb. Select the lightest available W section, using a factor of safety of 2.5, and calculate its slenderness ratio $L/k$. The proportional limit of this steel is 31,000 psi.

*Solution:* The design load $P$ now is $(2.5)(75,000)$, and

$$\left[ I = \frac{PL^2}{\pi^2 E} \right] \qquad \text{Required } I = \frac{(2.5)(75,000)(300)^2}{\pi^2(29,000,000)} = 59 \text{ in.}^4$$

$$\text{Use W 8} \times 48 \ (I_Y = 60.9 \text{ in.}^4). \qquad Ans.$$

For this section, $k = 2.08$. Hence

$$\frac{L}{k} = \frac{300}{2.08} = 144 \qquad (> 96, \text{ hence OK})$$

## PROBLEMS

NOTE: Column ends shall be presumed to be "hinged" or "pinned" unless otherwise stated.

**12-10.** A bronze rod 20 in. long with hinged ends is subjected to an axial compressive load $P$ of 366 lb. Calculate its required diameter $d$, if a factor of safety of 3 is used, and its slenderness ratio $L/k$. *Ans.* $d = \frac{1}{2}$ in.; $L/k = 160$

**12-11.** A round wood column of southern pine 8 ft 4 in. long must support an axial compressive load $P$ of 2,300 lb. Calculate (a) its required diameter $d$ if a factor of safety of 3 is used and (b) its slenderness ratio $L/k$.

**12-12.** A square wood column of Douglas fir S4S 12 ft long must support an axial compressive load $P$ of 4,000 lb. Determine the approximate size of column if a factor of safety of 3 is used. Calculate its slenderness ratio $L/k$.

**12-13.** An axial load of 45,000 lb must be supported by a standard steel pipe column 18 ft long. What diameter $d$ is appropriate if a factor of safety of 2 is used?
*Ans.* $d = 5$ in.; $L/k = 115$

**12-14.** A main member in a large truss is 16 ft long and is to be made of two structural-steel angles placed back to back with a $\frac{3}{8}$-in. gusset plate between. It is to support an axial compressive load of 20,000 lb. If a factor of safety of 2 is appropriate, what size angles will be most economical? What is $L/k$?

**12-15.** A structural-steel equal-leg angle 10 ft long is to be used as a brace capable of withstanding an axial compressive load of 3,300 lb. Using a factor of safety of 3, select the lightest available angle. Calculate $L/k$. (NOTE: A selection can be made by solving for the required value of $k$.)

## 12-10 Intermediate Columns

As stated in Art. 12-5, no single column-design formula has yet been devised which has proved to be satisfactory for all lengths of columns. For hinged-end columns in the "long" range, where only the *stiffness* $E$ of the material, and not its length, determines the buckling load, Euler's formula will predict closely the actual buckling load. In the "intermediate" range, however, the *length* of the material as well as its stiffness determines the buckling load.

Since many different materials are used, such as steel, brass, bronze, aluminum, wood, and others, with working strengths varying perhaps from 1,000 to 30,000 psi, it is not surprising that literally dozens of formulas have been devised. One edition of the "Structural Aluminum Handbook" lists 19 different column formulas for as many different aluminum alloys in the intermediate range. Despite this, the problem of column design is not difficult, since each industry usually confines itself to a few well-tested and proved formulas.

In order to simplify matters, only the most commonly used column formulas are presented here, grouped according to kind of material. Because all of them are of empirical origin, arrived at after much authoritative experimental testing, they are merely presented here along with explanations as to their proper use. Unlike Euler's formula, in which the required moment of inertia may be solved for directly, all formulas in the intermediate range, with $L/k$ up to 126.1 for A36 steel, give merely the allowable average unit load $P/A$ on a given column; this load in turn is dependent upon the slenderness ratio $L/k$. Because there are two variables, $A$ and $k$, in each formula, an appropriate column section can be selected only by the "trial-and-error" method. Two or three trials at the most are usually sufficient.

Since all structural design, whether it applies to a building, a bridge, a machine, or an airplane, demands absolute safety as to strength, the column formulas to be used in different geographical areas are specified by the code-making authorities within those areas, such as municipal, state, Federal, Army, Navy, and others, who are then guided by the special conditions within their areas.

Rolled steel sections are now available in several grades, designated according to yield strength and other properties as A7, A36, A242, A373, A440, and A441 by the American Society for Testing Materials (ASTM). Only A36 steel, the most commonly used, will be considered here.

## 12-11 Column Formulas, Structural Steel
Analysis here implies the determination of the allowable unit load $P/A$ on a given column, or of its allowable total load $P$.

The **Gordon-Rankine formula,** developed more than a century ago, is

$$\frac{P}{A} = \frac{18,000}{1 + (1/18,000)(L/k)^2} \tag{12-9}$$

The **American Institute of Steel Construction (AISC)** recommends the following formula for main and secondary compression members (columns, etc.) where the slenderness ratio $KL/k$ equals $C_c$ or less ($C_c = 126.1$ for A36 steel).

**For main and secondary members ($KL/k$ not over 120):**

$$\frac{P}{A} = \frac{\left[1 - \dfrac{(KL/k)^2}{2C_c{}^2}\right] s_y}{FS} \qquad \text{(12-10)}$$

where

$$C_c = \sqrt{\frac{2\pi^2 E}{s_y}}$$

and     Factor of safety $FS = \dfrac{5}{3} + \dfrac{3(KL/k)}{8C_c} - \dfrac{(KL/k)^3}{8C_c{}^3}$

(For A36 steel, $FS = 1.92$ when $C_c = KL/k$, or 126.1)

For main members with slenderness ratios greater than $C_c$ (the so-called "long" columns), the AISC recommends

**For main members when $KL/k$ exceeds $C_c$:**

$$\frac{P}{A} = \frac{12\pi^2 E}{23(KL/k)^2} \qquad \text{(12-11)}$$

For braces and secondary members, subjected only to temporary wind loads or other temporary forces, when $L/k$ exceeds $C_c$ (126.1 for A36 steel), the AISC recommends

**For braces and secondary members when $L/k$ exceeds 120:**

$$\frac{P}{A} = \frac{P/A \text{ by Eq. (12-10) or (12-11)}}{1.6 - L/200k} \qquad \text{(12-12)}$$

For Eq. (12-12), $K$ is always taken as unity. The modification (in the denominator) of Eq. (12-12) has the effect of increasing the allowable $P/A$ for braces and secondary members.

Allowable $P/A$ values for all three formulas are tabulated in Table E-14 in the Appendix. The maximum allowable $L/k$ for any member is 200.

These empirical AISC column formulas (12-10) to (12-12) apply to compression members in structures that are braced against sidesway (lateral displacement) by diagonal bracing, shear walls, or other adequate methods of preventing sidesway. When these conditions are met, as they are in the great majority of buildings, the effective-length factor $K$ shall be taken as unity, that is, $K = 1$. However, in such structures, further analysis may indicate that some degree of end restraint is present which may permit a value of $K$ less than unity and which then would permit a higher value of unit load $P/A$ to be used. In other design problems, such as building frames in which the rigidity of the joints connecting beams and columns offers the only resistance to sidesway, the end-restraint factor $K$ will lie between 1 and 2, which then has the effect of decreasing the allowable $P/A$.

FIG. 12-5 Allowable unit loads $P/A$ on axially loaded column of ASTM A36 steel.

Obviously, therefore, only the structural designer who is fully aware of all the requirements of the complete design can determine a safe and appropriate value of $K$, either by some rational analysis or by the exercise of his design experience and sound engineering judgment.

For the reasons stated above, the effective-length factor $K$, as it appears in Eqs. (12-10) to (12-12), shall be taken as unity ($K = 1$) in all problems, unless otherwise specified.

When values of $P/A$ are calculated for various ratios of $L/k$ up to 200 and are then plotted, as in Fig. 12-5, formula (12-10) gives the $AB$ portion of the curve $ABC$ and formula (12-11) the $BC$ portion. The curve $ABC$ is for *main* members and curve $BD$ for *secondary* members. Allowable AISC stresses may then be obtained by solving the applicable formulas, from the curves in Fig. 12-5, or from Table E-14. These allowable stresses are based on a factor of safety of about 2 based on the elastic limit and of about 4 based on ultimate strength.

The **American Railway Engineering Association (AREA)** has adopted the following formulas for main compression members in railway bridges:

**AREA:**
$$\frac{P}{A} = 15{,}000 - \frac{1}{3}\left(\frac{L}{k}\right)^2 \quad \text{for pinned ends} \qquad (12\text{-}13)$$

and
$$\frac{P}{A} = 15{,}000 - \frac{1}{4}\left(\frac{L}{k}\right)^2 \quad \text{for riveted ends} \qquad (12\text{-}14)$$

for $L/k$ values not greater than 140.

The **American Association of State Highway Officials (AASHO)** has adopted the same formulas for compression members in highway bridges, with a limit of $L/k$ of 120 for main members and 140 for secondary members.

## 12-12 Some Practical Aspects of Column Design

In most cases, column ends are not hinged or pinned, as has been presumed in all formulas presented in this article with the exception of the AREA formula (12-14), which allows a slightly higher stress when the column ends are riveted instead of hinged. Most cases of hinged-end columns occur in machine work: connecting rods in automobile engines and the side bars connecting the driving wheels of a locomotive are examples of hinged-end columns.

In structural work, the ends of the steel columns are usually restrained to some degree at the bottom by being welded to a baseplate, which in turn is bolted to a concrete footing, and at the various floor levels by the usual beam connections. Steel-pipe columns generally have plates welded to their ends which, when bolted to other parts of a structure, also restrain the ends to some degree. Such restraints, however, vary greatly and are difficult to evaluate. They are generally not used to permit higher stresses and therefore serve to add to the factor of safety. On the other hand, such restraints also serve to offset some possible lack of straightness of the column or some small unintentional and often unnoticed eccentricity of loading which, in some degree, is almost certain to exist.

**Main members** are those that resist *permanent*, known, calculated, or closely estimated loads. **Secondary members** are those that resist *temporary* loads, such as braces that may become loaded only temporarily because of wind, earthquake, or other temporary conditions.

Tables E-6 to E-12 list a wide selection of *structural-steel shapes* and their properties, sufficient for use in solving problems given in this text. Acquisition of "Steel Construction," the manual of the American Institute of Steel Construction, is strongly recommended for all who wish to extend their work beyond the problems of this text.

## 12-13 Analysis of Structural-Steel Columns

Analysis here implies the determination of the allowable unit load $P/A$ on a given column, or of its allowable total load $P$, as illustrated in the following problems.

### ILLUSTRATIVE PROBLEM

**12-16.** Calculate the allowable load $P$ on an A36 structural-steel column, W 8 × 31, when its unsupported length is (a) 8 ft, (b) 16 ft, and (c) 24 ft. Use AISC formulas or Table E-14. (Since $K = 1$, it may be omitted.)

*Solution:* To obtain the allowable load $P$, we first determine the allowable unit load $P/A$, which in turn depends upon the value of $L/k$. From Table E-6, $A = 9.12$ sq in., and the *least* $k_y = 2.01$ in. Then

(a)
$$\frac{L}{k} = \frac{96}{2.01} = 47.7$$

Since $L/k < 126.1$, use Eq. (12-10) or see Table E-14. From Table E-14.

$$\text{Allowable } \frac{P}{A} = 18,550 \text{ psi}$$

and
$$P = (18,550)(9.12) = 169,000 \text{ lb.} \quad \textit{Ans.}$$

(b)
$$\frac{L}{k} = \frac{192}{2.01} = 95.5$$

Since $L/k < 126.1$, use Eq. (12-10) or see Table E-14. From Table E-14.

$$\text{Allowable } \frac{P}{A} = 13,550 \text{ psi}$$

and
$$P = (13,550)(9.12) = 123,300 \text{ lb.} \quad \textit{Ans.}$$

(c)
$$\frac{L}{k} = \frac{288}{2.01} = 143.3$$

Since $L/k > 126.1$, use Eq. (12-11) or see Table E-14.

$$\left[ \frac{P}{A} = \frac{12\pi^2 E}{23(KL/k)^2} \right] \qquad \frac{P}{A} = \frac{12\pi^2(29,000,000)}{23(143.3)^2} = 7,270 \text{ psi}$$

and
$$P = (7,270)(9.12) = 66,300 \text{ lb.} \quad \textit{Ans.}$$

Interpolating in Table E-14 will give $P/A = 7,270$ psi. By studying the answers obtained, we see clearly that the allowable load $P$ on this column decreases as its slenderness ratio increases.

## ANALYSIS PROBLEMS

NOTE: The allowable unit load $P/A$ may be obtained from Table E-14. All members are main members unless specifically designated as secondary or braces. Steel is A36. $K = 1$.

**12-17.** Determine the allowable unit load $P/A$ and the allowable load $P$ for a structural-steel column, W 10 × 49, when its unsupported length is (a) 14 ft, (b) 21 ft, and (c) 28 ft.

**12-18.** Determine the allowable unit load $P/A$ and allowable load $P$ for a 3-in. standard steel-pipe column when its length is (a) 8 ft, (b) 10 ft, and (c) 12 ft.

**12-19.** Determine the allowable unit load $P/A$ and allowable load $P$ for a 4-in. standard steel-pipe column when its length is (a) 8 ft, (b) 12 ft, and (c) 18 ft.
*Ans.* (a) $P/A = 17,100$ psi, $P = 54,300$ lb; (b) $P/A = 13,430$ psi, $P = 42,650$ lb; (c) $P/A = 7,300$ psi, $P = 23,200$ lb

**12-20.** A double-angle truss compression member 8 ft long is made up of two angles 4 by 3 by ¼ in. riveted together long legs back to back with a ⅜-in. gusset plate between. Determine the allowable load $P$ for this member.

**12-21.** A brace in a structural-steel roof system is 14 ft long and is made up of two angles $3\frac{1}{2}$ by $2\frac{1}{2}$ by $\frac{1}{4}$ in. riveted together long legs back to back with a $\frac{3}{8}$-in. gusset plate between. Determine the allowable unit load $P/A$ and the allowable load $P$ for this member.

**12-22.** The top-chord member in a flat structural-steel roof truss is made up of two angles 4 by 3 by $\frac{3}{8}$ in. short legs riveted back to back with a $\frac{3}{8}$-in. gusset plate between. The 3-in. legs lie in the vertical plane in which the chord is braced by web members at 6-ft intervals. Diagonal steel rods brace the chord at 12-ft intervals in a horizontal plane. Determine the allowable load $P$ on this member.

*Ans. $P = 75,200$ lb*

**12-23.** A structural-steel column, W 10 × 33, has a laterally unsupported length of 24 ft in its strongest direction but is laterally supported at mid-length in its weakest direction. Determine the allowable unit load $P/A$ and the allowable load $P$ for this column.

**12-24.** Two 6- by 2-in. channels, each weighing 10.5 lb per ft, are welded together to form a 4- by 6-in. box column, with an unsupported length of 20 ft. Calculate the allowable load $P$ on this column.

*Ans. $P = 40,200$ lb*

**12-25.** A box column is built up of two 4- by 4- by $\frac{1}{4}$-in. angles to form a square section. If its unsupported length is 18 ft, what is the allowable load $P$ for this column?

**12-26.** A construction-derrick boom is built up of four 3- by 3- by $\frac{3}{8}$-in. steel angles latticed together to form a section such as shown in Figs. 12-1*i* and 12-2. The boom will be 60 ft long and will measure 24 in. from back to back of corner angles across each of the four sides. Calculate the allowable axial load $P$.

## 12-14 Design of Structural-Steel Columns

As previously mentioned, the design of columns is not a straightforward process but must be accomplished by successive trials. If we assume that $L/k$ lies between 30 and 120, we find from Table E-14 that the corresponding allowable unit loads (stresses) range from 19,940 psi for $L/k = 30$ down to 10,280 psi for $L/k = 120$. The mid-point would be $L/k = 75$, for which value the allowable stress is 15,900 psi. These stresses were obtained by solving formula (12-10). If conditions of a design problem do not suggest a more appropriate allowable stress for the first trial, 16,000 psi, say, is probably as good as any to use. For braces and secondary members with $L/k$ between 120 and 200, an estimated allowable $P/A$ of 8,000 psi is recommended. A five-step procedure is outlined below.

### DESIGN PROCEDURE. STRUCTURAL-STEEL COLUMNS

1. Using an estimated allowable stress $s'$ (from 10,000 to 20,000 psi), calculate the probable required area $A' = P/s'$.
2. Select trial section having approximately the probable required area and a high *least* radius of gyration $k$ ($r$ in some handbooks). (If a "*most economical*" section is required, it will invariably be found to be one of the *lightest*

three of the "square" groups, that is, 8 by 8, 10 by 10, 12 by 12, 14 by 14.)

3. Compute $L/k$ and determine actual allowable stress $s$ from formula (12-10) or (12-11) or use Table E-14.
4. Calculate allowable load $P' = sA$ on trial section and percent to which it is stressed:

$$\text{Percent stressed} = \frac{\text{actual load}}{\text{allowable load}} \; 100 = \frac{P}{P'} \; 100$$

5. Repeat Steps 2, 3, and 4 if section is stressed less than 95 percent or more than 102 percent (if stressed less than 95 percent, a more economical section can usually be found; a maximum overstress of 2 percent is usually permitted in practice).

The fact that a column is 100 percent stressed is not in itself sufficient evidence that it is the *lightest* available. For example, under a load of 200 kips, a W 14 × 68 is 100 percent stressed when $L$ is 20 ft, while a W 12 × 64 is only 80 percent stressed.

## ILLUSTRATIVE PROBLEMS

**12-27.** A six-story building has a structural-steel beam-and-column frame, appropriately fireproofed. The columns are spaced 20 ft on centers in one direction and 25 ft on centers in the perpendicular direction. Hence an interior column supports a floor area of 500 sq ft. The governing building code specifies that the frame must be designed to withstand the actual dead weight of the structure plus a possible roof snow load of 40 psf and a live load on each floor of 125 psf. The dead weight of the roof is estimated to be 80 psf and of each floor 100 psf. The unsupported length of the ground-floor column is 20 ft and of the others 16 ft. Design the third-floor column, using the most economical W section.

*Solution:* The third-floor column supports the fourth, fifth, and sixth floors and the roof. The load on this column therefore is

$$\begin{aligned}
\text{Roof dead load} \;\; (500)(80) &= 40,000 \text{ lb} \\
\text{Snow load} \qquad (500)(40) &= 20,000
\end{aligned}$$

Sixth-, fifth-, and fourth-floor loads:

$$\begin{aligned}
\text{Dead load} \;\; (500)(100)(3) &= 150,000 \\
\text{Live load} \;\; (500)(125)(3) &= \underline{187,500} \\
\text{Design load, third-floor column} &= 397,500 \text{ lb}
\end{aligned}$$

Proceeding through the five recommended steps, we have

1. Let $s' = 16,000$ psi; then required $A = 397,500/16,000 = 24.85$ sq in. Consulting Table E-6 we find that a W section 14 by $14\frac{1}{2}$ in. by 87 lb per ft comes nearest to having the required area. Then

2. *First Trial:*   W 14 × 14½ × 87   $A = 25.56$ sq in.   least $k_y = 3.70$

3. $\dfrac{L}{k} = \dfrac{192}{3.70} = 51.9$   From Table E-14, allowable $s = 18,180$ psi.

4. Allowable load $P' = sA = (18,180)(25.56) = 465,000$ lb.

$$\text{Percent stressed} = \frac{P}{P'} \cdot 100 = \frac{397,500}{465,000} \cdot 100 = 85.5 \text{ percent} \qquad \text{NG}$$

The capacity of this column is more than 14 percent in excess of that required. A more economical section can undoubtedly be found.

We shall try a lighter section, W 12 × 79, which has about 9 percent less area. Hence

2. *Second Trial:*   W 12 × 12 × 79   $A = 23.22$ sq in.   least $k_y = 3.05$

3. $\dfrac{L}{k} = \dfrac{192}{3.05} = 62.9$   From Table E-14, allowable $s = 17,150$ psi.

4. Allowable load $P' = sA = (17,150)(23.22) = 398,500$ lb.

$$\text{Percent stressed} = \frac{397,500}{398,500} \cdot 100 = 99.8 \text{ percent} \qquad \text{OK}$$

Use W 12 × 79.   Ans.

Since this W 12 × 79 is stressed to its capacity, a lighter 12 by 12 wide-flange section will not carry the load. (A W 12 × 72 can carry only 362,000 lb.)

**12-28.** The top-chord member of a flat roof truss is supported vertically at 8-ft intervals by web members and at 16-ft intervals horizontally by a system of bracing rods. Select the most economical double-angle section to withstand a compressive load of 83,000 lb. Gusset plates are ⅜ in. thick. Use AISC specifications.

*Solution:* We shall proceed with an estimated allowable unit stress of 16,000 psi, since the problem data give little inkling as to what might be a better estimate.

Owing to the method of bracing this member, the effective column length in the vertical plane is 8 ft and in the horizontal plane is 16 ft. Clearly, therefore, the resistance of the member to buckling in the horizontal plane must be about twice its resistance in the vertical plane. This ratio can be obtained by selecting two angles with *short* legs back to back. Hence

1. Let $s' = 16,000$ psi; then required $A' = P/s' = 83,000/16,000 = 5.19$ sq in.

From Table E-12 we find that the area of two 4- by 3½- by ⅜-in. angles is 5.34 sq in.

2. *First Trial:*   Two 4- by 3½- by ⅜-in. angles   $A = 5.34$ sq in.

3. Vertical plane,

$$\frac{L}{k_X} = \frac{96}{1.06} = 90.5 \quad \text{and} \quad s = 14,150 \text{ psi} \qquad \text{(Table E-14)}$$

Horizontal plane,

$$\frac{L}{k_Y} = \frac{192}{1.88} = 102.1 \quad \text{and} \quad s = 12,700 \text{ psi} \qquad \text{(governs)}$$

4. Allowable load $P' = sA = (12,700)(5.34) = 67,900$ lb     NG

This allowable load is 18 percent short of the actual load, and another choice must be made. The most economical section is usually one of the two *lightest* sections of a group. Hence we should look to the 5- by 3-in. and the 5- by $3\frac{1}{2}$-in. groups. But we discover immediately that all the 5 by 3s have $k_X$ values lower than that used in the first trial, which means that all have lower allowable stresses. The 5 by $3\frac{1}{2}$ by $\frac{3}{8}$ has about the same $k_X$ value and about 14 percent higher area. Hence

2. *Second Trial:*   Two 5- by $3\frac{1}{2}$- by $\frac{3}{8}$-in. angles   $A = 6.10$ sq in.

3. Vertical plane,

$$\frac{L}{k_X} = \frac{96}{1.02} = 94 \quad \text{and} \quad s = 13,720 \text{ psi} \quad \text{(governs)}$$

Horizontal plane,

$$\frac{L}{k_Y} = \frac{192}{2.40} = 80 \quad \text{and} \quad s = 15,360 \text{ psi}$$

4. Allowable load $P' = sA = (13,720)(6.10) = 83,700$ lb

$$\text{Percent stressed} = \frac{P}{P'} = \frac{83,000}{83,700} = 99.2 \text{ percent} \quad \text{OK}$$

Use two 5- by $3\frac{1}{2}$- by $\frac{3}{8}$-in. angles.   *Ans.*

## PROBLEMS

NOTE: Unless otherwise stated, the AISC formulas shall be used. Allowable stresses may be obtained from Table E-14. All members are main members unless specifically designated as secondary or braces. $K = 1$.

**12-29.** Select the most economical structural-steel W column 16 ft long to support an axial compressive load of 225,000 lb.                    *Ans.* W 10 × 49

**12-30.** A structural-steel column 24 ft long must support an axial compressive load of 360,000 lb. Select the most economical W section to use.

**12-31.** A steel column with an unsupported length of 24 ft must support an axial load of 66,000 lb. Select the most economical W section to use.

**12-32.** Select the most economical structural-steel W column 20 ft long to support an axial load of 36,000 lb.                    *Ans.* W 8 × 24

**12-33.** Design the first-floor column in Prob. 12-27.

**12-34.** Design the most economical double-angle main truss member 12 ft long to resist a compressive force of 66,000 lb. The long legs are to be riveted together back to back with a $\frac{3}{8}$-in. gusset plate between.

**12-35.** A main truss member 10 ft long is made up of two angles riveted together back to back with a $\frac{3}{8}$-in. gusset plate between. Design the most economical section to withstand a compressive load of 60,000 lb.

**12-36.** A riveted double-angle truss chord member is to be supported vertically at 6-ft intervals by web members and horizontally at 18-ft intervals by a system of braces. Gusset plates are $\frac{1}{2}$ in. thick. Select the most economical angles to use to withstand a load of 120 kips.

**12-37.** A riveted double-angle main truss member 9 ft long must resist a compressive load of 50,000 lb. What size angles will give the most economical section? Gusset plates are $\frac{3}{8}$ in. thick.

**12-38.** Design a riveted double-angle brace (secondary member) 14 ft long capable of withstanding a maximum compressive load of 25,000 lb; $\frac{3}{8}$-in. gusset plates are used.

**12-39.** A standard steel-pipe column 14 ft long must support an axial compressive load of 80,000 lb. Select the lightest available suitable pipe section.

**12-40.** Select a standard steel-pipe column 10 ft long to support an axial load of 40,000 lb.

**12-41.** A standard steel-pipe column 17 ft long must support an axial compressive load of 25,000 lb. Select a suitable size.

**12-42.** A piece of standard steel pipe 8 ft long is used as a brace and must be capable of resisting a maximum compressive load of 5,000 lb. What minimum diameter of pipe can be used?

**12-43.** A latticed column of the type shown in Figs. 12-1g and 12-2 is to be 30 ft long and must support an axial compressive load of 330,000 lb. (*a*) Select the most economical American Standard channels to use. (*b*) Calculate the required distance *b* from back to back of these channels to give equal moments of inertia with respect to the two major axes.
                    *Ans.* (*a*) 15 by $3\frac{3}{8}$ in. by 33.9 lb per ft; (*b*) $d = 9\frac{1}{2}$ in.

**12-44.** Solve Prob. 12-43 for a load of 90 kips and a length of 40 ft.

**12-45.** Two equal-leg angles are to be welded together to form a box column measuring 3 by $3\frac{1}{4}$ in. The column will be 12 ft long and must support an axial load of 26,000 lb. Select the appropriate equal-leg angles to use.

**12-46.** A construction-derrick boom is to be built up of 4 equal-leg angles latticed together to form a section such as illustrated in Figs. 12-1i and 12-2. The boom will be 80 ft long and will measure 20 in. from back to back of corner angles along each of the four sides. It must be capable of withstanding an axial load of 115,600 lb. Select the most economical angles to use. (NOTE: For preliminary design purposes, assume that the radius of gyration *k* of the boom equals 45 percent of its side dimension of 20 in. Then calculate *L/k* of boom, solve for required area *A*, select trial angles, and determine *actual k*, *L/k*, *P/A*, and allowable load *P'*.)

**12-47.** A compression member in a railway bridge is 24 ft long and must withstand an axial load of 240 kips. Using the AREA formula for riveted ends, select the most economical W section to use.

## 12-15 Analysis and Design of Structural-Aluminum Columns

Aluminum alloys are widely used in aircraft construction and in stationary structures where lightweight materials are required. Many alloys and tempers are available. Yield strengths vary from 11,000 to 53,000 psi. Because the strength and other properties vary so greatly, we find many formulas for the design of aluminum columns. Some of these formulas give the *ultimate* unit load, and so an appropriate factor of safety, usually from 2 to 4 in stationary structures, must be used in calculating the allowable load. Other formulas or

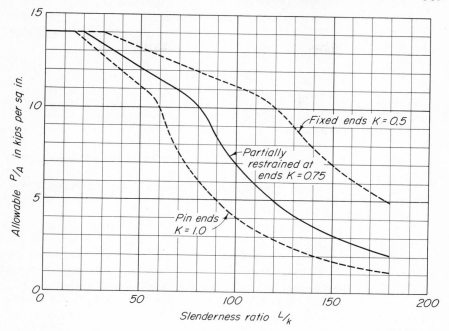

FIG. 12-6 Allowable unit loads $P/A$ on axially loaded columns of 6061-T6 aluminum.

graphs, such as those suggested by the American Society of Civil Engineers, give allowable average compressive stresses. The **ASCE** graph[1] for allowable compressive stresses in axially loaded columns of aluminum alloy 6061-T6 is shown in Fig. 12-6.

The **ASCE** specification suggests that, ordinarily, the curve for partial end restraint ($K = 0.75$) shall be used. The curves for pin-ended and fixed-ended columns are to be used as a guide in the selection of the allowable compressive stress for those cases in which the degree of end restraint is known to be different from that represented by partial restraint. The factor $K$ corresponds to the *effective-length* coefficients shown in Fig. 12-4 and given in Table 12-1.

NOTE: The cross sections of structural aluminum angles have properties similar to those of structural-steel angles. Properties of angles mentioned in the following problems may, therefore, be obtained from Tables E-10 to E-12. Aluminum pipe may be obtained with the same dimensions as steel pipe as given in Table E-15.

[1]Paper 970, *Journal of the Structural Division of the American Society of Civil Engineers,* vol. 82, no. ST3, May, 1956.

## ILLUSTRATIVE PROBLEM

**12-48.** Select an aluminum (6061-T6) pipe column to carry a 21,000-lb axial load. The ends of the 12-ft-long member are partially restrained against rotation.

*Solution:* An estimated allowable stress of 7,000 psi would require a cross-sectional area of 3 sq in. From Table E-15 choose for the first trial a 4-in. pipe with ANSI Schedule 40 dimensions. Note that $A = 3.174$ sq in. and $k = 1.51$ in. Therefore, $L/k = 144/1.51$, or 95.4. For Fig. 12-6, with $L/k = 95.4$ and using the curve for partially restrained ends, the allowable $P/A = 7,600$ psi. Hence,

$$P = (7,600)(3.174) = 24,100 \text{ lb} \qquad \text{OK}$$

Would a $3\frac{1}{2}$-in. pipe be satisfactory? From Table E-15, for $3\frac{1}{2}$-in. pipe, $A = 2.68$ sq in., and $k = 1.34$. Therefore, $L/k = 144/1.34 = 107.5$. From Fig. 12-6, $P/A = 6,200$ psi. $P = (6,200)(2.68) = 16,600$ lb, which is insufficient. Hence,

Use 4-in. Schedule 40 aluminum pipe. *Ans.*

## PROBLEMS

**12-49.** A round, hollow tube of aluminum alloy, 6061-T6, with outside diameter of 1.5 in. and inside diameter of 1.25 in., is subjected to an axial compressive load $P$. The section has the following properties: $A = 0.54$ sq in.; $I = 0.129$ in.$^4$; $k = 0.488$ in. If the length of this tube is 30 in. and its ends are assumed to be hinged, what maximum load $P$ may it safely support?

**12-50.** A rectangular hollow tube (see Fig. 12-1k) is formed from 6061-T6 aluminum. The outside dimensions are 6 by 4 in., and the wall thickness is $\frac{1}{4}$ in. Compute the allowable axial load if the column is 10 ft long and the ends are partially restrained against rotation. *Ans. P = 48,000 lb*

**12-51.** Solve Prob. 12-50 if wall thickness is $\frac{3}{8}$ in. and $L = 14$ ft.

**12-52.** Four 3- by 3- by $\frac{1}{4}$-in. structural-aluminum angles are riveted together, back to back, to form a symmetrical cross. Calculate the safe allowable axial load $P$ for this column if its unsupported length is 100 in. and the end-restraint factor $K$ is 0.75.

**12-53.** Select the most economical double-angle main truss member 11 ft long to resist a compressive force of 25,000 lb. The angles are attached to a $\frac{3}{8}$-in. gusset plate, they have an end-restraint factor $K = 0.75$, and they are of 6061-T6 aluminum.

**12-54.** A round aluminum-alloy tube, 6061-T6, 48 in. long, must withstand a calculated axial compressive force of 1,240 lb. Its ends are assumed to be hinged. If its inside diameter $d$ is to be 0.8 of its outside diameter $D$ ($d = 0.8D$), what should be the value of $D$?

## 12-16 Analysis and Design of Timber Columns

For design purposes, two sets of column formulas are in general use: (1) the Forest Products Laboratory (FPL) formulas, which have been widely used in the past, and (2) the less complicated National Forest Product Association (NFPA) formula. Still other column-design formulas are available. In practice, designers must, of course, always use the formula specified by the building code governing in the locality in which the column is to be used.

**Square timber columns** are generally used under normal design conditions. When axially loaded, the side dimensions *may* differ by 2 in.; when eccentrically loaded, they may differ by 2 in., 4 in., or more.

**Round timber columns** shall be presumed to have the same load-carrying capacity as square columns having the same area.

The **NFPA formula** applies to *all* columns having $L/d$ ratios of 50 or less. It is

**All Timber Columns** (NFPA formula for $L/d$ of 50 or less):

$$\frac{P}{A} = \frac{0.30E}{(L/d)^2} \qquad (12\text{-}15)$$

but in no case exceeding the allowable compressive stress $c$ parallel to grain, as given in Table 8-1. Hence this formula is valid only when it yields a $P/A$ value equal to or less than $c$. If a value of $P/A$ higher than $c$ is obtained, it must be disregarded, and the value of $c$ must then be used. Also

1. $P/A$ given by Eq. (12-15) is for "normal" load conditions, where maximum design load is *not* continuously applied.
2. $P/A$ must be reduced by 10 percent, if
   a. The maximum design load is continuously applied.
   b. The column is not in a dry location.

In effect, when so restricted and used, this formula allows the use of the full basic allowable stress $c$ for $L/d$ ratios up to *about* 20 and reduced values of $c$ for ratios from about 20 up to 50.

## PROCEDURE

Since $P/A$ and $L/d$ both are unknown, a solution is usually accomplished by the trial-and-error method as follows:

1. Calculate $P/L$ ($P$ in pounds and $L$ in feet) and obtain an **approximate** value of $P/A$, not, however, exceeding the value of $c$. Interpolate, if necessary.

| $P/L$: | 600 | 1,000 | 1,500 | 2,000 | 3,000 | 4,000 | 6,000 | 8,000 |
|---|---|---|---|---|---|---|---|---|
| Approx $P/A$: | 375 | 475 | 575 | 650 | 800 | 950 | 1,150 | 1,350 |

2. Calculate the approximate required area $A = P/(P/A)$ and select nearest size having that area. Ordinarily, the two side dimensions should not differ by more than 2 in.
3. Using the NFPA formula, calculate the true values $L/d$, $P/A$, and the allowable load $P'$ on the trial column. Compare $P'$ with actual load $P$. Make second trial if necessary.

## ILLUSTRATIVE PROBLEMS

**12-55.** Design a timber column 10 ft long of Douglas fir dense No. 1 S4S to support an axial load of 6,000 lb. Use the NFPA formula.

*Solution:* From Table 8-1, $c = 1,100$ psi and $E = 1,760,000$ psi. Following the procedure outlined above, $P/L = 6,000/10 = 600$ and the approximate value of $P/A$ then is 375 psi. Hence

$$\text{Approximate area required} = \frac{6,000}{375} = 13.4 \text{ sq in.}$$

Try 4- by 4-in. S4S (area = 12.25).

Since $L = 10$ ft $= 120$ in. and *least d* $= 3.5$ in., $L/d = 120/3.5 = 34.3$, and

$$\left[\frac{P}{A} = \frac{0.30E}{(L/d)^2}\right] \qquad \frac{P}{A} = \frac{(0.30)(1,760,000)}{(34.3)(34.3)} = 450 \text{ psi} \qquad \text{OK}$$

This value is *less* than the value of $c = 1,100$ psi given in Table 8-1 and will therefore be used. The allowable load on the trial column then is

Allowable load $P' = (450)(12.25) = 5,510$ lb     NG     ($P = 6,000$ lb)

Use 4- by 6-in. S4S Douglas fir (Allowable load = 450(19.25) = 8,670 lb)     *Ans.*

**12-56.** A timber column 12 ft long of southern pine No. 1 SR S4S must support an axial load of 30,000 lb. Using the NFPA formula, design this column.

*Solution:* From Table 8-1, $c = 925$ psi and $E = 1,760,000$ psi. Here, $P/L = 30,000/12 = 2,500$ and the approximate $P/A = 725$ psi. Consequently

$$\text{Approximate area required} = \frac{30,000}{725} = 41.3 \text{ sq in.}$$

Try 6- by 8-in. S4S ($A = 41.3$). Now, $L = 12$ ft $= 144$ in., *least d* $= 5.5$ in., $L/d = 144/5.5 = 26.2$, and

$$\left[\frac{P}{A} = \frac{0.30E}{(L/d)^2}\right] \qquad \frac{P}{A} = \frac{(0.30)(1,760,000)}{(26.2)(26.2)} = 769 \text{ psi} \qquad \text{OK} \qquad (<925)$$

Then

Allowable load $P' = (769)(41.3) = 31,700$ lb     OK     ($P = 30,000$)

Use 6- by 8-in. S4S southern pine.     *Ans.*

**12-57.** Using the NFPA formula, design a timber column 14 ft long of southern cypress common S4S to support an axial load of 90,000 lb.

*Solution:* From Table 8-1, $c = 1,125$ psi and $E = 1,132,000$ psi, $P/L = 90,000/14 = 6,430$, and approximate $P/A = 1,193$ psi. Since this exceeds $c = 1,125$ psi, the value of $c$ must be used.

$$\text{Approximate area required} = \frac{90,000}{1,125} = 80 \text{ sq in.}$$

Try 10- by 10-in. S4S ($A = 90.3$).

Then $L = 14$ ft $= 168$ in., *least d* $= 9.5$ in., $L/d = 168/9.5 = 17.7$, and

$$\left[\frac{P}{A} = \frac{0.30E}{(L/d)^2}\right] \qquad \frac{P}{A} = \frac{(0.30)(1,132,000)}{(17.7)(17.7)} = 1,084 \text{ psi}$$

Since the allowable stress is $c = 1,084$ psi, and the area $A = 90.3$ sq in., we find that

$$\text{Allowable load } P' = (1,084)(90.3) = 97,800 \text{ lb} \qquad \text{OK}$$

Use 10- by 10-in. S4S southern cypress common.    *Ans.*

## PROBLEMS

NOTE: Use the NFPA column-design formula in Probs. 12-58 to 12-65. Obtain values of $c$ and $E$ from Table 8-1. Specify only the most economical commercially available timber sizes.

**12-58.** Calculate the allowable unit load $P/A$ and the allowable load $P$ on each of three 6- by 6-in. S4S timber columns of Douglas fir select structural, respectively 5, 10, and 15 ft long.

*Ans.* 1,100 psi, 33,000 lb; 1,100 psi, 33,000 lb; 493 psi, 14,900 lb

**12-59.** Three 8- by 8-in. timber columns of southern pine No. 1 SR rough are respectively 6, 12, and 18 ft long. Calculate the allowable unit load $P/A$ and the allowable load $P$ for each column.

**12-60.** (*a*) Design a timber column 16 ft long of Douglas fir No. 1 S4S to support an axial load of 12,000 lb. (*b*) What size should be used if load is increased to 17,000 lb?

**12-61.** A timber column 16 ft long of southern pine No. 1 SR rough must support an axial load of 70,000 lb. Select the most economical available size.

**12-62.** Design a timber column 13 ft long of southern cypress common S4S to support an axial load of 115,000 lb.

**12-63.** Determine the required size of timber column of Douglas fir select structural rough 17 ft long to support an axial load of 175,000 lb.

**12-64.** (*a*) Design a timber column 12 ft long of Douglas fir No. 1 S4S to support an axial load of 60,000 lb. (*b*) What size should be used for a load of 80,000 lb?

**12-65.** Determine the required size of a timber column 16 ft long of southern pine No. 1 SR S4S to support an axial load of 135,000 lb.

**12-66.** (*a*) Design a timber column 15 ft long of southern cypress common rough to support an axial load of 50,000 lb. (*b*) What size would be required to support 60,000 lb?

**12-67.** Design a round timber column 10 ft long of Douglas fir No. 1 to support an axial load of 35,500 lb.

## 12-17 Eccentrically Loaded Columns

When the action line of the load on a column does not coincide with its longitudinal centroidal axis, the load is said to be **eccentric** and the column is **eccentrically loaded.** A complete treatment of the analysis (investigation) and design of eccentrically loaded columns would be very extensive and is beyond the intended scope of this text. Only the most commonly occurring cases can be discussed here, such as some eccentrically loaded structural-steel and timber columns that are usually found in structural frames of buildings.

The effects of eccentricity of load on the stresses in *short* members were discussed and illustrated in Chap. 11. The corresponding effects on the stresses in longer members, such as columns, are similar. However, a "long" column will necessarily *bend* slightly when subjected to an eccentric load, causing its longitudinal axis to deviate an amount $\Delta$ from a straight line, as illustrated in Fig. 11-12. This deviation $\Delta$ (a lateral deflection) will cause an additional moment $M = P\Delta$ and consequent internal stresses in addition to those obtained by Eq. (11-1). In comparatively long columns, these additional stresses due to $P\Delta$ may reach a sizable fraction of the total stress.

When it is necessary to consider the additional moment $P\Delta$, the maximum deflection $\Delta$ caused by the eccentric moment $M = P\Delta$ may be calculated, using the deflection equation in case 5, Table 9-1. The total combined stress in the column then becomes

$$s = \frac{P}{A} + \frac{Pe}{Z} + \frac{P\Delta}{Z} \qquad (12\text{-}16)$$

Eccentricity of loading should be avoided whenever possible because it leads to uneconomical design. It does, however, occur in several ways in normally designed structures.

Consider the stresses produced at a column cross section when an eccentric load is applied at the point of lateral support. As shown in Fig. 12-7$a$, a steel beam is fastened to the inside flange of an exterior steel column. We would assume the resultant load $P$ to be concentrated at the face of the inner column flange. The eccentric distance then is one-half the depth of the column section, and the eccentric moment is $Pe$. In Fig. 12-7$b$, a crane rail is supported by a bracket welded onto the flange of a steel column. The axial load $P_0$ may come from a roof load. The total direct load then is $P_0 + P$, and the eccentric moment is $Pe$. In Fig. 12-7$c$, the total direct load is $P_0 + P' + P''$, and the

Fig. 12-7 Eccentric loads on steel columns.

eccentric moment is $P''e'' - P'e'$ (assuming that $P''$ is greater than $P'$). The loads $P'$ and $P''$ may be assumed to act at the outer face of the column flange.

For the three cases in Fig. 12-7, the combined-stress formula becomes

Case *a*:
$$s = \frac{P}{A} + \frac{Pe}{Z} \tag{12-17}$$

Case *b*:
$$s = \frac{P_0 + P}{A} + \frac{Pe}{Z} \tag{12-18}$$

Case *c*:
$$s = \frac{P_0 + P' + P''}{A} + \frac{P''e'' - P'e'}{Z} \tag{12-19}$$

In all cases, $Z$ is the section modulus in the direction of bending.

When a beam-to-column connection is sufficiently rigid to transmit fully a moment from the beam to the column, the designer in practice will usually assume that the beam loads are carried in to the axis of the column and that no eccentricity of load exists. Completely welded connections would be in this category, as well as some types of riveted connections.

## 12-18 Design of Eccentrically Loaded Steel Columns

The possible lateral deflection $\Delta$ of structural-steel columns in building frames is usually small and is often disregarded. The total actual combined unit stress is then considered to be

$$s = \frac{\Sigma P}{A} + \frac{M}{Z} \tag{12-20}$$

If the actual axial stress $\Sigma P/A$ is designated by $f_a$ and the actual bending stress $M/Z$ by $f_b$, Eq. (12-20) becomes

$$s = f_a + f_b \tag{a}$$

If now we divide each term by $s$, we obtain

$$1 = \frac{f_a}{s} + \frac{f_b}{s} \tag{12-21}$$

For the design of structural-steel columns, the American Institute of Steel Construction (AISC) offers the following similar, but slightly changed, version of Eq. (12-21), which is now commonly used:

$$\frac{f_a}{F_a} + \frac{f_b}{F_b} \leqq 1 \tag{12-22}$$

in which $f_a$ = actual axial stress
$\quad\quad F_a$ = axial stress allowable if only axial column stress exists
$\quad\quad f_b$ = actual bending stress in column
$\quad\quad F_b$ = bending stress allowable if only bending stress exists

Equation (12-22) states, in effect, that the ratio of the actual direct axial stress to the allowable axial stress plus the ratio of the actual bending stress to the allowable bending stress shall not exceed unity.

Equation (12-22) may be referred to as a **stress factor** SF. A stress factor of 1.000 would indicate that the member is 100 percent stressed; 0.952 would indicate 95.2 percent stressed.

The quantities $f_a$, $F_a$, $f_b$, and $F_b$ in Eq. (12-22) obviously cannot be evaluated until a specific trial section has been selected. Clearly, therefore, the design of eccentrically loaded steel columns must be accomplished by the trial-and-error method.

To provide for the additional stress resulting from the lateral deflection $\Delta$ between points where a column is laterally supported, represented by the term $P\Delta/Z$ in Eq. (12-16), the latest AISC specification magnifies the computed bending stress $f_b$ given in Eq. (12-22) by an amplification factor.

The detailed design of an eccentrically loaded steel column is beyond the intended scope of this book. The interested reader is referred to the latest edition of the "Manual of Steel Construction" of the AISC for a worked-out design of a building column subjected to both an axial load and a bending moment.

### 12-19 Design of Eccentrically Loaded Timber Columns

Timber columns subjected to eccentric loads are most easily designed in a manner similar to that of steel columns, that is, by the trial-and-error method with unity as the limiting stress factor, as illustrated in Prob. 12-68. Since relatively few timber sections are commercially available (6 by 6, 6 by 8, 8 by 8, 8 by 10 in., etc.), the selection of an appropriate section can usually be accomplished in one or two trials. The additional bending stress due to lateral deflection caused by the eccentric moment will not be taken into account. Except in slender columns, this additional stress is comparatively small. For conservative design, however, it is suggested that the following stress factors not be exceeded: 1.000 for $L/d$ ratios up to 15; 0.900 for $L/d$ ratio of 25; and 0.800 for $L/d$ ratio of 35. Intermediate stress factors may be obtained by interpolation. Should greater refinement of design seem necessary, the formulas derived by J. A. Newlin,[2] of the Forest Products Laboratory, are recommended.

In Fig. 12-8, cases $a$ and $b$ should be solved as in Prob. 12-68. In case $c$, where the eccentric load $P$ is supported by a side bracket, which may be located at any point along the length of the column, the lateral forces $H$ ($H = P \tan \theta = Pe/b$) and the resulting reactions $R$ at the top and bottom of the column may readily be determined ($R = Pe/L$). The maximum bending moment $M$ in the column is then easily calculated if a free-body diagram

[2] See "National Design Specification for Stress-Grade Lumber and Its Fastenings," published by the National Forest Products Association.

FIG. 12-8  Eccentric loads on timber columns.

showing the $H$ and $R$ forces is used. This maximum bending moment will be $M = Pea/L$ when $a$ is greater than $c$ and $M = Pec/L$ when $c$ is greater than $a$. Unless specifically stated otherwise, the forces $P_O$ and $P$ shall be assumed to lie in a vertical plane lying parallel to the near face of the column and also containing the centroidal axis of the column.

## ILLUSTRATIVE PROBLEMS

**12-68.** Design a timber column 8 ft long of southern pine No. 1 dense SR rough to support an eccentric load $P$ of 15,000 lb in a manner similar to that illustrated in Fig. 12-8a. Assume that the eccentricity is 1.5 in. Use the NFPA formula.

*Solution:* From Table 8-1, we find that the allowable stresses in this southern pine are $c = 1,050$ psi if only axial stress exists and $F_b = 1,500$ psi if only bending stress exists. Let us assume that one-half of the combined unit stress in the column will be due to the direct load $P$ of 15,000 lb. The total unit stress that the column would stand could then be produced by a direct load of 30,000 lb, in which case $P/L$ would be 3,750, and the approximate $P/A$ would be 912 psi (see Procedure, Art. 12-16). This would call for a column with an area $A = 30,000/912 = 32.9$ sq in., leading to a choice of 6- by 6-in. column. That is,

$$\frac{1}{2}A = \frac{P}{s} = \frac{15,000}{900} = 16.7 \qquad \text{and} \qquad \text{required } A = 33.4 \text{ sq in.}$$

*First Trial:* 6 by 6 in., $A = 36$ sq in., $Z = bd^2/6 = 36$ in.$^3$, $P = 15,000$ lb,
$$M = (15,000)(1.5) = 22,500 \text{ lb-in.}, \quad L = 8 \text{ ft}, \quad L/d = 96/6 = 16$$

Then $\qquad$ Unit axial stress $f_a = \dfrac{P}{A} = \dfrac{15,000}{36} = 414$ psi

and $\qquad$ Unit bending stress $f_b = \dfrac{M}{Z} = \dfrac{22,500}{36} = 625$ psi

In order to calculate the stress factor, we must determine the allowable unit axial stress if only axial stress exists. Since $E = 1,760,000$ psi and $L/d = 16$,

$$\left[\frac{P}{A} = \frac{0.30E}{(L/d)^2}\right] \qquad F_a = \frac{(0.30)(1,760,000)}{(16)(16)} = 2,060 \text{ psi} \qquad \text{NG} \qquad \text{Use 1,050 psi.}$$

Because this value exceeds $c = 1,050$ psi from Table 8-1, the value $F_a = 1,050$ psi must be used. This is reasonably close to the 900 psi originally estimated. The stress factor then is

$$\text{SF} = \frac{414}{1,050} + \frac{625}{1,500} = 0.396 + 0.417 = 0.813 \qquad \text{OK}$$

Since this is less than the permissible stress factor of 0.990, obtained by interpolation for $L/d = 16$, the section is satisfactory.

Use 6- by 6-in. southern pine rough.   Ans.

**12-69.** Design a timber column 10 ft long of Douglas fir select structural rough to support an axial load $P_0$ of 62,000 lb plus an eccentric load $P$ of 10,000 lb applied on the major axis at a point 8 in. from the minor axis in a manner similar to that illustrated in Fig. 12-8b.

*Solution:* From Table 8-1, $E = 1,760,000$ psi, the maximum allowable direct stress $c = 1,100$ psi, and the maximum allowable bending stress is 1,600 psi. Let us assume that two-thirds of the actual combined unit stress in the column will be due to the direct load $P$ of $60,000 + 10,000 = 70,000$ lb. The total unit stress the column could withstand could then be produced by a direct load of $70,000 + 35,000 = 105,000$ lb. $P/L$ would then equal $105,000/10 = 10,500$, which indicates an approximate $P/A$ equal to $c = 1,450$ psi. Then

$$\tfrac{2}{3}A = \frac{P}{s} = \frac{72,000}{1,200} = 65.5 \qquad \text{and} \qquad \text{required } A = 98.2 \text{ sq in.}$$

An unequal-side column with greater resistance to bending in the plane of bending will be more economical than a square column. An 8 by 12 with an area of 96 sq in. seems most suitable. That is,

*First Trial:* 8 by 12 in., $A = 96$ sq in., $Z = bd^2/6 = (8)(12)(12)/6 = 192$ in.$^3$, $P = 72,000$ lb, $M = (10,000)(8) = 80,000$ lb-in., $L = 14$ ft, $L/d = 168/8 = 21$

Then    Unit axial stress $f_a = \dfrac{P}{A} = \dfrac{72,000}{96} = 750$ psi

and   Unit bending stress $f_b = \dfrac{M}{Z} = \dfrac{80,000}{192} = 417$ psi

Allowable axial stress $F_a = \dfrac{(0.30)(1,760,000)}{(21)(21)} = 1,195$ psi   NG   Use 1,100 psi.

Allowable bending stress $F_b = 1,600$ psi (from Table 8-1)

The stress factor then is

$$\text{SF} = \frac{f_a}{F_a} + \frac{f_b}{F_b} = \frac{750}{1,100} + \frac{417}{1,600} = 0.682 + 0.260 = 0.942 \qquad \text{OK}$$

By interpolation, maximum permissible stress factor for $L/d = 21$ is 0.940.

Use 8- by 12-in. Douglas fir. *Ans.*

**12-70.** Design a timber column 12 ft long of southern pine No. 1 dense SR rough to support an axial load $P_o$ of 25,000 lb plus an eccentric load $P$ of 6,000 lb applied on a bracket at a distance $e = 24$ in. from the column axis in a manner similar to that shown in Fig. 12-8c. The bracket is so placed that $a = 3$ ft, $b = 4$ ft, and $c = 5$ ft.

*Solution:* From Table 8-1, the allowable stresses are $c = 1,050$ psi if only axial stress exists and $F_b = 1,500$ psi if only bending stress exists. Assume that one-half of the combined unit stress in the column will be due to the direct load of 25,000 lb + 6,000 lb = 31,000 lb and that the probable allowable stress $s$, if only axial stress existed, would be 1,000 psi. Then

$$\tfrac{1}{2}A = \frac{\Sigma P}{s} = \frac{25,000 + 6,000}{1,000} = 31 \quad \text{and} \quad \text{required } A = 62 \text{ sq in.}$$

*First Trial:* 8 by 8 in., $A = 64$ sq in., $Z = (8)(8)^2/6 = 85.3$ in.³

$$\text{Max bending moment } M = \frac{Pec}{L} = \frac{(6,000)(24)(60)}{144} = 60,000 \text{ lb-in.}$$

$$\text{Unit axial stress } f_a = \frac{\Sigma P}{A} = \frac{25,000 + 6,000}{64} = 485 \text{ psi}$$

$$\text{Unit bending stress } f_b = \frac{M}{Z} = \frac{60,000}{85.3} = 703 \text{ psi}$$

From Table 8-1, $c = 1,050$ psi, and $E = 1,760,000$ psi. Since $L/d = 144/8 = 18$,

$$\left[\frac{P}{A} = \frac{0.30E}{(L/d)^2}\right] \quad F_a = \frac{(0.30)(1,760,000)}{(18)(18)} = 1,630 \text{ psi} \quad \text{NG} \quad \text{Use 1,050 psi.}$$

Because $F_a$ cannot exceed $c = 1,050$ psi, that value must be used. This is reasonably close to the 1,000 psi originally assumed. The stress factor then is

$$\text{SF} = \frac{f_a}{F_a} + \frac{f_b}{F_b} = \frac{485}{1,050} + \frac{703}{1,500} = 0.462 + 0.469 = 0.931 \quad \text{OK}$$

This is less than, but very close to, the permissible stress factor of 0.970, obtained by interpolation for $L/d = 18$. Hence

Use 8- by 8-in. southern pine No. 1 dense SR rough. *Ans.*

## PROBLEMS

NOTE: The side dimensions of timber columns subjected to eccentric loads may differ by 2 or 4 in.

**12-71.** Design a timber column 12 ft long of southern pine No. 1 SR rough to support an eccentric load $P$ of 42,000 lb in a manner similar to that illustrated in Fig. 12-8a. Assume that the eccentricity $e$ is 2 in.

**12-72.** A timber column 10 ft long of southern cypress select rough is subjected to an eccentric load $P$ of 45,000 lb similar to the column illustrated in Fig. 12-8$a$. Assume that the eccentricity is 3 in. Select the most economical column to use.

**12-73.** Design a timber column 12 ft long of Douglas fir select structural rough to support an axial load $P_O$ of 30,000 lb plus an eccentric load $P$ of 12,500 lb applied at a point 8 in. from the column axis, in a manner similar to that illustrated in Fig. 12-8$b$. *Ans.* 8 by 10 in. rough

**12-74.** A timber column 8 ft long of southern pine No. 1 SR rough is subjected to an axial load $P_O$ of 12,000 lb plus an eccentric load $P$ of 10,000 lb applied at a point 6 in. from the axis of the column in a manner similar to that illustrated in Fig. 12-8$b$. Select the most economical column.

**12-75.** Design a timber column 9 ft long of southern pine No. 2 SR S4S loaded as illustrated in Fig. 12-8$b$; $P_O = 10,000$ lb, $P = 2,500$ lb, and $e = 5.5$ in.

**12-76.** A mountain hemlock select structural column is loaded in a manner similar to that shown in Fig. 12-8$c$. The axial load $P_O = 23,000$ lb, the eccentric load $P = 10,000$ lb, $e = 2$ ft, $a = 6$ ft, $b = 4$ ft, and $c = 6$ ft. Select the most economical S4S column size.

**12-77.** A timber column of southern cypress common rough is loaded as illustrated in Fig. 12-8$c$; $P_O = 37,000$ lb, $P = 3,600$ lb, and $e = 2.5$ ft. The column is 10 ft long, and the bracket is so placed that $a = 5$ ft, $b = 3$ ft, and $c = 2$ ft. Design the most economical column to use.

**12-78.** Design a timber column 16 ft long of southern pine No. 1 dense SR rough to support an axial load $P_O$ of 44 kips and an eccentric load $P$ of 6 kips applied at a point 4 ft from the column axis in a manner similar to that illustrated in Fig. 12-8$c$. Let $a = 7$ ft, $b = 4$ ft, and $c = 5$ ft. *Ans.* 10 by 14 in. rough

## SUMMARY
### (By article number)

**12-1.** Under end loads, concentric or eccentric, columns tend to buckle laterally (sideways). This tendency increases with slenderness.

**12-3.** The **slenderness ratio** of a column, $KL/k$ for metal columns and $L/d$ for timber columns, determines its allowable load $P/A$ per unit (square inch) of cross-sectional area. $P/A$ decreases from a maximum when the slenderness ratio is quite low to approximately 20 percent of the maximum when the slenderness ratio is at the highest permissible in usual practice. Depending on its slenderness ratio, a column is classified as *short, intermediate,* or *long.* Formulas have been devised, nearly all derived from experimental data, giving the allowable unit load $P/A$ for the columns in each of these groups.

**12-5.** **Euler's formula,** the only one derived entirely mathematically, is

$$\frac{P}{A} = \frac{\pi^2 E}{(L/k)^2} \tag{12-2}$$

where $P$ is the critical or *buckling load.* To obtain the safe load, $P$ must be divided by an appropriate factor of safety.

**12-6.** **End restraint** has great influence on the tendency of a "long" column to buckle under load. A given "long" column whose ends are fully restrained against rotation has four times the load-carrying capacity of the same column with "hinged" ends (free to rotate), *provided, however, that in no case must P/A exceed the value permissible on the "simple-column length" of the column* in question. Columns in the "intermediate" range are much less affected by end restraint, and "short" columns are not affected at all,

**12-11.** For the **design of structural-steel columns and other compression members** in buildings and similar structures, most building codes specify the use of the three AISC formulas, Eqs. (12-10) to (12-12). Values of $P/A$ for these formulas are listed in Table E-14.

**12-15.** The ASCE specifications for aluminum columns give graphically the allowable stress $P/A$ as a function of the slenderness ratio $L/k$.

**12-16.** The NFPA formula for long timber columns is

$$\frac{P}{A} = \frac{0.30E}{(L/d)^2} \tag{12-15}$$

**12-17. Eccentrically loaded columns** are those in which the action line of the load is parallel to but does not coincide with the axis of the column. Such columns are subjected both to direct axial stress and bending stress. Formulas giving the allowable unit load $P/A$ on such columns become rather complex. A simple rule is that eccentrically loaded columns shall be so proportioned that the **stress factor**

$$\frac{f_a}{F_a} + \frac{f_b}{F_b} \text{ shall not exceed 1} \tag{12-22}$$

For timber columns, this permissible stress factor decreases from 1 when $L/d$ is about 15 to 0.8 when $L/d = 35$. These are arbitrary limits.

## REVIEW PROBLEMS

**12-79.** A piece of hard-steel shafting having a proportional limit of 60,000 psi is used as a column supporting an axial load. If its diameter is 1 in. and $E = 30,000,000$ psi, calculate (a) the lowest limit of length $L$ for which Euler's formula applies and (b) the buckling load $P$ when $L = 40$ in.

*Ans.* (a) $L = 17.5$ in.; (b) $P = 9,080$ lb

**12-80.** A strut in an airplane frame is a 1-in. square, hollow tube of aluminum, 14S-T4, with walls 0.1 in. thick. Using Euler's formula, calculate the buckling load $P$ on this strut if $L = 50$ in.

**12-81.** Using Euler's formula, calculate the safe axial compressive load $P$ on a 2-in.-diameter standard steel-pipe column 8 ft 4 in. long if a factor of safety of $2\frac{1}{2}$ is used. *Ans.* $P = 5,290$ lb

**12-82.** The radius of gyration $k$ of a thin-walled square tube, $d$ by $d$ in., is given by $d/\sqrt{6}$. Develop the equations (a) giving the lower limit of the length $L$ for which Euler's formula applies if the proportional limit is $s_p$, and (b) the corresponding length $L$ for a solid square section with the same outside dimensions.

**12-83.** A 7-ft-long piece of standard steel pipe is to be used to withstand a compressive axial load of 4,000 lb. Its ends are presumed to be hinged. $E$ is 29,000,000 psi and a factor of safety of 2 is considered appropriate. Determine the required diameter $d$ of pipe, using Euler's formula. Calculate also its slenderness ratio $L/k$ to determine if Euler's formula holds. See Table 12-2.

**12-84.** A round rod of hard steel having a proportional limit of 40,000 psi is to be used as a machine part capable of transmitting an estimated axial compressive load $P$ of 284 lb. If $E = 30,000,000$ psi and a factor of safety of 2 is used, what minimum diameter $d$ of rod may be used, using Euler's formula? $L = 10$ in.

**12-85.** A square, hollow tube of structural aluminum, 14S-T6, is to be used as a strut in an airplane frame to withstand a calculated axial compressive load of 9,320 lb. Its ends are presumed to be hinged. The proportional limit is 30,000 psi and $E$ is 10,300,000 psi. If the length of the strut is 50 in. and the inside dimension of the tube is 0.9 of the outside dimension $d$, what should be the dimension $d$? Use a factor of safety of 2.

**12-86.** A box column is built up of two 9- by 2½-in. standard structural-steel channels, each weighing 13.4 lb per ft, to form a section 5 by 9 in. similar to the one illustrated in Fig. 12-1$k$. If the unsupported length of this column is 24 ft, what is its allowable load $P$? Use AISC specifications.          *Ans. $P = 87,000$ lb*

**12-87.** Select the most economical rolled structural-steel W section to use as a column 10 ft 6 in. long to support an axial compressive load $P$ of 60,000 lb. Use AISC specifications.

**12-88.** A double-angle truss chord member 13 ft long must resist a compressive load of 60,000 lb. Angles are ⅜ in. back to back. Select the most economical angles using AISC specifications.

**12-89.** A structural T-section is used for the upper (compressive) chord of a steel truss. The tee is formed by shearing the web of an W 8 × 20 into two identical tees. Calculate the allowable axial compressive load using AISC specifications if the member is braced at 6-ft intervals in the plane of the web and at 12-ft intervals in the plane of the flange.

**12-90.** A single-angle structural-steel brace 70 in. long must be capable of withstanding a compressive load of 4,600 lb. Select the most economical equal-leg angle to use. Use AISC specifications.          *Ans. 2 by 2 by $\frac{3}{16}$ in.*

**12-91.** A latticed column of the type shown in Figs. 12-1$h$ and 12-2 is to be 40 ft long and must support an axial load of 172,000 lb. (*a*) Select the most economical standard structural-steel channels to use. (*b*) Calculate the required distance $b$ from back to back of channels to give equal moments of inertia with respect to both major axes.

**12-92.** Solve Prob. 12-47 for a length of 25 ft and a load of 265,000 lb.

**12-93.** A round bar of aluminum alloy, 6061-T6, 30 in. long in an aircraft frame is to be subjected to an axial compressive load $P$ of 1,200 lb. Its ends will be free to rotate so that $K = 1$. Determine the required diameter $d$ of this bar.

**12-94.** A square tube of aluminum alloy, 6061-T6, 56 in. long is to be subjected to an axial compressive load $P$. Determine the allowable load $P$ if the tube is 1¼- by 1¼-in. outside dimensions and the wall thickness is ⅛ in. Ends are partially restrained.

**12-95.** Design a timber column 10 ft long of Douglas fir No. 1 S4S to support an axial compressive load of 45,000 lb.

**12-96.** Select the most economical timber column 8 ft 4 in. long of mountain hemlock select structural S4S to support an axial load $P$ of 10 kips.

*Ans. 4 by 6 in. S4S*

**12-97.** Design a round timber column 12 ft long of southern cypress select to support an axial load $P$ of 60,400 lb.

**12-98.** Design a round timber column 8 ft long of southern pine No. 1 SR to support an axial load $P$ of 42,300 lb.

*Ans. D* $= 7.65$ in.

# REVIEW QUESTIONS

**12-1.** Describe a column and its function and the different types of columns.

**12-2.** In what manner does a column usually fail?

**12-3.** What imperfections of material, of manufacture, and of loading would hasten failure?

**12-4.** Which type of column cross section is most efficient?

**12-5.** What is meant by the slenderness ratio of a column?

**12-6.** What slenderness-ratio limits determine (approximately) the group in which a given column belongs?

**12-7.** What is meant by the critical or buckling load on a column?

**12-8.** To what specific group of columns does Euler's formula apply? What load does $P$ represent? Does this formula contain a factor of safety?

**12-9.** In connection with columns, what is meant by end restraint?

**12-10.** Discuss the effect of end restraint on the load-carrying capacity of a column.

**12-11.** In column formulas, why is $P/A$ referred to as the allowable unit load rather than the allowable unit stress?

**12-12.** In structural-steel columns, what distinguishes main members from secondary members?

**12-13.** Do the structural-steel column formulas contain a factor of safety? What load does $P$ represent?

**12-14.** Why must steel columns be designed by a trial-and-error method?

**12-15.** In determining the slenderness ratio of a structural-steel or aluminum column, which of its two radii of gyration must always be used?

**12-16.** When the unsupported column length in one plane differs from that in a perpendicular plane, which $L/k$ governs the allowable load?

**12-17.** Do the curves in Fig. 12-6 for axially loaded aluminum columns include a factor of safety?

**12-18.** Describe the internal stress picture in an eccentrically loaded column.

**12-19.** In what kind of column does the lateral deflection $\Delta$ become an important factor?

**12-20.** What is the stress factor and what is its function?

# CHAPTER 13

# Special Topics

## 13-1 Impact or Dynamic Loading

In the preceding chapters all loads and forces acting on structural members, connections, or parts have been considered to be *static;* that is, they have been *gradually* applied until the maximum was reached and then have remained static.

However, many problems arise, especially in machine design, in which loads are caused by impacts or other dynamic conditions. The solutions of such problems differ radically from those in which only static forces are present, as is shown in the following examples.

## 13-2 Strain Energy

The work-energy concept is developed in engineering mechanics.[1] A force moving through a distance does work. In the case of a gradually applied load, as in Fig. 13-1, the work done by the varying force is the product of the *average* force and the displacement $\delta$. In symbols,

$$U = \frac{P\delta}{2} = \frac{P}{2}\frac{PL}{AE} \quad \text{or} \quad U = \frac{P^2LA}{2A^2E}$$

Since $P/A$ is the unit stress $s$ and $LA$ is the volume of the member,

$$U = \frac{s^2}{2E} \text{ (volume)} \tag{13-1}$$

---

[1]See, for example, the authors' "Applied Engineering Mechanics," 3d ed., McGraw-Hill Book Company, New York, 1972.

FIG. 13-1 Gradually applied axial load $P$.

The work done on the block is stored as *strain energy* in the material. The block behaves as a very stiff spring, provided the stress $s$ is below the proportional limit of the material. When the load is removed, the stored energy is recovered.

The strain energy per unit volume is

$$\frac{U}{\text{volume}} = \frac{s^2}{2E} \tag{13-2}$$

The amount of energy transmitted to a structural member under impact is sometimes referred to as an *energy load*. The stresses caused by such energy loads can be computed from Eq. (13-2).

### ILLUSTRATIVE PROBLEMS

**13-1.** The elastic limit of a certain steel is 32,000 psi, and its modulus of elasticity $E$ is 30,000,000 psi. What maximum amount of recoverable energy can be stored in each cubic inch of this material?

*Solution:* If a bar of this material is stressed beyond its elastic limit, it will be permanently deformed, and some of the applied work cannot be recovered. If all the stored energy is to be recovered, $s$ in Eq. (13-2) is the elastic limit, and we then have

$$\left[\frac{U}{\text{volume}} = \frac{s^2}{2E}\right] \qquad \frac{U}{\text{volume}} = \frac{(32,000)(32,000)}{(2)(30,000,000)} = 17.05 \ \frac{\text{lb-in.}}{\text{cu in.}} \qquad Ans.$$

This maximum recoverable amount of stored energy for a material is called the *modulus of resilience* of that material. While the recoverable strain energy is determined by the elastic limit of the material, the expression $s^2/2E$ for the strain energy per unit volume is applicable only to the proportional limit. The error in applying $s^2/2E$ to the elastic limit is small.

FIG. 13-2 Weight dropped on spring.

**13-2.** A 10-lb weight is dropped from a height of 3 in. onto a spring, as shown in Fig. 13-2. The spring constant $k$ is 100 lb per in.; that is, a 100-lb force will compress the spring 1 in., and a 300-lb force will compress it 3 in. Calculate the maximum spring deflection $\delta$.

*Solution:* The potential energy lost by the falling weight is $U = (10)(3 + \delta)$.

The work done on the spring is stored in it as strain energy and is evaluated from $U = \frac{1}{2}k\delta^2$. Equating the work done on the spring to its strain energy, we obtain

$$(10)(3 + \delta) = \frac{1}{2}k\delta^2$$

Since $k = 100$ lb per in.,

$$30 + 10\delta = 50\delta^2 \quad \text{or} \quad \delta^2 - 0.2\delta - 0.6 = 0$$

When this quadratic equation is solved, either by formula or by completing the square, we obtain

$$\delta = 0.88 \text{ in.} \quad \textit{Ans.}$$

Note that the static spring deflection under the 10-lb weight would be

$$[\text{W} = k\delta] \qquad 10 = 100\delta \quad \text{or} \quad \delta = 0.1 \text{ in.}$$

## PROBLEMS

**13-3.** A 1-in.-diameter aluminum rod is 9 ft long. Its yield stress $s$ is 16,000 psi, and its modulus of elasticity $E$ is 10,000,000 psi. What energy load may be applied to the rod without exceeding the yield stress?     *Ans.* $U = 1,085$ lb.-in.

**13-4.** The weight $W$ in Fig. 13-3 is 10 lb and drops $h = 3$ in. onto the stop. The rod is 1 in. in diameter and is 20 ft long. Calculate the maximum unit rod stress $s$ if $E = 3,000,000$ psi.     *Ans.* $s = 990$ psi ($s = 977$ psi if $\delta$ neglected)

**13-5.** If the weight $W$ in Fig. 13-3 is 10 lb, from what height $h$ may it be dropped if the unit stress $s$ in the steel rod is not to exceed 20,000 psi? The rod is $1\frac{1}{4}$ in. in diameter, $L$ is 60 in., and $E$ is 30,000,000 psi. Neglect the elongation $\delta$.

**13-6.** The cross-sectional area of the upper portion of each of the rods in Fig. 13-4 is 3 sq in. and of the bottom portion is 1 sq in. Determine the maximum unit stress $s$ in each rod if $E = 10,000,000$ psi. Neglect the elongation $\delta$. Note that the

FIG. 13-3 Probs. 13-4 and 13-5.    FIG. 13-4 Probs. 13-6 and 13-19.

unit stress in the lower portion is three times that in the upper portion because
$T = S_1A_1 = S_2A_2$.

## 13-3 Impact Loading

Consider the simple beam shown in Fig. 13-5a. The impact load $W$ is
allowed to fall freely onto the mid-span of the beam. The beam deflects an
amount $\delta$ under this condition, as indicated in Fig. 13-5b. This same deflection
$\delta$ could be produced by a gradually applied load of sufficient magnitude. Call
this equivalent static load $P_{EQ}$ as indicated in Fig. 13-5d.

Because the deflections and stresses under the equivalent load $P_{EQ}$ are the

FIG. 13-5 Derivation of impact factor. (a) Impact load. (b) Deflection under impact
load. (c) Deflection under gradually applied load $W$. (d) Equivalent gradually
applied load to give some deflection as impact load $W$.

same as those under the impact load $W$, the internal strain energies are the same. The external work done by the loads, then, must also be the same. Putting this statement in equation form, we obtain

$$W(h + \delta) = \frac{P_{EQ}\delta}{2} \tag{a}$$

Since the deflection of the beam is proportional to the load that produces that deflection,

$$\frac{P_{EQ}}{\delta} = \frac{W}{\delta_{ST}} \qquad \text{or} \qquad P_{EQ} = W\frac{\delta}{\delta_{ST}} \tag{b}$$

where $\delta_{ST}$ is the static deflection under a gradually applied load $W$, as in Fig. 13-5c. Substituting the value of $P_{EQ}$ from Eq. $(b)$ in Eq. $(a)$ gives

$$W(h + \delta) = \frac{W\delta^2}{2\delta_{ST}} \qquad \text{or} \qquad h + \delta = \frac{\delta^2}{2\delta_{ST}}$$

Rearranging gives

$$\delta^2 - 2\delta_{ST}\delta = 2h\delta_{ST}$$

Completing the square yields

$$\delta^2 - 2\delta_{ST}\delta + \delta_{ST}{}^2 = \delta_{ST}{}^2 + 2h\delta_{ST}$$

or

$$(\delta - \delta_{ST})^2 = \delta_{ST}{}^2\left(1 + \frac{2h}{\delta_{ST}}\right)$$

Taking the square root of each side gives

$$\delta - \delta_{ST} = \delta_{ST}\sqrt{1 + \frac{2h}{\delta_{ST}}}$$

or

$$\boldsymbol{\delta = \delta_{ST}\left(1 + \sqrt{1 + \frac{2h}{\delta_{ST}}}\right)} \tag{13-3}$$

We conclude from Eq. (13-3) that the deflection $\delta$ under impact loading is considerably greater than the static deflection $\delta_{ST}$ for the same load gradually applied. The factor $1 + \sqrt{1 + (2h/\delta_{ST})}$ can be thought of as an **impact factor,** which always exceeds unity.

Since the stresses in a beam are proportional to the load that produces them, we may state

$$\frac{s_{ST}}{W} = \frac{s}{P_{EQ}} \tag{c}$$

where $s_{ST}$ is the stress due to the gradually applied load $W$ and $s$ is the stress due to the equivalent load $P_{EQ}$.

Recall from Eq. $(b)$ that $P_{EQ} = W(\delta/\delta_{ST})$. Substituting for $P_{EQ}$ in Eq. $(c)$, we obtain

$$\frac{s_{ST}}{W} = \frac{s\delta_{ST}}{W\delta} \quad \text{or} \quad s = \frac{\delta}{\delta_{ST}}\,s_{ST}$$

From Eq. (13-3), $\delta/\delta_{ST} = 1 + \sqrt{1 + (2h/\delta_{ST})}$. Hence

$$s = \left(1 + \sqrt{1 + \frac{2h}{\delta_{ST}}}\right)s_{ST} \qquad \text{(13-4)}$$

The stress $s$ under impact loading is, then, the impact factor multiplied by the stress $s_{ST}$ under the same loading gradually applied, that is, the static load.

## ILLUSTRATIVE PROBLEMS

**13-7.** The beam shown in Fig. 13-5$a$ is a $W\ 10 \times 33$ section. Compute the maximum bending stress $s$ if a 1,000-lb weight is dropped from a height $h = 4$ in. at mid-span. The length of the beam is 12 ft, and $E = 30,000,000$ psi.

*Solution:* The maximum moment under a concentrated static load of 1,000 lb is $PL/4$ (see Fig. 9-31). Therefore $M = (1,000)(144)/4 = 36,000$ lb-in. The maximum bending stress is

$$\left[s = \frac{M}{Z}\right] \qquad s = \frac{36,000}{35} = 1,027 \text{ psi}$$

The maximum deflection under the concentrated static load, from Fig. 9-31, is

$$\left[\delta = \frac{PL^3}{48EI}\right] \qquad \delta_{ST} = \frac{(1,000)(144)^3}{(48)(30,000,000)(170.9)} = 0.012\ 14 \text{ in.}$$

The maximum stress under the impact load is

$$\left[s = \left(1 + \sqrt{1 + \frac{2h}{\delta_{ST}}}\right)s_{ST}\right] \qquad s = \left[1 + \sqrt{1 + \frac{(2)(4)}{0.012\ 14}}\right](1,027)$$

or $\qquad s = (26.75)(1,027) = 27,400 \text{ psi.} \qquad Ans.$

**13-8.** The shaft in Fig. 13-6 is subjected to a torsional impact load as shown. The shaft is 6 ft long and its diameter is 2 in. Calculate the maximum torsional shearing stress $s_s$ if $W = 10$ lb, $e = 6$ in., $h = 4$ in., and $G = 10,000,000$ psi.

*Solution:* The polar moment of inertia $J$ of the shaft is

$$\left[J = \frac{\pi R^4}{2}\right] \qquad J = \frac{\pi(1)^4}{2} = 1.57 \text{ in.}^4$$

If the 10-lb load were *gradually* applied, the shearing stress would be

$$\left[s_s = \frac{TD}{2J}\right] \qquad s_s = \frac{(10)(6)(2)}{(2)(1.57)} = 38.2 \text{ psi}$$

FIG. 13-6 Torsional impact load on shaft.

The angle of twist $\theta$ under this *gradually* applied load would be

$$\left[\theta = \frac{TL}{GJ}\right] \qquad \theta = \frac{(10)(6)(72)}{(10,000,000)(1.57)} = 0.000\ 275 \text{ radian}$$

The deflection $\delta$ at the load center would then be

$$[\delta = e\theta] \qquad \delta = (6)(0.000\ 275) = 0.001\ 65 \text{ in.}$$

The maximum stress $s_s$ under the impact load is given by Eq. (13-4):

$$\left[s_s = \left(1 + \sqrt{1 + \frac{2h}{\delta_{ST}}}\right)s_{ST}\right] \qquad s_s = \left[1 + \sqrt{1 + \frac{(2)(4)}{0.001\ 65}}\right](38.2)$$

$$s_s = (70.7)(38.2) = 2,700 \text{ psi} \qquad Ans.$$

## PROBLEMS

**13-9.** From what height $h$ may a 100-lb weight be dropped on a 4- by 10-in. rough timber beam if the bending stress must not exceed 2,000 psi? The beam span is 10 ft, $E = 1,800,000$ psi, and the load is applied at mid-span.     *Ans. h = 5.67 in.*

**13-10.** Solve Prob. 13-7 if the weight $W = 500$ lb, $h = 3$ in., $L = 10$ ft, and the beam is of Douglas fir with a 6- by 10-in. section (S4S).

**13-11.** Calculate the total angle of twist $\theta$ of the shaft shown in Fig. 13-6 under the impact load if the data are as given in Prob. 13-8.

**13-12.** Check the spring deflection of Prob. 13-2 using Eq. (13-3).

**13-13.** Check the rod stress of Prob. 13-4 using Eq. (13-4).

## 13-4 Prestressed Beams

A beam material that is relatively weak in tension may be placed in compression by the pull of longitudinal tendons. If this prestressing is done before the application of the usual bending loads, the tendency of the bending moments to place one side of the beam in tension simply results in the reduction of the compressive prestress on that side of the beam. Of course, the bending compressive stress on the other side also adds to the initial compressive stress there. The prestressing tendon may be placed off-center, or may be draped, so that the initial compressive stress will be high on that side of the beam where later bending stresses will tend to be tensile. The off-center tendon will cause small compressive prestress on the side of the beam that will later be compressed still further by the applied bending moments.

Consider the beam shown in Fig. 13-7. The beam is composed of concrete blocks 6 in. wide by 12 in. deep. A loose-fitting threaded steel rod runs longitudinally as shown. The nuts on the end of the rod are tightened until the maximum stress in the beam is 1,000 psi as shown in Fig. 13-7d. The rod pull $P$ is given by

$$s = \frac{P}{A} + \frac{Mc}{I} = \frac{P}{A} + \frac{Pec}{I}$$

$$1,000 = \frac{P}{72} + \frac{P(2)(6)}{6(12)^3/12} = 0.0139P + 0.0139P$$

FIG. 13-7 Prestressed beam under uniform load. (a) Beam section. (b) Prestressed beam. (c) Moment diagram. (d) Beam stress due to rod tension. (e) Stress due to bending load. (f) Final stress diagram.

or

$$P = \frac{1,000}{0.0278} = 36,000 \text{ lb}$$

The stress at the top of the beam due to the tendon force will be

$$\left[ s = \frac{P}{A} + \frac{Pec}{I} \right] \qquad s = \frac{36,000}{72} - \frac{36,000(2)(6)}{6(12)^3/12} = 0$$

Let us now add a bending moment to the beam that will cause a compressive bending stress of 1,000 psi at the top of the beam and a tensile bending stress of 1,000 psi at the bottom of the beam. (See Fig. 13-7e.) If the stresses due to the tendon force are added to those caused by the later applied bending moment, the result will be the stresses shown in Fig. 13-7f. The stresses without the applied bending load are those in Fig. 13-7d and stresses with the applied bending load are shown in Fig. 13-7f. In neither case are there any tensile stresses in the beam. Therefore, the beam material need not be capable of withstanding large tensile stresses.

What is the magnitude of the maximum bending moment that may be applied to this prestressed beam? We assume that tension is undesirable. The bending moment that will cause the bending stresses shown in Fig. 13-7e is

$$\left[ M = \frac{sI}{c} \right] \qquad M = \frac{1,000(6)(12)^3/12}{6} = 144,000 \text{ lb-in.}$$

If this moment is the result of a uniformly distributed load, the bending moment diagram will be that shown in Fig. 13-7c and the maximum moment will be $wL^2/8$. The allowable uniform load (dead plus live) is

$$\left[ w = \frac{8 \, M_{max}}{L^2} \right] \qquad w = \frac{8(144,000/12)}{10^2} = 960 \text{ lb/ft}$$

## PROBLEMS

**13-14.** The beam shown in Fig. 13-7 is subjected simultaneously to a 36,000-lb tendon pull and a downward uniform load of 960 lb/ft. Calculate the resultant maximum stresses at the top and bottom of the beam at a section 2 ft 6 in. from the left support. *Ans.* $s_{top} = 750$ psi; $s_{btm} = 250$ psi

**13-15.** Consider the beam shown in Fig. 13-7, but let the tendon be located at the mid-depth of the beam. (a) Calculate the required tendon force to cause a uniform compressive stress of 500 psi over each section. (b) What uniform vertical load, in pounds per foot, may be placed on the prestressed beam without causing a resultant tensile stress? *Ans.* (a) $F = 36,000$ lb; (b) $w = 480$ lb/ft

**13-16.** The beam shown in Fig. 13-8 has a rectangular cross section with a width of 6 in. and a depth of 18 in. (a) Calculate the maximum flexural bending stress

FIG. 13-8 Prob. 13-16.

FIG. 13-9 Prob. 13-17. ($a$) Beam section. ($b$) Prestress. ($c$) Final stresses at mid-span.

in the beam due to the concentrated load shown. ($b$) Calculate the required tendon force to give sufficient compressive prestress to the beam so that the combined prestress and bending stress at any cross section will be everywhere compressive. Assume that the tendon will be located 6 in. above the bottom of the beam. Neglect the weight of the beam.

**13-17.** A beam is composed of concrete blocks held together by a tensioned tendon as shown in Fig. 13-9a. The allowable compressive stress in the concrete is 1,000 psi, tensile stress in the concrete is to be avoided, and all concrete thicknesses are 1 in. ($a$) Calculate the maximum permissible tendon force and ($b$) the maximum allowable total uniform load on a 10-ft-long simple span.

## SUMMARY
### (By article number)

**13-2.** The **strain energy** in a uniformly stressed member is

$$U = \frac{s^2}{2E}(\text{volume}) \tag{13-1}$$

**13-3.** The deflection under impact loading is

$$\delta = \delta_{ST}\left(1 + \sqrt{1 + \frac{2h}{\delta_{ST}}}\right) \tag{13-3}$$

where $\delta_{ST}$ is the deflection if the load were gradually applied. The parenthetical part of the equation is an impact factor.

The maximum stress under an impact load is

$$s = \left(1 + \sqrt{1 + \frac{2h}{\delta_{ST}}}\right)s_{ST} \tag{13-4}$$

FIG. 13-10 Prob. 13-18.

FIG. 13-11 Probs. 13-20 and 13-21.

## REVIEW PROBLEMS

**13-18.** The weight $W$ falls freely 1 in. onto the spring shown in Fig. 13-10. The spring constant $k$ is 2,000 lb per in. and the modulus of elasticity $E$ of the rod is 10,000,000 psi. Calculate the maximum stress in the rod. The elongation of the rod may be neglected, since it is extremely small compared to the deformation of the spring. (HINT: The maximum force in the rod equals the maximum force in the spring.)

*Ans.* $s = 204$ psi

**13-19.** Solve Prob. 13-6 if $W = 10$ lb and $E = 30,000,000$ psi.

*Ans.* $s_a = 4,750$ psi; $s_b = 6,710$ psi

**13-20.** A weight $W$ is dropped at mid-span on a beam as shown in Fig. 13-11. Calculate the maximum bending stress $s$ if $W = 100$ lb, $h = 4$ in., $k = 200$ lb per in., $E = 1,000,000$ psi, $I = 48$ in.$^4$, $L = 48$ in., and the beam is 4 in. deep.

*Ans.* $s = 456$ psi

**13-21.** Solve Prob. 13-20 if $W = 120$ lb, $h = 3$ in., $k = 60$ lb per in., $E = 1,600,000$ psi, $I = 32$ in.$^4$, $L = 60$ in., and the depth of the beam is 3 in.

**13-22.** Solve Prob. 13-16 if the beam carries a uniform load of 240 lb per ft in addition to the two concentrated loads. Let the beam width be 12 in.

## REVIEW QUESTIONS

**13-1.** Distinguish between static and dynamic loads.

**13-2.** Define strain energy.

**13-3.** Define modulus of resilience of a material.

**13-4.** What is meant by the term "impact factor"?

**13-5.** How do stresses due to dynamic loading compare to those caused by static loading?

**13-6.** Show by simple sketches the typical stress distribution in a prestressed beam due to ($a$) the tendon force only, ($b$) bending due to lateral loads only, and ($c$) both causes.

# APPENDIX A

# Centroids of Areas by Integration

To locate the centroid of an area, as was stated in Art. 5-1, we need only equate the moment $A\bar{x}$ of the entire area, with respect to some reference axis, to the sum of the moments of the component areas $\Sigma ax$. When each component area is infinitesimally small, or $dA$, the equations with respect to the $Y$ and $X$ axes become

$$A\bar{x} = \int x \, dA \qquad (a)$$

and

$$A\bar{y} = \int y \, dA \qquad (b)$$

**Rules.** In selecting the differential element $dA$, the following rules must be observed:

1. All points of the element must be the same distance from the moment axis.
2. The position of the centroid of the element must be known, so that the moment of the element with respect to the moment axis is the product of the element and the distance of its centroid from the axis. The locations of the centroids of a triangle, a parabola, and a sector of a circle are shown in Probs. A-1 to A-3.

## ILLUSTRATIVE PROBLEMS

**A-1.** Locate the centroid of the triangle shown in Fig. A-1 with respect to the $Y$ axis.

*Solution:* Let the differential area $dA$ be a vertical strip parallel to the side $a$. This

333

FIG. A-1 Centroid of a triangle.

FIG. A-2 Centroid of a parabolic segment.

area $dA = y\ dx$. From Eq. $(a)$, we have

$$\left[ A\bar{x} = \int x\ dA \right] \qquad \tfrac{1}{2}ba \cdot \bar{x} = \int_0^b xy\ dx \qquad (c)$$

But by similar triangles

$$\frac{y}{x} = \frac{a}{b} \qquad \text{or} \qquad y = \frac{ax}{b} \qquad (d)$$

When this value of $y$ is substituted in Eq. $(c)$, we obtain

$$\tfrac{1}{2}ba \cdot \bar{x} = \frac{a}{b} \int_0^b x^2\ dx \qquad \text{and} \qquad \bar{x} = \tfrac{2}{3}b \qquad Ans.$$

**A-2.** Locate the centroid with respect to the $Y$ axis of the parabolic segment shown in Fig. A-2 and bounded by the $X$ axis, the vertical side $b$, and the parabola $y = kx^2$.

*Solution:* As the differential area $dA$ we select a vertical strip parallel to side $b$. This area $dA = y\ dx$. But $y = kx^2$, and when $x = a$, $b = ka^2$. Hence

$$\left[ A = \int dA \right] \qquad A = \int_0^a y\ dx = \int_0^a kx^2\ dx = \tfrac{1}{3}ka^3 \qquad (e)$$

To find $\bar{x}$ from Eq. $(a)$, we have

$$\left[ A\bar{x} = \int x\ dA \right] \qquad \tfrac{1}{3}ka^3 \cdot \bar{x} = \int_0^a xy\ dx = k \int_0^a x^3\ dx \qquad (f)$$

from which $\qquad \tfrac{1}{3}ka^3 \cdot \bar{x} = \tfrac{1}{4}ka^4 \qquad \text{and} \qquad \bar{x} = \tfrac{3}{4}a \qquad Ans.$

To find $\bar{y}$, we may use the same elemental strip. But our moment axis is now the $X$ axis, and all points of the element are not at the same distance from the axis. Hence rule 2 applies, and the moment of each elemental strip with respect to the $X$ axis then is the product of its area $y\ dx$ and its centroidal distance $\tfrac{1}{2}y$, or $\tfrac{1}{2} \cdot y\ dx$. Then, since $y = kx^2$ and $b = ka^2$,

$$\left[ A\bar{y} = \int y\ dA \right] \qquad \tfrac{1}{3}ka^3 \cdot \bar{y} = \int_0^a \tfrac{1}{2}y \cdot y\ dx = \tfrac{1}{2}k^2 \int_0^a x^4\ dx \qquad (g)$$

or          $\frac{1}{3}ka^3 \cdot \bar{y} = \dfrac{k^2 a^5}{10}$     and     $\bar{y} = \frac{3}{10}ka^2 = \frac{3}{10}b$     *Ans.*

**A-3.** Locate the centroid of the area of the sector of a circle shown in Fig. A-3.
*Solution:* Let the area of the sector be symmetrical about the $X$ axis. The radius is $r$ and the subtended angle is $2\alpha$. The differential area $dA$ here selected is a triangle whose area is $\frac{1}{2}r^2\,d\theta$. The centroid of this triangle lies at a distance $x = \frac{2}{3}r\cos\theta$ from the $Y$ axis. The area of the sector is

$$\left[A = \int dA\right] \qquad A = \int_{-\alpha}^{+\alpha} \frac{1}{2}r^2\,d\theta = r^2\alpha \tag{h}$$

Then from Eq. $(a)$

$$\left[A\bar{x} = \int x\,dA\right] \qquad r^2\alpha \cdot \bar{x} = \int_{-\alpha}^{+\alpha} \frac{2}{3}r\cos\theta\,\frac{1}{2}r^2\,d\theta \tag{i}$$

or          $r^2\alpha \cdot \bar{x} = \dfrac{1}{3}r^3 \displaystyle\int_{-\alpha}^{+\alpha} \cos\theta\,d\theta = \dfrac{2}{3}r^3\sin\alpha$ \hfill $(j)$

and          $\bar{x} = \dfrac{2}{3}\dfrac{r\sin\alpha}{\alpha}$     *Ans.*

When $\alpha = 90°$, as in the semicircle in Fig. A-4, $\bar{x} = 4r/3\pi$.

Fig. A-3 Centroid of a circular sector.          Fig. A-4  Centroid of a semicircle.

## PROBLEMS

**A-4.** Find the distance $\bar{y}$ from the $X$ axis to the centroid of the triangle shown in Fig. A-1 using $(a)$ rule 2 and a strip parallel to the $Y$ axis and $(b)$ rule 1 and a strip parallel to the $X$ axis. \hfill *Ans.* $\bar{y} = \frac{1}{3}a$

**A-5.** Locate the position of the centroid of the shaded area shown in Fig. A-5.

*Ans.* $\bar{x} = \frac{4}{5}a$; $\bar{y} = \frac{2}{7}b$

**A-6.** Determine the position of the centroid of the shaded area shown in Fig. A-6.

*Ans.* $\bar{x} = \frac{3}{5}a$; $\bar{y} = \frac{3}{8}b$

**A-7.** Locate the centroid of the shaded area shown in Fig. A-7.

*Ans.* $\bar{x} = \frac{3}{10}a$; $\bar{y} = \frac{3}{4}b$

Fig. A-5 Prob. A-5.          Fig. A-6 Prob. A-6.          Fig. A-7 Prob. A-7.

# APPENDIX B

# Moments of Inertia
# of Plane Areas

**Rectangular Moment of Inertia.** As stated in Art. 5-4, the moment of inertia of an area with respect to an axis in the plane of the area is required in the design of beams and columns. The exact moments of inertia of simple plane areas, such as a rectangle, a triangle, or a circle, must be determined by integration. The basic expression for the moment of inertia $I$ of any plane area with respect to an axis in its plane, as was mentioned in Art. 5-4, is

$$I = \int y^2 \, dA \qquad\qquad (a)$$

According to this equation, *the moment of inertia of a plane area with respect to an axis in the plane is the sum of the products of the elemental areas, each multiplied by the square of the distance from the axis to its centroid.*

**Rules.** In selecting the elemental area, the following rules must be observed:

1. All points in the elemental area must be the same distance from the axis.
2. The moment of inertia of the elemental area with respect to the inertia axis must be known. The moment of inertia of the entire area is then the summation of the moments of inertia of all the elements.

### ILLUSTRATIVE PROBLEMS

**B-1.** Find the moment of inertia of the rectangular area shown in Fig. B-1 with respect to $(a)$ a centroidal axis parallel to the base $b$ and $(b)$ an axis coinciding with the base.

*Solution: (a) Axis through centroid.* Let the width of the area be $b$ and its depth $h$, and let the elemental area be $b \, dy$, a strip parallel to the inertia axis $XX$. The

FIG. B-1 Moment of inertia of rectangular area. (*a*) Axis through centroid. (*b*) Axis coinciding with base.

moment of inertia of a single strip with respect to this axis is $y^2\,dA$ and of the entire area between the limits of $h/2$ and $-h/2$ then is

$$\left[I = \int y^2\,dA\right] \qquad I_{X_C} = \int_{-h/2}^{+h/2} y^2 b\,dy = b\left[\frac{y^3}{3}\right]_{-h/2}^{+h/2}$$

or
$$I_{X_C} = \frac{b}{3}\left(\frac{h^3}{8} + \frac{h^3}{8}\right) = \frac{bh^3}{12} \qquad Ans. \qquad (b)$$

(*b*) *Axis coinciding with base.* The summation is now made between the limits of 0 and $h$. That is,

$$\left[I = \int y^2\,dA\right] \qquad I_X = \int_0^h y^2 b\,dy = b\left[\frac{y^3}{3}\right]_0^h = \frac{bh^3}{3} \qquad Ans.$$

**B-2.** Determine the moment of inertia of the triangular area shown in Fig. B-2 with respect to (*a*) a centroidal axis parallel to the base $b$ and (*b*) an axis coinciding with the base.

FIG. B-2 Moment of inertia of a triangular area. (*a*) Axis through centroid. (*b*) Axis coinciding with base.

*Solution:* (*a*) *Axis through centroid.* The elemental area selected is a strip parallel to the centroidal inertia axis $X_C$. The moment of inertia of this strip with respect to axis $X_C$ is $y^2\,dA$, or $y^2x\,dy$, and the entire triangle then is

$$\left[I = \int y^2\,dA\right] \qquad\qquad I_{X_C} = \int_{-h/3}^{+2h/3} y^2 x\,dy \qquad\qquad (c)$$

By similar triangles

$$\frac{x}{\tfrac{2}{3}h - y} = \frac{b}{h} \qquad \text{or} \qquad x = \frac{b}{h}(\tfrac{2}{3}h - y) \qquad\qquad (d)$$

The moment of inertia of the entire area then is

$$\left[I = \int y^2\,dA\right] \quad I_{X_C} = \int_{-h/3}^{+2h/3} y^2 \frac{b}{h}(\tfrac{2}{3}h - y)\,dy$$

$$= \frac{2}{3}b \int_{-h/3}^{+2h/3} y^2\,dy - \frac{b}{h}\int_{-h/3}^{+2h/3} y^3\,dy$$

$$= \frac{2}{3}b\left[\frac{y^3}{3}\right]_{-h/3}^{+2h/3} - \frac{b}{h}\left[\frac{y^4}{4}\right]_{-h/3}^{+2h/3}$$

$$= \frac{2}{9}b\left(\frac{8}{27}h^3 + \frac{1}{27}h^3\right) - \frac{b}{4h}\left(\frac{16}{81}h^4 - \frac{1}{81}h^4\right)$$

or $$I_{X_C} = \frac{bh^3}{36} \qquad Ans. \qquad\qquad (e)$$

(*b*) *Axis coinciding with base.* When the same elemental strip is used,

$$\left[I = \int y^2\,dA\right] \qquad\qquad I_X = \int_0^h y^2 x\,dy \qquad\qquad (f)$$

But by similar triangles

$$\frac{x}{h - y} = \frac{b}{h} \qquad \text{and} \qquad x = \frac{b}{h}(h - y) \qquad\qquad (g)$$

The moment of inertia of the entire area then is

$$\left[I = \int y^2\,dA\right] \qquad I_X = \int_0^h y^2 \cdot \frac{b}{h}(h - y)\,dy$$

$$= b\int_0^h y^2\,dy - \frac{b}{h}\int_0^h y^3\,dy$$

$$= b\left[\frac{y^3}{3}\right]_0^h - \frac{b}{h}\left[\frac{y^4}{4}\right]_0^h$$

$$= \frac{bh^3}{3} - \frac{bh^3}{4} = \frac{bh^3}{12} \qquad Ans. \qquad\qquad (h)$$

**B-3.** Find the rectangular moment of inertia of the circular area shown in Fig. B-3, with respect to a diameter.

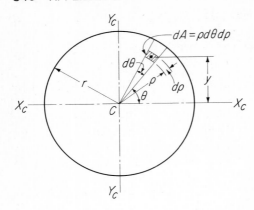

FIG. B-3 Rectangular moment of inertia of a circular area.

*Solution:* The elemental area is selected as shown in Fig. B-3. Its moment of inertia with respect to the $X_C$ axis is $y^2\, dA$, or $\rho^2 \sin^2 \theta\, \rho\, d\theta\, d\rho$, since $y = \rho \sin \theta$. A double integration is now necessary. The moment of inertia of the elemental sector is

$$\int_0^r \rho^2 \sin^2 \theta\, \rho\, d\theta\, d\rho$$

When this quantity is then again integrated between the limits of $\theta = 0$ and $\theta = 2\pi$ radians, we obtain the moment of inertia of the entire circle. That is,

$$\left[ I = \int y^2\, dA \right] \qquad I_{X_C} = \int_0^r \int_0^{2\pi} \rho^3\, d\rho \sin^2 \theta\, d\theta$$

$$= \frac{r^4}{4} \int_0^{2\pi} \sin^2 \theta\, d\theta = \frac{\pi r^4}{4} \qquad Ans. \qquad (i)$$

**The Transfer Formula for Parallel Axes.** In Art. 5-7 was described the method of computing the moment of inertia of an area with respect to a noncentroidal axis in the plane of the area. This was accomplished by use of Eq. (5-5), the so-called **transfer formula,** the derivation of which is shown below.

The moment of inertia of the elemental area $dA$ in Fig. B-4 with respect to axis $XX$ is

FIG. B-4 Moment of inertia of an area with respect to a noncentroidal axis.

$$I_X = \int y_1{}^2 \, dA = \int (y + d)^2 \, dA \qquad (j)$$

Expanding this, we obtain

$$I_X = \int y^2 \, dA + 2d \int y \, dA + d^2 \int dA \qquad (k)$$

The first term is recognized as the moment of inertia of the area with respect to its own centroidal axis $X_C$. The second term, when integrated, becomes $\bar{y}A$, the first moment of the area with respect to axis $X_C$. When this axis passes through the centroid of the area, as it does, $\bar{y}$ becomes zero, since the first moment of an area with respect to a centroidal axis is zero. Hence the second term becomes zero. The third term, when integrated, becomes $Ad^2$. Accordingly

$$I_X = I_{X_C} + Ad^2 \qquad \textbf{(5-4)}$$

As an example of the use of this equation, let us obtain the moment of inertia of the triangle in Prob. B-2 with respect to the centroidal axis $X_C$, knowing its moment of inertia with respect to the $X$ axis to be $bh^3/12$. Now $d$, the distance between the two parallel axes, is $\frac{1}{3}h$. Hence

$$\frac{bh^3}{12} = I_{X_C} + \frac{bh}{2}\left(\frac{h}{3}\right)^2 \qquad (l)$$

and

$$I_{X_C} = \frac{bh^3}{12} - \frac{bh^3}{18} = \frac{bh^3}{36} \qquad (m)$$

**B-4.** In Prob. B-3 the moment of inertia of the circular area with respect to a diameter was found to be $\frac{1}{4}\pi r^4$. Consequently the moment of inertia of a semicircle with respect to the same axis would be one-half of that, which is $\frac{1}{8}\pi r^4$, or $0.3927R^4$, as listed in Table 5-1. By use of the transfer formula, find the moment of inertia of the semicircular area with respect to the centroidal axis $X_C$, which is given as $0.110R^4$ $(r = R)$.

*Solution:* Since $A = \pi R^2/2$ and $d = 0.424R$ (Art. 5-2),

$$[I_X = I_{X_C} + Ad^2] \qquad 0.3927R^4 = I_{X_C} + \frac{1}{2}\pi R^2(0.424R)^2$$

or

$$I_{X_C} = 0.3927R^4 - 0.2827R^4 = 0.1100R^4 \qquad Ans. \qquad (n)$$

**Polar Moment of Inertia.** In the design of shafts and other members subjected to twisting, we need to know the moment of inertia of the cross-sectional area of the member with respect to an axis *perpendicular* to the plane of the area. This is called the *polar moment of inertia* of the area and is denoted by the symbol $J$.

**B-5.** Find the polar moment of inertia of the circular area shown in Fig. B-5 with respect to an axis perpendicular to the plane of the area and passing through its centroid $C$.

FIG. B-5 Polar moment of inertia of a circular area.

*Solution:* The elemental area selected is circular, as shown in Fig. B-5. Its moment of inertia with respect to an axis through $C$ is $\rho^2\ dA$. Integrating this quantity between the limits of $\rho = 0$ and $\rho = r$ gives the moment of inertia of the entire area. That is, since $dA = 2\pi\rho\ d\rho$,

$$\left[ J = \int \rho^2\ dA \right] \qquad J = \int_0^r \rho^2\ 2\pi\rho\ d\rho$$

$$= 2\pi \int_0^r \rho^3\ d\rho = 2\pi \left[ \frac{\rho^4}{4} \right]_0^r = \frac{\pi r^4}{2} \qquad Ans. \tag{o}$$

# APPENDIX C

# Derivation of the Laws of Beam Diagrams

**Relationship between Load and Shear and between Shear and Moment in Beams.** In Fig. C-1 is shown a loaded, simple beam. Below is an enlarged view of a small vertical segment of length $dx$ of this beam. Let the shear and moment on the left side of this segment be $V$ and $M$. If the length of the segment is infinitesimally small and is denoted by $dx$, the resultant of the load

FIG. C-1 Relationship between load and shear and between shear and moment.

on the segment is $w\,dx$ concentrated at a distance $dx/2$ from either vertical face of the segment.

On the right face of the segment the shear has been decreased by an infinitesimally small amount $dV$ equal to the load $w\,dx$ on the segment. We may now write an equation expressing the relationship between load and shear at any section in a beam. That is,

$$[\Sigma V = 0] \qquad\qquad V = w\,dx + (V - dV) \qquad\qquad (a)$$

or

$$dV = w\,dx \qquad \text{and} \qquad w = \frac{dV}{dx} \qquad\qquad (b)$$

Also, the moment of the stresses on the right face of the segment has been increased by an infinitesimally small amount $dM$. Since $\Sigma M = 0$ about any point such as $c$, for example, an equation may now be written expressing the relationship between shear and moment at any section in the beam. That is,

$$[\Sigma M_c = 0] \qquad\qquad (M + dM) + w\,dx\,\frac{dx}{2} = M + V\,dx$$

The term $w\,dx\,(dx/2)$ may be disregarded, since it becomes $w(dx)^2/2$, which is an infinitesimal of the second order. Then,

$$dM = V\,dx \qquad \text{and} \qquad V = \frac{dM}{dx} \qquad\qquad (c)$$

Equation $(b)$ is interpreted as follows: At any section along a beam, the equation expressing the load $w$ is the first derivative of the equation expressing the shear $V$. Conversely, *the shear equation is the first integral of the load equation.*

In similar manner, from Eq. $(c)$, we deduce that *the moment equation is the first integral of the shear equation.*

Expressing these relationships mathematically, we have

$$V = \textstyle\int w\,dx \qquad\qquad \textbf{(C-1)}$$

and

$$M = \textstyle\int V\,dx \qquad\qquad \textbf{(C-2)}$$

**Proof of First and Second Laws of Beam Diagrams.** The **first law of beam diagrams,** as stated in Chap. 6, is

$$\begin{bmatrix} \textbf{Slope of curve at any point} \\ \textbf{in any diagram} \end{bmatrix} = \begin{bmatrix} \textbf{length of ordinate at corresponding} \\ \textbf{point in next higher diagram} \end{bmatrix}$$

*Proof:* In the shear diagram in Fig. C-2, the slope of the shear curve at any point is $dV/dx = w$, the ordinate in the load (next higher) diagram, also

FIG. C-2 Relationship between load and
shear diagrams and between shear and
moment diagrams.

verified by Eq. ($b$). Similarly, in the moment diagram the slope of the moment curve at any point is $dM/dx = V$, the ordinate in the shear (next higher) diagram, also verified by Eq. ($c$).

The **second law of beam diagrams,** as stated in Chap. 6, is

$$\begin{bmatrix}\textbf{Difference in length of any} \\ \textbf{two ordinates in any diagram}\end{bmatrix} = \begin{bmatrix}\textbf{area between corresponding ordi-} \\ \textbf{nates in next higher diagram}\end{bmatrix}$$

*Proof:* In the shear diagram in Fig. C-2, the difference in length of any two shear ordinates is $dV$, which equals $w\,dx$ [see Eq. ($b$)], the area between corresponding ordinates in the load (next higher) diagram. Thus the *difference* in length of any two shear ordinates $V_1$ and $V_2$ equals the entire area under the load curve between load ordinates $w_1$ and $w_2$. Similarly, in the moment diagram the difference in length of any two moment ordinates is $dM$, which equals $V\,dx$ [see Eq. ($c$)], the area between corresponding ordinates in the shear (next higher) diagram. Thus the *difference* in length of any two moment ordinates $M_1$ and $M_2$ equals the entire area under the shear curve between shear ordinates $V_1$ and $V_2$.

**Location of Section of Maximum Moment.** Since the lengths of the shear ordinates are equal to the slope of the moment-diagram curve at corresponding points along the beam, it follows, then, that the shear will be zero or will pass through zero where the bending moment is a maximum. The principle is illustrated in Figs. 6-20, 6-21, and 6-25.

# APPENDIX D

# Derivation of the Flexure Formula. Location of the Neutral Axis

In Fig. D-1 is shown a side view and a cross-sectional view of a prismatic beam subjected to bending loads. The beam is of irregular cross section but is symmetrical about its vertical axis. Because of the shape of the beam, the neutral axis clearly lies nearer the bottom of the beam. The unit stress on any fiber is proportional to its distance from the neutral axis. Let $s$ be the maximum unit stress in the extreme fiber at $c$ distance from the neutral axis and $f$ the fiber stress at $y$ distance from the neutral axis. Then by similarity

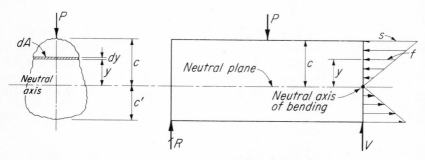

Fig. D-1 Derivation of the general flexure formula.

of triangles

$$\frac{s}{c} = \frac{f}{y} \qquad \text{and} \qquad f = \frac{s}{c} y \qquad (a)$$

The total force exerted on the infinitesimal area $dA$ is the product of the unit stress $f$ and the area $dA$. Then, substituting $(s/c)y$ for $f$, we obtain

$$f \, dA = \left(\frac{s}{c} y\right) dA \qquad (b)$$

The internal resisting moment RM of this force about the neutral axis is

$$\text{RM} = \left(\frac{s}{c} y\right) dA \, y = \frac{s}{c} y^2 \, dA \qquad (c)$$

The total internal resisting moment RM about the neutral axis is then obtained by a summation of the moments of the forces on all of the elemental areas. That is,

$$\text{RM} = \frac{s}{c} \int_{-c'}^{+c} y^2 \, dA \qquad (d)$$

This latter term, $\int y^2 \, dA$, is called the **static moment of inertia,** or second moment of area, of the cross section of the beam and is represented by the symbol $I$. Then, substituting $I$ in Eq. $(d)$, we obtain

$$\textbf{RM} = \boldsymbol{M} = \boldsymbol{s}\frac{\boldsymbol{I}}{\boldsymbol{c}} \qquad (7\text{-}1)$$

where $M$ is the external bending moment which at any section equals the internal resisting moment RM.

**Location of the Neutral Axis.** In connection with the design of a beam, the required moment of inertia is that with respect to the neutral axis of its cross-sectional area. Consequently it is necessary to locate this neutral axis before a determination of the moment of inertia can be made. The following development is proof that **the neutral axis of the cross-sectional area of a beam passes through the centroid of that area.**

At any section of a bent beam the resultants $C$ and $T$ of the compressive and tensile stresses, respectively, are equal. That is, the total compressive stress on one side of the neutral axis equals the total tensile stress on the other side.

Reference to Fig. D-1 and to Eq. $(b)$ indicates that the force on any elemental area $dA$ on either side of the neutral axis is $(s/c)y \, dA$. The algebraic sum of all forces on the entire cross-sectional area is then $(s/c) \int_{-c'}^{+c} y \, dA$,

since the total sum $C$ of the compressive stresses must always equal the total sum $T$ of the tensile stresses. Since $s/c$ obviously cannot equal zero, it follows that the term $\int_{-c'}^{+c} y\,dA$ must equal zero. But the general term $\int y\,dA$ expresses the first moment of the area of the cross section about the neutral axis, and this moment can be zero only if the neutral axis passes through the centroid of the section. The above statement that the neutral axis always passes through the centroid of the cross-sectional area of a beam has thus been proved.

# APPENDIX E

# Tables

The following tables, with the exception of Tables E-1 and E-2, have been taken from "Steel Construction," the manual of the American Institute of Steel Construction, and are reprinted here by permission.

**SYMBOLS USED TO DESIGNATE STRUCTURAL-STEEL SHAPES**

| New Designation | Type of Shape | Old Designation |
|---|---|---|
| W 24 × 76<br>W 14 × 26 | W shape | 24 WF 76<br>14 B 26 |
| S 24 × 100 | S shape | 24 I 100 |
| M 8 × 18.5<br>M 10 × 9<br>M 8 × 34.3 | M shape | 8 M 18.5<br>10 JR 9.0<br>8 × 8 M 34.3 |
| C 12 × 20.7 | American Standard Channel | 12 [ 20.7 |
| MC 12 × 45<br>MC 12 × 10.6 | Miscellaneous Channel | 12 × 4 [ 45.0<br>12 JR [ 10.6 |
| HP 14 × 73 | HP shape | 14 BP 73 |
| L 6 × 6 × ¾<br>L 6 × 4 × ⅝ | Equal Leg Angle<br>Unequal Leg Angle | ∠ 6 × 6 × ¾<br>∠ 6 × 4 × ⅝ |
| WT 12 × 38<br>WT 7 × 13 | Structural Tee cut from W shape | ST 12 WF 38<br>ST 7 B 13 |
| ST 12 × 50 | Structural Tee cut from S shape | ST 12 I 50 |
| MT 4 × 9.25<br>MT 5 × 4.5<br>MT 4 × 17.15 | Structural Tee cut from M shape | ST 4 M 9.25<br>ST 5 JR 4.5<br>ST 4 M 17.15 |
| PL ½ × 18 | Plate | PL 18 × ½ |
| Bar 1 ⊡<br>Bar 1¼ ⌀<br>Bar 2½ × ½ | Square Bar<br>Round Bar<br>Flat Bar | Bar 1 ⊡<br>Bar 1¼ ⌀<br>Bar 2½ × ½ |
| Pipe 4 Std.<br>Pipe 4 X-Strong<br>Pipe 4 XX-Strong | Pipe | Pipe 4 Std.<br>Pipe 4 X-Strong<br>Pipe 4 XX-Strong |
| TS 4 × 4 × .375<br>TS 5 × 3 × .375<br>TS 3 OD × .250 | Structural Tubing: Square<br>Structural Tubing: Rectangular<br>Structural Tubing: Circular | Tube 4 × 4 × .375<br>Tube 5 × 3 × .375<br>Tube 3 OD × .250 |

# E-1

## AVERAGE MECHANICAL PROPERTIES OF SOME ENGINEERING MATERIALS

| Material | Weight, pounds per cubic foot | Allowable stresses, psi | | | Elastic limit, psi | | Ultimate strength, psi | | | Modulus of elasticity, psi |
|---|---|---|---|---|---|---|---|---|---|---|
| | | Tension | Com-pression | Shear | Tension | Com-pression | Tension | Com-pression | Shear | |
| Steel, carbon, ASTM-A36 | 490 | 22,000 | 22,000 | 14,500 | 36,000[b] | .... | 58,000–80,000 | .... | .... | 29,000,000 |
| low alloy, ASTM-A441 | 490 | 30,000 | 30,000 | 20,000 | 50,000[b] | .... | 70,000 | .... | .... | 29,000,000 |
| cast, soft | 480 | 15,000 | 20,000 | 9,250 | 27,000 | 27,000 | 60,000 | 60,000 | 45,000 | 29,000,000 |
| Iron, wrought | 480 | 16,000 | 16,000 | 10,000 | 26,000 | 26,000 | 48,000 | 48,000 | 40,000 | 28,000,000 |
| cast, gray | 450 | 5,000–8,000 | 20,000–30,000 | 7,500–12,500 | .... | .... | .... | .... | .... | 14,000,000–18,300,000 |
| malleable | 475 | 6,000 | 16,000 | 6,000 | 18,000 | 18,000 | 32,000 | 46,000 | 40,000 | 20,000,000 |
| Aluminum alloy, 14S-T4 | 174 | 16,000 | 16,000 | 10,000 | 41,000 | 41,000 | 62,000 | .... | 38,000 | 10,000,000 |
| 6061-T6 | 174 | 15,000 | 14,000 | 10,000 | 30,000 | 30,000 | 42,000 | .... | 27,000 | 10,000,000 |
| Brass, cast | 520 | 4,000 | 4,000 | 4,000 | 6,000 | 6,000 | 21,000 | 30,000 | 36,000 | 9,000,000 |
| Bronze | 535 | 25,000 | 25,000 | .... | 40,000 | 40,000 | 75,000 | 120,000 | .... | 14,500,000 |
| Copper, rods, bolts | 555 | 8,000 | 8,000 | .... | 10,000 | 10,000 | 32,000 | 32,000 | .... | 16,000,000 |
| Brick, lime mortar | 120 | .... | 250 | 20 | .... | .... | .... | .... | .... | 2,000,000 |
| cement mortar | 130 | .... | 300 | 25 | .... | .... | .... | .... | .... | 2,000,000 |
| Concrete, 1:2:4 mix | 150 | .... | 1,000 | .... | .... | .... | .... | .... | .... | 2,000,000 |

[a] The values given in this table are average, and conform in general to accepted practice.

[b] Minimum yield point.

# TIMBER
## AMERICAN STANDARD SIZES
### PROPERTIES FOR DESIGNING

# E-2

| Nominal Size | American Standard Dressed Size | Area of Section | Weight per Foot | Moment of Inertia | Section Modulus | Nominal Size | American Standard Dressed Size | Area of Section | Weight per Foot | Moment of Inertia | Section Modulus |
|---|---|---|---|---|---|---|---|---|---|---|---|
| In. | In. | In.$^2$ | Lb. | In.$^4$ | In.$^3$ | In. | In. | In.$^2$ | Lb. | In.$^4$ | In.$^3$ |
| 2 x 4 | 1½x3½ | 5.25 | 1.46 | 5.36 | 3.06 | 10x10 | 9½x9½ | 90.3 | 25.0 | 679 | 143 |
| 6 | 5½ | 8.25 | 2.29 | 20.80 | 7.56 | 12 | 11½ | 109 | 30.3 | 1204 | 209 |
| 8 | 7¼ | 10.88 | 3.02 | 47.6 | 13.14 | 14 | 13½ | 128 | 35.6 | 1948 | 289 |
| 10 | 9¼ | 13.88 | 3.85 | 98.9 | 21.4 | 16 | 15½ | 147 | 40.9 | 2948 | 380 |
| 12 | 11¼ | 16.88 | 4.68 | 178 | 31.6 | 18 | 17½ | 166 | 46.1 | 4243 | 485 |
| 14 | 13¼ | 19.88 | 5.52 | 291 | 43.9 | 20 | 19½ | 185 | 51.4 | 5870 | 602 |
| 16 | 15¼ | 22.88 | 6.35 | 443 | 58.2 | 22 | 21½ | 204 | 56.7 | 7868 | 732 |
| 18 | 17¼ | 25.88 | 7.17 | 642 | 74.5 | 24 | 23½ | 223 | 62.0 | 10274 | 874 |
| 3 x 4 | 2½x3½ | 8.75 | 2.43 | 8.93 | 5.10 | 12x12 | 11½x11½ | 132 | 36.7 | 1458 | 253 |
| 6 | 5½ | 13.75 | 3.82 | 34.7 | 12.6 | 14 | 13½ | 155 | 43.1 | 2358 | 349 |
| 8 | 7¼ | 18.12 | 5.04 | 79.4 | 21.9 | 16 | 15½ | 178 | 49.5 | 3569 | 460 |
| 10 | 9¼ | 23.12 | 6.42 | 165 | 35.6 | 18 | 17½ | 201 | 55.9 | 5136 | 587 |
| 12 | 11¼ | 28.12 | 7.81 | 297 | 52.7 | 20 | 19½ | 224 | 62.3 | 7106 | 729 |
| 14 | 13¼ | 33.12 | 9.20 | 485 | 73.2 | 22 | 21½ | 247 | 68.7 | 9524 | 886 |
| 16 | 15¼ | 38.12 | 10.59 | 739 | 96.9 | 24 | 23½ | 270 | 75.0 | 12437 | 1058 |
| 18 | 17¼ | 43.12 | 11.98 | 1070 | 124 | | | | | | |
| 4 x 4 | 3½x3½ | 12.25 | 3.40 | 12.5 | 7.15 | 14x14 | 13½x13½ | 182 | 50.6 | 2768 | 410 |
| 6 | 5½ | 19.25 | 5.35 | 48.5 | 17.65 | 16 | 15½ | 209 | 58.1 | 4189 | 541 |
| 8 | 7¼ | 25.38 | 7.05 | 111.1 | 30.66 | 18 | 17½ | 236 | 65.6 | 6029 | 689 |
| 10 | 9¼ | 32.38 | 8.93 | 231 | 49.9 | 20 | 19½ | 263 | 73.1 | 8342 | 856 |
| 12 | 11¼ | 39.38 | 10.94 | 415 | 73.8 | 22 | 21½ | 290 | 80.6 | 11181 | 1040 |
| 14 | 13¼ | 46.38 | 12.89 | 678 | 102.4 | 24 | 23½ | 317 | 88.1 | 14600 | 1243 |
| 16 | 15¼ | 53.38 | 14.93 | 1034 | 136.0 | 16x16 | 15½x15½ | 240 | 66.7 | 4810 | 621 |
| 18 | 17¼ | 60.38 | 16.75 | 1498 | 173.5 | 18 | 17½ | 271 | 75.3 | 6923 | 791 |
| 6 x 6 | 5½x5½ | 30.3 | 8.40 | 76.3 | 27.7 | 20 | 19½ | 302 | 83.9 | 9578 | 982 |
| 8 | 7½ | 41.3 | 11.4 | 193 | 51.6 | 22 | 21½ | 333 | 92.5 | 12837 | 1194 |
| 10 | 9½ | 52.3 | 14.5 | 393 | 82.7 | 24 | 23½ | 364 | 101 | 16763 | 1427 |
| 12 | 11½ | 63.3 | 17.5 | 697 | 121 | 18x18 | 17½x17½ | 306 | 85.0 | 7816 | 893 |
| 14 | 13½ | 74.3 | 20.6 | 1128 | 167 | 20 | 19½ | 341 | 94.8 | 10813 | 1109 |
| 16 | 15½ | 85.3 | 23.6 | 1707 | 220 | 22 | 21½ | 376 | 105 | 14493 | 1348 |
| 18 | 17½ | 96.3 | 26.7 | 2456 | 281 | 24 | 23½ | 411 | 114 | 18926 | 1611 |
| 20 | 19½ | 107.3 | 29.8 | 3398 | 349 | 26 | 25½ | 446 | 124 | 24181 | 1897 |
| 8 x 8 | 7½x7½ | 56.3 | 15.6 | 264 | 70.3 | 20x20 | 19½x19½ | 380 | 106 | 12049 | 1236 |
| 10 | 9½ | 71.3 | 19.8 | 536 | 113 | 22 | 21½ | 419 | 116 | 16150 | 1502 |
| 12 | 11½ | 86.3 | 23.9 | 951 | 165 | 24 | 23½ | 458 | 127 | 21089 | 1795 |
| 14 | 13½ | 101.3 | 28.0 | 1538 | 228 | 26 | 25½ | 497 | 138 | 26945 | 2113 |
| 16 | 15½ | 116.3 | 32.0 | 2327 | 300 | 28 | 27½ | 536 | 149 | 33795 | 2458 |
| 18 | 17½ | 131.3 | 36.4 | 3350 | 383 | 24x24 | 23½x23½ | 552 | 153 | 25415 | 2163 |
| 20 | 19½ | 146.3 | 40.6 | 4634 | 475 | 26 | 25½ | 599 | 166 | 32472 | 2547 |
| 22 | 21½ | 161.3 | 44.8 | 6211 | 578 | 28 | 27½ | 646 | 180 | 40727 | 2962 |
| | | | | | | 30 | 29½ | 693 | 193 | 50275 | 3408 |

**All properties and weights given are for dressed size only.**
**The weights given above are based on assumed average weight of 40 pounds per cubic foot.**

# E-3

## SCREW THREADS
### Unified Standard Series—UNC and 4 UN
#### A.S.A. B1.1—1960

**Thread Dimensions**

Nominal size (basic major dia.)
No. threads per inch ($n$)
Thread series symbol
$^c$Thread class symbol
Left hand thread. No symbol req'd for right hand thread.

$\frac{3}{4}$–**10 UNC 2A LH**

**Standard Designations**

| Diameter | | Area | | | $^b$Th'ds per Inch $n$ | Diameter | | Area | | | $^b$Th'ds per Inch $n$ |
|---|---|---|---|---|---|---|---|---|---|---|---|
| Basic Major $D$ | Root $K$ | Gross $A_D$ | Root $A_K$ | $^a$Tensile Stress | | Basic Major $D$ | Root $K$ | Gross $A_D$ | Root $A_K$ | $^a$Tensile Stress | |
| In. | In. | In.² | In.² | In.² | | In. | In. | In.² | In.² | In.² | |
| ¼ | .185 | .049 | .027 | .032 | 20 | 2¾ | 2.425 | 5.940 | 4.62 | 4.93 | 4 |
| ⅜ | .294 | .110 | .068 | .078 | 16 | 3 | 2.675 | 7.069 | 5.62 | 5.97 | 4 |
| ½ | .400 | .196 | .126 | .142 | 13 | 3¼ | 2.925 | 8.296 | 6.72 | 7.10 | 4 |
| ⅝ | .507 | .307 | .202 | .226 | 11 | 3½ | 3.175 | 9.621 | 7.92 | 8.33 | 4 |
| ¾ | .620 | .442 | .302 | .334 | 10 | 3¾ | 3.425 | 11.045 | 9.21 | 9.66 | 4 |
| ⅞ | .731 | .601 | .419 | .462 | 9 | 4 | 3.675 | 12.566 | 10.6 | 11.1 | 4 |
| 1 | .838 | .785 | .551 | .606 | 8 | 4¼ | 3.925 | 14.186 | 12.1 | 12.6 | 4 |
| 1⅛ | .939 | .994 | .693 | .763 | 7 | 4½ | 4.175 | 15.904 | 13.7 | 14.2 | 4 |
| 1¼ | 1.064 | 1.227 | .890 | .969 | 7 | 4¾ | 4.425 | 17.721 | 15.4 | 16.0 | 4 |
| 1⅜ | 1.158 | 1.485 | 1.05 | 1.16 | 6 | 5 | 4.675 | 19.635 | 17.2 | 17.8 | 4 |
| 1½ | 1.283 | 1.767 | 1.29 | 1.41 | 6 | 5¼ | 4.925 | 21.648 | 19.1 | 19.7 | 4 |
| 1¾ | 1.490 | 2.405 | 1.74 | 1.90 | 5 | 5½ | 5.175 | 23.758 | 21.0 | 21.7 | 4 |
| 2 | 1.711 | 3.142 | 2.30 | 2.50 | 4½ | 5¾ | 5.425 | 25.967 | 23.1 | 23.8 | 4 |
| 2¼ | 1.961 | 3.976 | 3.02 | 3.25 | 4½ | 6 | 5.675 | 28.274 | 25.3 | 26.0 | 4 |
| 2½ | 2.175 | 4.909 | 3.72 | 4.00 | 4 | | | | | | |

$^a$ Tensile stress area $= 0.7854 \left( D - \dfrac{.9743}{n} \right)^2$.

$^b$ For basic major diameters of ¼ to 4 in. incl., thread series is UNC (coarse); for 4¼ in. diameter and larger, thread series is 4UN.

$^c$ 2A denotes Class 2 fit external thread. 2B denotes Class 2 fit internal thread.

---

## LENGTH OF THREAD ON BOLTS
#### A.S.A. B18.2—1960

| Length of Bolt | Diameter of Bolt, $D$, Inches | | | | | | | | | | | | | | | | |
|---|---|---|---|---|---|---|---|---|---|---|---|---|---|---|---|---|---|
| | ¼ | ⅜ | ½ | ⅝ | ¾ | ⅞ | 1 | 1⅛ | 1¼ | 1⅜ | 1½ | 1¾ | 2 | 2¼ | 2½ | 2¾ | 3 |
| To 6 in. Incl. | ¾ | 1 | 1¼ | 1½ | 1¾ | 2 | 2¼ | 2½ | 2¾ | 3 | 3¼ | 3¾ | 4¼ | 4¾ | 5¼ | 5¾ | 6 |
| Over 6 in. | 1 | 1¼ | 1½ | 1¾ | 2 | 2¼ | 2½ | 2¾ | 3 | 3¼ | 3½ | 4 | 4½ | 5 | 5½ | 6 | 6½ |

**Note 1.** Thread length for bolts up to 6 in. long is $2D + ¼$. For bolts over 6 in. long, thread length is $2D + ½$. These proportions may be used to compute thread length for diameters not shown in the table. Bolts which are too short for listed or computed thread lengths are threaded as close to the head as possible.

**Note 2.** For thread lengths for high strength bolts, refer to "Specifications for Structural Joints Using ASTM A325 Bolts."

## RIVET GAGES   E-4

| | Leg | 8 | 7 | 6 | 5 | 4 | 3½ | 3 | 2½ | 2 | 1¾ | 1½ | 1⅜ | 1¼ | 1 | CRIMPS |
|---|---|---|---|---|---|---|---|---|---|---|---|---|---|---|---|---|
| | g | 4½ | 4 | 3½ | 3 | | 2½ | 2 | 1¾ | 1⅜ | 1⅛ | 1 | ⅞ | ⅞ | ¾ ⅝ | $b = t + 1½$ |
| | g₁ | 3 | 2½ | 2¼ | 2 | | | | | | | | | | | Min. = 2 |
| | g₂ | 3 | 3 | 2½ | 1¾ | | | | | | | | | | | |

CRIMPS: $b = t + 1\tfrac{1}{2}$   Min. $= 2$

## SQUARE AND ROUND BARS   E-5
### WEIGHT AND AREA

| Size Inches | Weight Lb. per Foot ■ | Weight Lb. per Foot ● | Area Square Inches ◨ | Area Square Inches ◉ | Size Inches | Weight Lb. per Foot ■ | Weight Lb. per Foot ● | Area Square Inches ◨ | Area Square Inches ◉ |
|---|---|---|---|---|---|---|---|---|---|
| 0 | | | | | 2 | 13.600 | 10.681 | 4.0000 | 3.1416 |
| 1/16 | .013 | .010 | .0039 | .0031 | 1/16 | 14.463 | 11.359 | 4.2539 | 3.3410 |
| 1/8 | .053 | .042 | .0156 | .0123 | 1/8 | 15.353 | 12.058 | 4.5156 | 3.5466 |
| 3/16 | .120 | .094 | .0352 | .0276 | 3/16 | 16.270 | 12.778 | 4.7852 | 3.7583 |
| 1/4 | .213 | .167 | .0625 | .0491 | 1/4 | 17.213 | 13.519 | 5.0625 | 3.9761 |
| 5/16 | .332 | .261 | .0977 | .0767 | 5/16 | 18.182 | 14.280 | 5.3477 | 4.2000 |
| 3/8 | .478 | .376 | .1406 | .1105 | 3/8 | 19.178 | 15.062 | 5.6406 | 4.4301 |
| 7/16 | .651 | .511 | .1914 | .1503 | 7/16 | 20.201 | 15.866 | 5.9414 | 4.6664 |
| 1/2 | .850 | .668 | .2500 | .1963 | 1/2 | 21.250 | 16.690 | 6.2500 | 4.9087 |
| 9/16 | 1.076 | .845 | .3164 | .2485 | 9/16 | 22.326 | 17.534 | 6.5664 | 5.1572 |
| 5/8 | 1.328 | 1.043 | .3906 | .3068 | 5/8 | 23.428 | 18.400 | 6.8906 | 5.4119 |
| 11/16 | 1.607 | 1.262 | .4727 | .3712 | 11/16 | 24.557 | 19.287 | 7.2227 | 5.6727 |
| 3/4 | 1.913 | 1.502 | .5625 | .4418 | 3/4 | 25.713 | 20.195 | 7.5625 | 5.9396 |
| 13/16 | 2.245 | 1.763 | .6602 | .5185 | 13/16 | 26.895 | 21.123 | 7.9102 | 6.2126 |
| 7/8 | 2.603 | 2.044 | .7656 | .6013 | 7/8 | 28.103 | 22.072 | 8.2656 | 6.4918 |
| 15/16 | 2.988 | 2.347 | .8789 | .6903 | 15/16 | 29.338 | 23.042 | 8.6289 | 6.7771 |
| 1 | 3.400 | 2.670 | 1.0000 | .7854 | 3 | 30.60 | 24.03 | 9.000 | 7.069 |
| 1/16 | 3.838 | 3.015 | 1.1289 | .8866 | 1/16 | 31.89 | 25.05 | 9.379 | 7.366 |
| 1/8 | 4.303 | 3.380 | 1.2656 | .9940 | 1/8 | 33.20 | 26.08 | 9.766 | 7.670 |
| 3/16 | 4.795 | 3.766 | 1.4102 | 1.1075 | 3/16 | 34.54 | 27.13 | 10.160 | 7.980 |
| 1/4 | 5.313 | 4.172 | 1.5625 | 1.2272 | 1/4 | 35.91 | 28.21 | 10.563 | 8.296 |
| 5/16 | 5.857 | 4.600 | 1.7227 | 1.3530 | 5/16 | 37.31 | 29.30 | 10.973 | 8.618 |
| 3/8 | 6.428 | 5.049 | 1.8906 | 1.4849 | 3/8 | 38.73 | 30.42 | 11.391 | 8.946 |
| 7/16 | 7.026 | 5.518 | 2.0664 | 1.6230 | 7/16 | 40.18 | 31.55 | 11.816 | 9.281 |
| 1/2 | 7.650 | 6.008 | 2.2500 | 1.7671 | 1/2 | 41.65 | 32.71 | 12.250 | 9.621 |
| 9/16 | 8.301 | 6.519 | 2.4414 | 1.9175 | 9/16 | 43.15 | 33.89 | 12.691 | 9.968 |
| 5/8 | 8.978 | 7.051 | 2.6406 | 2.0739 | 5/8 | 44.68 | 35.09 | 13.141 | 10.321 |
| 11/16 | 9.682 | 7.604 | 2.8477 | 2.2365 | 11/16 | 46.23 | 36.31 | 13.598 | 10.680 |
| 3/4 | 10.413 | 8.178 | 3.0625 | 2.4053 | 3/4 | 47.81 | 37.55 | 14.063 | 11.045 |
| 13/16 | 11.170 | 8.773 | 3.2852 | 2.5802 | 13/16 | 49.42 | 38.81 | 14.535 | 11.416 |
| 7/8 | 11.953 | 9.388 | 3.5156 | 2.7612 | 7/8 | 51.05 | 40.10 | 15.016 | 11.793 |
| 15/16 | 12.763 | 10.024 | 3.7539 | 2.9483 | 15/16 | 52.71 | 41.40 | 15.504 | 12.177 |
| 2 | 13.600 | 10.681 | 4.0000 | 3.1416 | 4 | 54.40 | 42.73 | 16.000 | 12.566 |

# E-6

**W SHAPES**

**PROPERTIES FOR DESIGNING**

| Nominal Size | Weight per Foot | Area | Depth | Flange Width | Flange Thickness | Web Thickness | AXIS X-X I | AXIS X-X Z | AXIS X-X k | AXIS Y-Y I | AXIS Y-Y Z | AXIS Y-Y k |
|---|---|---|---|---|---|---|---|---|---|---|---|---|
| In. | Lb. | In.² | In. | In. | In. | In. | In.⁴ | In.³ | In. | In.⁴ | In.³ | In. |
| 27 x 10 | 94 | 27.65 | 26.91 | 9.990 | .747 | .490 | 3266.7 | 242.8 | 10.87 | 115.1 | 23.0 | 2.04 |
| 24 x 12 | 120 | 35.29 | 24.31 | 12.088 | .930 | .556 | 3635.3 | 299.1 | 10.15 | 254.0 | 42.0 | 2.68 |
|  | 110 | 32.36 | 24.16 | 12.042 | :855 | .510 | 3315.0 | 274.4 | 10.12 | 229.1 | 38.0 | 2.66 |
|  | 100 | 29.43 | 24.00 | 12.000 | .775 | .468 | 2987.3 | 248.9 | 10.08 | 203.5 | 33.9 | 2.63 |
| 24 x 9 | 94 | 27.63 | 24.29 | 9.061 | .872 | .516 | 2683.0 | 220.9 | 9.85 | 102.2 | 22.6 | 1.92 |
|  | 84 | 24.71 | 24.09 | 9.015 | .772 | .470 | 2364.3 | 196.3 | 9.78 | 88.3 | 19.6 | 1.89 |
|  | 76 | 22.37 | 23.91 | 8.985 | .682 | .440 | 2096.4 | 175.4 | 9.68 | 76.5 | 17.0 | 1.85 |
| 21 x 13 | 142 | 41.76 | 21.46 | 13.132 | 1.095 | .659 | 3403.1 | 317.2 | 9.03 | 385.9 | 58.8 | 3.04 |
|  | 127 | 37.34 | 21.24 | 13.061 | .985 | .588 | 3017.2 | 284.1 | 8.99 | 338.6 | 51.8 | 3.01 |
|  | 112 | 32.93 | 21.00 | 13.000 | .865 | .527 | 2620.6 | 249.6 | 8.92 | 289.7 | 44.6 | 2.96 |
| 21 x 9 | 96 | 28.21 | 21.14 | 9.038 | .935 | .575 | 2088.9 | 197.6 | 8.60 | 109.3 | 24.2 | 1.97 |
|  | 82 | 24.10 | 20.86 | 8.962 | .795 | .499 | 1752.4 | 168.0 | 8.53 | 89.6 | 20.0 | 1.93 |
| 21 x 8¼ | 73 | 21.46 | 21.24 | 8.295 | .740 | .455 | 1600.3 | 150.7 | 8.64 | 66.2 | 16.0 | 1.76 |
|  | 68 | 20.02 | 21.13 | 8.270 | .685 | .430 | 1478.3 | 139.9 | 8.59 | 60.4 | 14.6 | 1.74 |
|  | 62 | 18.23 | 20.99 | 8.240 | .615 | .400 | 1326.8 | 126.4 | 8.53 | 53.1 | 12.9 | 1.71 |
| 18 x 11¾ | 114 | 33.51 | 18.48 | 11.833 | .991 | .595 | 2033.8 | 220.1 | 7.79 | 255.6 | 43.2 | 2.76 |
|  | 105 | 30.86 | 18.32 | 11.792 | .911 | .554 | 1852.5 | 202.2 | 7.75 | 231.0 | 39.2 | 2.73 |
|  | 96 | 28.22 | 18.16 | 11.750 | .831 | .512 | 1674.7 | 184.4 | 7.70 | 206.8 | 35.2 | 2.71 |
| 18 x 8¾ | 85 | 24.97 | 18.32 | 8.838 | .911 | .526 | 1429.9 | 156.1 | 7.57 | 99.4 | 22.5 | 2.00 |
|  | 77 | 22.63 | 18.16 | 8.787 | .831 | .475 | 1286.8 | 141.7 | 7.54 | 88.6 | 20.2 | 1.98 |
|  | 70 | 20.56 | 18.00 | 8.750 | .751 | .438 | 1153.9 | 128.2 | 7.49 | 78.5 | 17.9 | 1.95 |
|  | 64 | 18.80 | 17.87 | 8.715 | .686 | .403 | 1045.8 | 117.0 | 7.46 | 70.3 | 16.1 | 1.93 |
| 18 x 7½ | 60 | 17.64 | 18.25 | 7.558 | .695 | .416 | 984.0 | 107.8 | 7.47 | 47.1 | 12.5 | 1.63 |
|  | 55 | 16.19 | 18.12 | 7.532 | .630 | .390 | 889.9 | 98.2 | 7.41 | 42.0 | 11.1 | 1.61 |
|  | 50 | 14.71 | 18.00 | 7.500 | .570 | .358 | 800.6 | 89.0 | 7.38 | 37.2 | 9.9 | 1.59 |

# E-6

# W SHAPES

## PROPERTIES FOR DESIGNING

### (continued)

| Nominal Size | Weight per Foot | Area | Depth | Flange Width | Flange Thickness | Web Thickness | AXIS X-X I | AXIS X-X Z | AXIS X-X k | AXIS Y-Y I | AXIS Y-Y Z | AXIS Y-Y k |
|---|---|---|---|---|---|---|---|---|---|---|---|---|
| In. | Lb. | In.² | In. | In. | In. | In. | In.⁴ | In.³ | In. | In.⁴ | In.³ | In. |
| 16 x 11½ | 96 | 28.22 | 16.32 | 11.533 | .875 | .535 | 1355.1 | 166.1 | 6.93 | 207.2 | 35.9 | 2.71 |
|  | 88 | 25.87 | 16.16 | 11.502 | .795 | .504 | 1222.6 | 151.3 | 6.87 | 185.2 | 32.2 | 2.67 |
| 16 x 8½ | 78 | 22.92 | 16.32 | 8.586 | .875 | .529 | 1042.6 | 127.8 | 6.74 | 87.5 | 20.4 | 1.95 |
|  | 71 | 20.86 | 16.16 | 8.543 | .795 | .486 | 936.9 | 115.9 | 6.70 | 77.9 | 18.2 | 1.93 |
|  | 64 | 18.80 | 16.00 | 8.500 | .715 | .443 | 833.8 | 104.2 | 6.66 | 68.4 | 16.1 | 1.91 |
|  | 58 | 17.04 | 15.86 | 8.464 | .645 | .407 | 746.4 | 94.1 | 6.62 | 60.5 | 14.3 | 1.88 |
| 16 x 7 | 50 | 14.70 | 16.25 | 7.073 | .628 | .380 | 655.4 | 80.7 | 6.68 | 34.8 | 9.8 | 1.54 |
|  | 45 | 13.24 | 16.12 | 7.039 | .563 | .346 | 583.3 | 72.4 | 6.64 | 30.5 | 8.7 | 1.52 |
|  | 40 | 11.77 | 16.00 | 7.000 | .503 | .307 | 515.5 | 64.4 | 6.62 | 26.5 | 7.6 | 1.50 |
|  | 36 | 10.59 | 15.85 | 6.992 | .428 | .299 | 446.3 | 56.3 | 6.49 | 22.1 | 6.3 | 1.45 |
| 14 x 14½ | 95 | 27.94 | 14.12 | 14.545 | .748 | .465 | 1063.5 | 150.6 | 6.17 | 383.7 | 52.8 | 3.71 |
|  | 87 | 25.56 | 14.00 | 14.500 | .688 | .420 | 966.9 | 138.1 | 6.15 | 349.7 | 48.2 | 3.70 |
| 14 x 12 | 84 | 24.71 | 14.18 | 12.023 | .778 | .451 | 928.4 | 130.9 | 6.13 | 225.5 | 37.5 | 3.02 |
|  | 78 | 22.94 | 14.06 | 12.000 | .718 | .428 | 851.2 | 121.1 | 6.09 | 206.9 | 34.5 | 3.00 |
| 14 x 10 | 74 | 21.76 | 14.19 | 10.072 | .783 | .450 | 796.8 | 112.3 | 6.05 | 133.5 | 26.5 | 2.48 |
|  | 68 | 20.00 | 14.06 | 10.040 | .718 | .418 | 724.1 | 103.0 | 6.02 | 121.2 | 24.1 | 2.46 |
|  | 61 | 17.94 | 13.91 | 10.000 | .643 | .378 | 641.5 | 92.2 | 5.98 | 107.3 | 21.5 | 2.45 |
| 14 x 8 | 53 | 15.59 | 13.94 | 8.062 | .658 | .370 | 542.1 | 77.8 | 5.90 | 57.5 | 14.3 | 1.92 |
|  | 48 | 14.11 | 13.81 | 8.031 | .593 | .339 | 484.9 | 70.2 | 5.86 | 51.3 | 12.8 | 1.91 |
|  | 43 | 12.65 | 13.68 | 8.000 | .528 | .308 | 429.0 | 62.7 | 5.82 | 45.1 | 11.3 | 1.89 |
| 14 x 6¾ | 38 | 11.17 | 14.12 | 6.776 | .513 | .313 | 385.3 | 54.6 | 5.87 | 24.6 | 7.3 | 1.49 |
|  | 34 | 10.00 | 14.00 | 6.750 | .453 | .287 | 339.2 | 48.5 | 5.83 | 21.3 | 6.3 | 1.46 |
|  | 30 | 8.81 | 13.86 | 6.733 | .383 | .270 | 289.6 | 41.8 | 5.73 | 17.5 | 5.2 | 1.41 |
| 12 x 12 | 106 | 31.19 | 12.88 | 12.230 | .986 | .620 | 930.7 | 144.5 | 5.46 | 300.9 | 49.2 | 3.11 |
|  | 99 | 29.09 | 12.75 | 12.190 | .921 | .580 | 858.5 | 134.7 | 5.43 | 278.2 | 45.7 | 3.09 |
|  | 92 | 27.06 | 12.62 | 12.155 | .856 | .545 | 788.9 | 125.0 | 5.40 | 256.4 | 42.2 | 3.08 |
|  | 85 | 24.98 | 12.50 | 12.105 | .796 | .495 | 723.3 | 115.7 | 5.38 | 235.5 | 38.9 | 3.07 |
|  | 79 | 23.22 | 12.38 | 12.080 | .736 | .470 | 663.0 | 107.1 | 5.34 | 216.4 | 35.8 | 3.05 |
|  | 72 | 21.16 | 12.25 | 12.040 | .671 | .430 | 597.4 | 97.5 | 5.31 | 195.3 | 32.4 | 3.04 |
|  | 65 | 19.11 | 12.12 | 12.000 | .606 | .390 | 533.4 | 88.0 | 5.28 | 174.6 | 29.1 | 3.02 |

# E-6

## W SHAPES

### PROPERTIES FOR DESIGNING
#### (continued)

| Nominal Size | Weight per Foot | Area | Depth | Flange | | Web Thick-ness | AXIS X-X | | | AXIS Y-Y | | |
|---|---|---|---|---|---|---|---|---|---|---|---|---|
| | | | | Width | Thick-ness | | I | Z | k | I | Z | k |
| In. | Lb. | In.² | In. | In. | In. | In. | In.⁴ | In.³ | In. | In.⁴ | In.³ | In. |
| 12 x 10 | 58 | 17.06 | 12.19 | 10.014 | .641 | .359 | 476.1 | 78.1 | 5.28 | 107.4 | 21.4 | 2.51 |
| | 53 | 15.59 | 12.06 | 10.000 | .576 | .345 | 426.2 | 70.7 | 5.23 | 96.1 | 19.2 | 2.48 |
| 12 x 8 | 50 | 14.71 | 12.19 | 8.077 | .641 | .371 | 394.5 | 64.7 | 5.18 | 56.4 | 14.0 | 1.96 |
| | 45 | 13.24 | 12.06 | 8.042 | .576 | .336 | 350.8 | 58.2 | 5.15 | 50.0 | 12.4 | 1.94 |
| | 40 | 11.77 | 11.94 | 8.000 | .516 | .294 | 310.1 | 51.9 | 5.13 | 44.1 | 11.0 | 1.94 |
| 12 x 6½ | 36 | 10.59 | 12.24 | 6.565 | .540 | .305 | 280.8 | 45.9 | 5.15 | 23.7 | 7.2 | 1.50 |
| | 31 | 9.12 | 12.09 | 6.525 | .465 | .265 | 238.4 | 39.4 | 5.11 | 19.8 | 6.1 | 1.47 |
| | 27 | 7.97 | 11.95 | 6.500 | .400 | .240 | 204.1 | 34.1 | 5.06 | 16.6 | 5.1 | 1.44 |
| 10 x 10 | 112 | 32.92 | 11.38 | 10.415 | 1.248 | .755 | 718.7 | 126.3 | 4.67 | 235.4 | 45.2 | 2.67 |
| | 100 | 29.43 | 11.12 | 10.345 | 1.118 | .685 | 625.0 | 112.4 | 4.61 | 206.6 | 39.9 | 2.65 |
| | 89 | 26.19 | 10.88 | 10.275 | .998 | .615 | 542.4 | 99.7 | 4.55 | 180.6 | 35.2 | 2.63 |
| | 77 | 22.67 | 10.62 | 10.195 | .868 | .535 | 457.2 | 86.1 | 4.49 | 153.4 | 30.1 | 2.60 |
| | 72 | 21.18 | 10.50 | 10.170 | .808 | .510 | 420.7 | 80.1 | 4.46 | 141.8 | 27.9 | 2.59 |
| | 66 | 19.41 | 10.38 | 10.117 | .748 | .457 | 382.5 | 73.7 | 4.44 | 129.2 | 25.5 | 2.58 |
| | 60 | 17.66 | 10.25 | 10.075 | .683 | .415 | 343.7 | 67.1 | 4.41 | 116.5 | 23.1 | 2.57 |
| | 54 | 15.88 | 10.12 | 10.028 | .618 | .368 | 305.7 | 60.4 | 4.39 | 103.9 | 20.7 | 2.56 |
| | 49 | 14.40 | 10.00 | 10.000 | .558 | .340 | 272.9 | 54.6 | 4.35 | 93.0 | 18.6 | 2.54 |
| 10 x 8 | 45 | 13.24 | 10.12 | 8.022 | .618 | .350 | 248.6 | 49.1 | 4.33 | 53.2 | 13.3 | 2.00 |
| | 39 | 11.48 | 9.94 | 7.990 | .528 | .318 | 209.7 | 42.2 | 4.27 | 44.9 | 11.2 | 1.98 |
| | 33 | 9.71 | 9.75 | 7.964 | .433 | .292 | 170.9 | 35.0 | 4.20 | 36.5 | 9.2 | 1.94 |
| 10 x 5¾ | 29 | 8.53 | 10.22 | 5.799 | .500 | .289 | 157.3 | 30.8 | 4.29 | 15.2 | 5.2 | 1.34 |
| | 25 | 7.35 | 10.08 | 5.762 | .430 | .252 | 133.2 | 26.4 | 4.26 | 12.7 | 4.4 | 1.31 |
| | 21 | 6.19 | 9.90 | 5.750 | .340 | .240 | 106.3 | 21.5 | 4.14 | 9.7 | 3.4 | 1.25 |
| 8 x 8 | 67 | 19.70 | 9.00 | 8.287 | .933 | .575 | 271.8 | 60.4 | 3.71 | 88.6 | 21.4 | 2.12 |
| | 58 | 17.06 | 8.75 | 8.222 | .808 | .510 | 227.3 | 52.0 | 3.65 | 74.9 | 18.2 | 2.10 |
| | 48 | 14.11 | 8.50 | 8.117 | .683 | .405 | 183.7 | 43.2 | 3.61 | 60.9 | 15.0 | 2.08 |
| | 40 | 11.76 | 8.25 | 8.077 | .558 | .365 | 146.3 | 35.5 | 3.53 | 49.0 | 12.1 | 2.04 |
| | 35 | 10.30 | 8.12 | 8.027 | .493 | .315 | 126.5 | 31.1 | 3.50 | 42.5 | 10.6 | 2.03 |
| | 31 | 9.12 | 8.00 | 8.000 | .433 | .288 | 109.7 | 27.4 | 3.47 | 37.0 | 9.2 | 2.01 |
| 8 x 6½ | 28 | 8.23 | 8.06 | 6.540 | .463 | .285 | 97.8 | 24.3 | 3.45 | 21.6 | 6.6 | 1.62 |
| | 24 | 7.06 | 7.93 | 6.500 | .398 | .245 | 82.5 | 20.8 | 3.42 | 18.2 | 5.6 | 1.61 |
| 8 x 5¼ | 20 | 5.88 | 8.14 | 5.268 | .378 | .248 | 69.2 | 17.0 | 3.43 | 8.5 | 3.2 | 1.20 |
| | 17 | 5.00 | 8.00 | 5.250 | .308 | .230 | 56.4 | 14.1 | 3.36 | 6.7 | 2.6 | 1.16 |

# W SHAPES
# MISCELLANEOUS
# COLUMNS AND BEAMS

**E-7**

## PROPERTIES FOR DESIGNING

| Nominal Size | Weight per Foot | Area | Depth | Flange Width | Flange Thickness | Web Thickness | AXIS X-X I | AXIS X-X Z | AXIS X-X k | AXIS Y-Y I | AXIS Y-Y Z | AXIS Y-Y k |
|---|---|---|---|---|---|---|---|---|---|---|---|---|
| In. | Lb. | In.² | In. | In. | In. | In. | In.⁴ | In.³ | In. | In.⁴ | In.³ | In. |
| **W SHAPES AND LIGHT COLUMNS** | | | | | | | | | | | | |
| 6 x 6 | 25 | 7 37 | 6.37 | 6.080 | .456 | .320 | 53.5 | 16.8 | 2.69 | 17.1 | 5.6 | 1.52 |
|  | 20 | 5.90 | 6.20 | 6.018 | .367 | .258 | 41.7 | 13.4 | 2.66 | 13.3 | 4.4 | 1.50 |
|  | 15.5 | 4.62 | 6.00 | 6.000 | .269 | .240 | 30.3 | 10.1 | 2.56 | 9.69 | 3.2 | 1.45 |
| 5 x 5 | 18.5 | 5.45 | 5.12 | 5.025 | .420 | .265 | 25.4 | 9.94 | 2.16 | 8.89 | 3.54 | 1.28 |
|  | 16 | 4.70 | 5.00 | 5.000 | .360 | .240 | 21.3 | 8.53 | 2.13 | 7.51 | 3.00 | 1.26 |
| 4 x 4 | 13 | 3.82 | 4.16 | 4.060 | .345 | .280 | 11.3 | 5.45 | 1.72 | 3.76 | 1.85 | .99 |
| **LIGHT BEAMS** | | | | | | | | | | | | |
| 12 x 4 | 22 | 6.47 | 12.31 | 4.030 | .424 | .260 | 155.7 | 25.3 | 4.91 | 4.55 | 2.26 | .84 |
|  | 19 | 5.62 | 12.16 | 4.010 | .349 | .240 | 130.1 | 21.4 | 4.81 | 3.67 | 1.83 | .81 |
|  | 16½ | 4.86 | 12.00 | 4.000 | .269 | .230 | 105.3 | 17.5 | 4.65 | 2.79 | 1.39 | .76 |
|  | 14 | 4.14 | 11.91 | 3.970 | .224 | .200 | 88.2 | 14.8 | 4.61 | 2.25 | 1.13 | .74 |
| 10 x 4 | 19 | 5.61 | 10.25 | 4.020 | .394 | .250 | 96.2 | 18.8 | 4.14 | 4.19 | 2.08 | .86 |
|  | 17 | 4.98 | 10.12 | 4.010 | .329 | .240 | 81.8 | 16.2 | 4.05 | 3.45 | 1.72 | .83 |
|  | 15 | 4.40 | 10.00 | 4.000 | .269 | .230 | 68.8 | 13.8 | 3.95 | 2.79 | 1.39 | .80 |
|  | 11½ | 3.39 | 9.87 | 3.950 | .204 | .180 | 51.9 | 10.5 | 3.92 | 2.01 | 1.02 | .77 |
| 8 x 4 | 15 | 4.43 | 8.12 | 4.015 | .314 | .245 | 48.0 | 11.8 | 3.29 | 3.30 | 1.65 | .86 |
|  | 13 | 3.83 | 8.00 | 4.000 | .254 | .230 | 39.5 | 9.88 | 3.21 | 2.62 | 1.31 | .83 |
|  | 10 | 2.95 | 7.90 | 3.940 | .204 | .170 | 30.8 | 7.79 | 3.23 | 1.99 | 1.01 | .82 |
| 6 x 4 | 16 | 4.72 | 6.25 | 4.030 | .404 | .260 | 31.7 | 10.1 | 2.59 | 4.32 | 2.14 | .96 |
|  | 12 | 3.53 | 6.00 | 4.000 | .279 | .230 | 21.7 | 7.24 | 2.48 | 2.89 | 1.44 | .90 |

# E-7

## MISCELLANEOUS SHAPES (M)

## PROPERTIES FOR DESIGNING
### (continued)

| Nominal Size | Weight per Foot | Area | Depth | Width of Flange | Web Thick-ness | AXIS X-X | | | AXIS Y-Y | | |
|---|---|---|---|---|---|---|---|---|---|---|---|
| | | | | | | I | Z | k | I | Z | k |
| In. | Lb. | In.² | In. | In. | In. | In.⁴ | In.³ | In. | In.⁴ | In.³ | In. |

### LIGHT COLUMNS

| Nominal Size | Weight per Foot | Area | Depth | Width of Flange | Web Thick-ness | I | Z | k | I | Z | k |
|---|---|---|---|---|---|---|---|---|---|---|---|
| * 8 x 8 | 34.3 | 10.09 | 8.00 | 8.000 | .375 | 115.5 | 28.9 | 3.40 | 35.1 | 8.8 | 1.87 |
| * 6 x 6 | 25.0 | 7.35 | 6.00 | 5.938 | .313 | 47.0 | 15.7 | 2.53 | 14.9 | 5.0 | 1.43 |
| | 20.0 | 5.88 | 6.00 | 5.938 | .250 | 38.8 | 12.9 | 2.57 | 11.4 | 3.8 | 1.39 |
| § 5 x 5 | 18.9 | 5.56 | 5.00 | 5.000 | .313 | 23.8 | 9.5 | 2.08 | 7.8 | 3.1 | 1.20 |
| † 4 x 4 | 13.0 | 3.82 | 4.00 | 3.937 | .250 | 10.4 | 5.2 | 1.65 | 3.4 | 1.7 | .94 |

### STANDARD MILL BEAMS

| Nominal Size | Weight per Foot | Area | Depth | Width of Flange | Web Thick-ness | I | Z | k | I | Z | k |
|---|---|---|---|---|---|---|---|---|---|---|---|
| ‡8 x 6½ | 24 | 7.06 | 8.00 | 6.50 | .24 | 83.8 | 21.0 | 3.45 | 16.52 | 5.08 | 1.53 |
| ‡8 x 5¼ | 20 | 5.88 | 8.00 | 5.36 | .35 | 60.7 | 15.2 | 3.22 | 6.60 | 2.46 | 1.06 |
| | 17 | 5.00 | 8.00 | 5.25 | .24 | 56.0 | 14.0 | 3.35 | 6.16 | 2.35 | 1.11 |

### JUNIOR BEAMS

| Nominal Size | Weight per Foot | Area | Depth | Width of Flange | Web Thick-ness | I | Z | k | I | Z | k |
|---|---|---|---|---|---|---|---|---|---|---|---|
| ⊕12 x 3 | 11.8 | 3.45 | 12.00 | 3.063 | .175 | 72.2 | 12.0 | 4.57 | .98 | .64 | .53 |
| ⊕10 x 2¾ | 9.0 | 2.64 | 10.00 | 2.688 | .155 | 39.0 | 7.8 | 3.85 | .61 | .45 | .48 |
| ⊕ 8 x 2¼ | 6.5 | 1.92 | 8.00 | 2.281 | .135 | 18.7 | 4.7 | 3.12 | .34 | .30 | .42 |
| ⊕ 7 x 2⅛ | 5.5 | 1.61 | 7.00 | 2.078 | .126 | 12.1 | 3.5 | 2.74 | .25 | .24 | .39 |
| ⊕ 6 x 1⅞ | 4.4 | 1.30 | 6.00 | 1.844 | .114 | 7.3 | 2.4 | 2.37 | .17 | .18 | .36 |

## JUNIOR CHANNELS (MC)

| Nominal Size | Weight per Foot | Area | Depth | Width of Flange | Web Thick-ness | AXIS X-X | | | AXIS Y-Y | | | |
|---|---|---|---|---|---|---|---|---|---|---|---|---|
| | | | | | | I | S | r | I | S | r | x |
| In. | Lb. | In.² | In. | In. | In. | In.⁴ | In.³ | In. | In.⁴ | In.³ | In. | In. |
| ⊕12 x 1½ | 10.6 | 3.12 | 12.00 | 1.500 | .190 | 55.8 | 9.3 | 4.23 | .39 | .32 | .35 | .27 |
| ⊕10 x 1½ | 8.4 | 2.47 | 10.00 | 1.500 | .170 | 32.3 | 6.5 | 3.61 | .33 | .28 | .37 | .29 |
| ⊕10 x 1⅛ | 6.5 | 1.91 | 10.00 | 1.125 | .150 | 22.1 | 4.4 | 3.47 | .12 | .13 | .25 | .19 |

*Rolled by Carnegie-Illinois Steel Corp. and Inland Steel Co.
§Rolled by Carnegie-Illinois Steel Corp. and Bethlehem Steel Co.-M.
†Rolled by Carnegie-Illinois Steel Corp.-M.
‡Rolled by The Phoenix Iron Co.-B.
⊕Rolled by Jones & Laughlin Steel Corp.-Jr.

# E-8

## AMERICAN STANDARD
## BEAMS (S)
### PROPERTIES FOR DESIGNING

| Nominal Size | Weight per Foot | Area | Depth | Flange | | Web Thick-ness | AXIS X-X | | | AXIS Y-Y | | |
|---|---|---|---|---|---|---|---|---|---|---|---|---|
| | | | | Width | Thick-ness | | I | Z | k | I | Z | k |
| In. | Lb. | In.² | In. | In. | In. | In. | In.⁴ | In.³ | In. | In.⁴ | In.³ | In. |
| 24 x 7⅞ | 120.0 | 35.13 | 24.00 | 8.048 | 1.102 | .798 | 3010.8 | 250.9 | 9.26 | 84.9 | 21.1 | 1.56 |
| | 105.9 | 30.98 | 24.00 | 7.875 | 1.102 | .625 | 2811.5 | 234.3 | 9.53 | 78.9 | 20.0 | 1.60 |
| 24 x 7 | 100.0 | 29.25 | 24.00 | 7.247 | .871 | .747 | 2371.8 | 197.6 | 9.05 | 48.4 | 13.4 | 1.29 |
| | 90.0 | 26.30 | 24.00 | 7.124 | .871 | .624 | 2230.1 | 185.8 | 9.21 | 45.5 | 12.8 | 1.32 |
| | 79.9 | 23.33 | 24.00 | 7.000 | .871 | .500 | 2087.2 | 173.9 | 9.46 | 42.9 | 12.2 | 1.36 |
| 20 x 7 | 95.0 | 27.74 | 20.00 | 7.200 | .916 | .800 | 1599.7 | 160.0 | 7.59 | 50.5 | 14.0 | 1.35 |
| | 85.0 | 24.80 | 20.00 | 7.053 | .916 | .653 | 1501.7 | 150.2 | 7.78 | 47.0 | 13.3 | 1.38 |
| 20 x 6¼ | 75.0 | 21.90 | 20.00 | 6.391 | .789 | .641 | 1263.5 | 126.3 | 7.60 | 30.1 | 9.4 | 1.17 |
| | 65.4 | 19.08 | 20.00 | 6.250 | .789 | .500 | 1169.5 | 116.9 | 7.83 | 27.9 | 8.9 | 1.21 |
| 18 x 6 | 70.0 | 20.46 | 18.00 | 6.251 | .691 | .711 | 917.5 | 101.9 | 6.70 | 24.5 | 7.8 | 1.09 |
| | 54.7 | 15.94 | 18.00 | 6.000 | .691 | .460 | 795.5 | 88.4 | 7.07 | 21.2 | 7.1 | 1.15 |
| 15 x 5½ | 50.0 | 14.59 | 15.00 | 5.640 | .622 | .550 | 481.1 | 64.2 | 5.74 | 16.0 | 5.7 | 1.05 |
| | 42.9 | 12.49 | 15.00 | 5.500 | .622 | .410 | 441.8 | 58.9 | 5.95 | 14.6 | 5.3 | 1.08 |
| 12 x 5¼ | 50.0 | 14.57 | 12.00 | 5.477 | .659 | .687 | 301.6 | 50.3 | 4.55 | 16.0 | 5.8 | 1.05 |
| | 40.8 | 11.84 | 12.00 | 5.250 | .659 | .460 | 268.9 | 44.8 | 4.77 | 13.8 | 5.3 | 1.08 |
| 12 x 5 | 35.0 | 10.20 | 12.00 | 5.078 | .544 | .428 | 227.0 | 37.8 | 4.72 | 10.0 | 3.9 | .99 |
| | 31.8 | 9.26 | 12.00 | 5.000 | .544 | .350 | 215.8 | 36.0 | 4.83 | 9.5 | 3.8 | 1.01 |
| 10 x 4⅝ | 35.0 | 10.22 | 10.00 | 4.944 | .491 | .594 | 145.8 | 29.2 | 3.78 | 8.5 | 3.4 | .91 |
| | 25.4 | 7.38 | 10.00 | 4.660 | .491 | .310 | 122.1 | 24.4 | 4.07 | 6.9 | 3.0 | .97 |
| 8 x 4 | 23.0 | 6.71 | 8.00 | 4.171 | .425 | .441 | 64.2 | 16.0 | 3.09 | 4.4 | 2.1 | .81 |
| | 18.4 | 5.34 | 8.00 | 4.000 | .425 | .270 | 56.9 | 14.2 | 3.26 | 3.8 | 1.9 | .84 |
| 7 x 3⅝ | 20.0 | 5.83 | 7.00 | 3.860 | .392 | .450 | 41.9 | 12.0 | 2.68 | 3.1 | 1.6 | .74 |
| | 15.3 | 4.43 | 7.00 | 3.660 | .392 | .250 | 36.2 | 10.4 | 2.86 | 2.7 | 1.5 | .78 |
| 6 x 3⅜ | 17.25 | 5.02 | 6.00 | 3.565 | .359 | .465 | 26.0 | 8.7 | 2.28 | 2.3 | 1.3 | .68 |
| | 12.5 | 3.61 | 6.00 | 3.330 | .359 | .230 | 21.8 | 7.3 | 2.46 | 1.8 | 1.1 | .72 |
| 5 x 3 | 14.75 | 4.29 | 5.00 | 3.284 | .326 | .494 | 15.0 | 6.0 | 1.87 | 1.7 | 1.0 | .63 |
| | 10.0 | 2.87 | 5.00 | 3.000 | .326 | .210 | 12.1 | 4.8 | 2.05 | 1.2 | .82 | .65 |
| 4 x 2⅝ | 9.5 | 2.76 | 4.00 | 2.796 | .293 | .326 | 6.7 | 3.3 | 1.56 | .91 | .65 | .58 |
| | 7.7 | 2.21 | 4.00 | 2.660 | .293 | .190 | 6.0 | 3.0 | 1.64 | .77 | .58 | .59 |
| 3 x 2⅜ | 7.5 | 2.17 | 3.00 | 2.509 | .260 | .349 | 2.9 | 1.9 | 1.15 | .59 | .47 | .52 |
| | 5.7 | 1.64 | 3.00 | 2.330 | .260 | .170 | 2.5 | 1.7 | 1.23 | .46 | .40 | .53 |

# E-9

[

## AMERICAN STANDARD
## CHANNELS (C)

## PROPERTIES FOR DESIGNING

| Nominal Size | Weight per Foot | Area | Depth | Flange | | Web Thickness | AXIS X-X | | | AXIS Y-Y | | | |
|---|---|---|---|---|---|---|---|---|---|---|---|---|---|
| | | | | Width | Average Thickness | | I | Z | k | I | Z | k | x |
| In. | Lb. | In.² | In. | In. | In. | In. | In.⁴ | In.³ | In. | In.⁴ | In.³ | In. | In. |
| *18 x 4 | 58.0 | 16.98 | 18.00 | 4.200 | .625 | .700 | 670.7 | 74.5 | 6.29 | 18.5 | 5.6 | 1.04 | .88 |
| | 51.9 | 15.18 | 18.00 | 4.100 | .625 | .600 | 622.1 | 69.1 | 6.40 | 17.1 | 5.3 | 1.06 | .87 |
| | 45.8 | 13.38 | 18.00 | 4.000 | .625 | .500 | 573.5 | 63.7 | 6.55 | 15.8 | 5.1 | 1.09 | .89 |
| | 42.7 | 12.48 | 18.00 | 3.950 | .625 | .450 | 549.2 | 61.0 | 6.64 | 15.0 | 4.9 | 1.10 | .90 |
| 15 x 3⅜ | 50.0 | 14.64 | 15.00 | 3.716 | .650 | .716 | 401.4 | 53.6 | 5.24 | 11.2 | 3.8 | .87 | .80 |
| | 40.0 | 11.70 | 15.00 | 3.520 | .650 | .520 | 346.3 | 46.2 | 5.44 | 9.3 | 3.4 | .89 | .78 |
| | 33.9 | 9.90 | 15.00 | 3.400 | .650 | .400 | 312.6 | 41.7 | 5.62 | 8.2 | 3.2 | .91 | .79 |
| 12 x 3 | 30.0 | 8.79 | 12.00 | 3.170 | .501 | .510 | 161.2 | 26.9 | 4.28 | 5.2 | 2.1 | .77 | .68 |
| | 25.0 | 7.32 | 12.00 | 3.047 | .501 | .387 | 143.5 | 23.9 | 4.43 | 4.5 | 1.9 | .79 | .68 |
| | 20.7 | 6.03 | 12.00 | 2.940 | .501 | .280 | 128.1 | 21.4 | 4.61 | 3.9 | 1.7 | .81 | .70 |
| 10 x 2⅝ | 30.0 | 8.80 | 10.00 | 3.033 | .436 | .673 | 103.0 | 20.6 | 3.42 | 4.0 | 1.7 | .67 | .65 |
| | 25.0 | 7.33 | 10.00 | 2.886 | .436 | .526 | 90.7 | 18.1 | 3.52 | 3.4 | 1.5 | .68 | .62 |
| | 20.0 | 5.86 | 10.00 | 2.739 | .436 | .379 | 78.5 | 15.7 | 3.66 | 2.8 | 1.3 | .70 | .61 |
| | 15.3 | 4.47 | 10.00 | 2.600 | .436 | .240 | 66.9 | 13.4 | 3.87 | 2.3 | 1.2 | .72 | .64 |
| 9 x 2½ | 20.0 | 5.86 | 9.00 | 2.648 | .413 | .448 | 60.6 | 13.5 | 3.22 | 2.4 | 1.2 | .65 | .59 |
| | 15.0 | 4.39 | 9.00 | 2.485 | .413 | .285 | 50.7 | 11.3 | 3.40 | 1.9 | 1.0 | .67 | .59 |
| | 13.4 | 3.89 | 9.00 | 2.430 | .413 | .230 | 47.3 | 10.5 | 3.49 | 1.8 | .97 | .67 | .61 |
| 8 x 2¼ | 18.75 | 5.49 | 8.00 | 2.527 | .390 | .487 | 43.7 | 10.9 | 2.82 | 2.0 | 1.0 | .60 | .57 |
| | 13.75 | 4.02 | 8.00 | 2.343 | .390 | .303 | 35.8 | 9.0 | 2.99 | 1.5 | .86 | .62 | .56 |
| | 11.5 | 3.36 | 8.00 | 2.260 | .390 | .220 | 32.3 | 8.1 | 3.10 | 1.3 | .79 | .63 | .58 |
| 7 x 2⅛ | 14.75 | 4.32 | 7.00 | 2.299 | .366 | .419 | 27.1 | 7.7 | 2.51 | 1.4 | .79 | .57 | .53 |
| | 12.25 | 3.58 | 7.00 | 2.194 | .366 | .314 | 24.1 | 6.9 | 2.59 | 1.2 | .71 | .58 | .53 |
| | 9.8 | 2.85 | 7.00 | 2.090 | .366 | .210 | 21.1 | 6.0 | 2.72 | .98 | .63 | .59 | .55 |
| 6 x 2 | 13.0 | 3.81 | 6.00 | 2.157 | .343 | .437 | 17.3 | 5.8 | 2.13 | 1.1 | .65 | .53 | .52 |
| | 10.5 | 3.07 | 6.00 | 2.034 | .343 | .314 | 15.1 | 5.0 | 2.22 | .87 | .57 | .53 | .50 |
| | 8.2 | 2.39 | 6.00 | 1.920 | .343 | .200 | 13.0 | 4.3 | 2.34 | .70 | .50 | .54 | .52 |
| 5 x 1¾ | 9.0 | 2.63 | 5.00 | 1.885 | .320 | .325 | 8.8 | 3.5 | 1.83 | .64 | .45 | .49 | .48 |
| | 6.7 | 1.95 | 5.00 | 1.750 | .320 | .190 | 7.4 | 3.0 | 1.95 | .48 | .38 | .50 | .49 |
| 4 x 1⅝ | 7.25 | 2.12 | 4.00 | 1.720 | .296 | .320 | 4.5 | 2.3 | 1.47 | .44 | .35 | .46 | .46 |
| | 5.4 | 1.56 | 4.00 | 1.580 | .296 | .180 | 3.8 | 1.9 | 1.56 | .32 | .29 | .45 | .46 |
| 3 x 1½ | 6.0 | 1.75 | 3.00 | 1.596 | .273 | .356 | 2.1 | 1.4 | 1.08 | .31 | .27 | .42 | .46 |
| | 5.0 | 1.46 | 3.00 | 1.498 | .273 | .258 | 1.8 | 1.2 | 1.12 | .25 | .24 | .41 | .44 |
| | 4.1 | 1.19 | 3.00 | 1.410 | .273 | .170 | 1.6 | 1.1 | 1.17 | .20 | .21 | .41 | .44 |

*Car and Shipbuilding Channel; not an American Standard.

# E-10

## ANGLES
## EQUAL LEGS

### PROPERTIES FOR DESIGNING

| Size | Thickness | Weight per Foot | Area | AXIS X-X AND AXIS Y-Y | | | | AXIS Z-Z |
|---|---|---|---|---|---|---|---|---|
| | | | | I | Z | k | x or y | k |
| In. | In. | Lb. | In.² | In.⁴ | In.³ | In. | In. | In. |
| 6 x 6 | ⅝ | 24.2 | 7.11 | 24.2 | 5.7 | 1.84 | 1.73 | 1.18 |
| | ⁹⁄₁₆ | 21.9 | 6.43 | 22.1 | 5.1 | 1.85 | 1.71 | 1.18 |
| | ½ | 19.6 | 5.75 | 19.9 | 4.6 | 1.86 | 1.68 | 1.18 |
| | ⁷⁄₁₆ | 17.2 | 5.06 | 17.7 | 4.1 | 1.87 | 1.66 | 1.19 |
| | ⅜ | 14.9 | 4.36 | 15.4 | 3.5 | 1.88 | 1.64 | 1.19 |
| | ⁵⁄₁₆ | 12.5 | 3.66 | 13.0 | 3.0 | 1.89 | 1.61 | 1.19 |
| 5 x 5 | ⅞ | 27.2 | 7.98 | 17.8 | 5.2 | 1.49 | 1.57 | .97 |
| | ¾ | 23.6 | 6.94 | 15.7 | 4.5 | 1.51 | 1.52 | .97 |
| | ⅝ | 20.0 | 5.86 | 13.6 | 3.9 | 1.52 | 1.48 | .98 |
| | ½ | 16.2 | 4.75 | 11.3 | 3.2 | 1.54 | 1.43 | .98 |
| | ⁷⁄₁₆ | 14.3 | 4.18 | 10.0 | 2.8 | 1.55 | 1.41 | .98 |
| | ⅜ | 12.3 | 3.61 | 8.7 | 2.4 | 1.56 | 1.39 | .99 |
| | ⁵⁄₁₆ | 10.3 | 3.03 | 7.4 | 2.0 | 1.57 | 1.37 | .99 |
| 4 x 4 | ¾ | 18.5 | 5.44 | 7.7 | 2.8 | 1.19 | 1.27 | .78 |
| | ⅝ | 15.7 | 4.61 | 6.7 | 2.4 | 1.20 | 1.23 | .78 |
| | ½ | 12.8 | 3.75 | 5.6 | 2.0 | 1.22 | 1.18 | .78 |
| | ⁷⁄₁₆ | 11.3 | 3.31 | 5.0 | 1.8 | 1.23 | 1.16 | .78 |
| | ⅜ | 9.8 | 2.86 | 4.4 | 1.5 | 1.23 | 1.14 | .79 |
| | ⁵⁄₁₆ | 8.2 | 2.40 | 3.7 | 1.3 | 1.24 | 1.12 | .79 |
| | ¼ | 6.6 | 1.94 | 3.0 | 1.1 | 1.25 | 1.09 | .80 |
| 3½ x 3½ | ½ | 11.1 | 3.25 | 3.6 | 1.5 | 1.06 | 1.06 | .68 |
| | ⁷⁄₁₆ | 9.8 | 2.87 | 3.3 | 1.3 | 1.07 | 1.04 | .68 |
| | ⅜ | 8.5 | 2.48 | 2.9 | 1.2 | 1.07 | 1.01 | .69 |
| | ⁵⁄₁₆ | 7.2 | 2.09 | 2.5 | .98 | 1.08 | .99 | .69 |
| | ¼ | 5.8 | 1.69 | 2.0 | .79 | 1.09 | .97 | .69 |
| 3 x 3 | ½ | 9.4 | 2.75 | 2.2 | 1.1 | .90 | .93 | .58 |
| | ⁷⁄₁₆ | 8.3 | 2.43 | 2.0 | .95 | .91 | .91 | .58 |
| | ⅜ | 7.2 | 2.11 | 1.8 | .83 | .91 | .89 | .58 |
| | ⁵⁄₁₆ | 6.1 | 1.78 | 1.5 | .71 | .92 | .87 | .59 |
| | ¼ | 4.9 | 1.44 | 1.2 | .58 | .93 | .84 | .59 |
| | ³⁄₁₆ | 3.71 | 1.09 | .96 | .44 | .94 | .82 | .59 |
| 2½ x 2½ | ½ | 7.7 | 2.25 | 1.2 | .72 | .74 | .81 | .49 |
| | ⅜ | 5.9 | 1.73 | .98 | .57 | .75 | .76 | .49 |
| | ⁵⁄₁₆ | 5.0 | 1.47 | .85 | .48 | .76 | .74 | .49 |
| | ¼ | 4.1 | 1.19 | .70 | .39 | .77 | .72 | .49 |
| | ³⁄₁₆ | 3.07 | .90 | .55 | .30 | .78 | .69 | .49 |
| 2 x 2 | ⅜ | 4.7 | 1.36 | .48 | .35 | .59 | .64 | .39 |
| | ⁵⁄₁₆ | 3.92 | 1.15 | .42 | .30 | .60 | .61 | .39 |
| | ¼ | 3.19 | .94 | .35 | .25 | .61 | .59 | .39 |
| | ³⁄₁₆ | 2.44 | .71 | .27 | .19 | .62 | .57 | .39 |
| | ⅛ | 1.65 | .48 | .19 | .13 | .63 | .55 | .40 |

# E-11

## ANGLES
## UNEQUAL LEGS
### PROPERTIES FOR DESIGNING

| Size | Thickness | Weight per Foot | Area | AXIS X-X | | | | AXIS Y-Y | | | | AXIS Z-Z | |
|------|-----------|-----------------|------|-----|-----|-----|-----|-----|-----|-----|-----|-----|-------|
| | | | | I | Z | k | y | I | Z | k | x | k | Tan $\alpha$ |
| In. | In. | Lb. | In.² | In.⁴ | In.³ | In. | In. | In.⁴ | In.³ | In. | In. | In. | |
| 6 x 4 | ⅝ | 20.0 | 5.86 | 21.1 | 5.3 | 1.90 | 2.03 | 7.5 | 2.5 | 1.13 | 1.03 | .86 | .435 |
| | 9/16 | 18.1 | 5.31 | 19.3 | 4.8 | 1.90 | 2.01 | 6.9 | 2.3 | 1.14 | 1.01 | .87 | .438 |
| | ½ | 16.2 | 4.75 | 17.4 | 4.3 | 1.91 | 1.99 | 6.3 | 2.1 | 1.15 | .99 | .87 | .440 |
| | 7/16 | 14.3 | 4.18 | 15.5 | 3.8 | 1.92 | 1.96 | 5.6 | 1.9 | 1.16 | .96 | .87 | .443 |
| | ⅜ | 12.3 | 3.61 | 13.5 | 3.3 | 1.93 | 1.94 | 4.9 | 1.6 | 1.17 | .94 | .88 | .446 |
| | 5/16 | 10.3 | 3.03 | 11.4 | 2.8 | 1.94 | 1.92 | 4.2 | 1.4 | 1.17 | .92 | .88 | .449 |
| 6 x 3½ | ½ | 15.3 | 4.50 | 16.6 | 4.2 | 1.92 | 2.08 | 4.3 | 1.6 | .97 | .83 | .76 | .344 |
| | ⅜ | 11.7 | 3.42 | 12.9 | 3.2 | 1.94 | 2.04 | 3.3 | 1.2 | .99 | .79 | .77 | .350 |
| | 5/16 | 9.8 | 2.87 | 10.9 | 2.7 | 1.95 | 2.01 | 2.9 | 1.0 | 1.00 | .76 | .77 | .352 |
| | ¼ | 7.9 | 2.31 | 8.9 | 2.2 | 1.96 | 1.99 | 2.3 | 0.85 | 1.01 | .74 | .78 | .355 |
| 5 x 3½ | ¾ | 19.8 | 5.81 | 13.9 | 4.3 | 1.55 | 1.75 | 5.6 | 2.2 | .98 | 1.00 | .75 | .464 |
| | ⅝ | 16.8 | 4.92 | 12.0 | 3.7 | 1.56 | 1.70 | 4.8 | 1.9 | .99 | .95 | .75 | .472 |
| | ½ | 13.6 | 4.00 | 10.0 | 3.0 | 1.58 | 1.66 | 4.1 | 1.6 | 1.01 | .91 | .75 | .479 |
| | 7/16 | 12.0 | 3.53 | 8.9 | 2.6 | 1.59 | 1.63 | 3.6 | 1.4 | 1.01 | .88 | .76 | .482 |
| | ⅜ | 10.4 | 3.05 | 7.8 | 2.3 | 1.60 | 1.61 | 3.2 | 1.2 | 1.02 | .86 | .76 | .486 |
| | 5/16 | 8.7 | 2.56 | 6.6 | 1.9 | 1.61 | 1.59 | 2.7 | 1.0 | 1.03 | .84 | .76 | .489 |
| | ¼ | 7.0 | 2.06 | 5.4 | 1.6 | 1.61 | 1.56 | 2.2 | .83 | 1.04 | .81 | .76 | .492 |
| 5 x 3 | ½ | 12.8 | 3.75 | 9.5 | 2.9 | 1.59 | 1.75 | 2.6 | 1.1 | .83 | .75 | .65 | .357 |
| | 7/16 | 11.3 | 3.31 | 8.4 | 2.6 | 1.60 | 1.73 | 2.3 | 1.0 | .84 | .73 | .65 | .361 |
| | ⅜ | 9.8 | 2.86 | 7.4 | 2.2 | 1.61 | 1.70 | 2.0 | .89 | .84 | .70 | .65 | .364 |
| | 5/16 | 8.2 | 2.40 | 6.3 | 1.9 | 1.61 | 1.68 | 1.8 | .75 | .85 | .68 | .66 | .368 |
| | ¼ | 6.6 | 1.94 | 5.1 | 1.5 | 1.62 | 1.66 | 1.4 | .61 | .86 | .66 | .66 | .371 |
| 4 x 3½ | ⅝ | 14.7 | 4.30 | 6.4 | 2.4 | 1.22 | 1.29 | 4.5 | 1.8 | 1.03 | 1.04 | .72 | .745 |
| | ½ | 11.9 | 3.50 | 5.3 | 1.9 | 1.23 | 1.25 | 3.8 | 1.5 | 1.04 | 1.00 | .72 | .750 |
| | 7/16 | 10.6 | 3.09 | 4.8 | 1.7 | 1.24 | 1.23 | 3.4 | 1.4 | 1.05 | .98 | .72 | .753 |
| | ⅜ | 9.1 | 2.67 | 4.2 | 1.5 | 1.25 | 1.21 | 3.0 | 1.2 | 1.06 | .96 | .73 | .755 |
| | 5/16 | 7.7 | 2.25 | 3.6 | 1.3 | 1.26 | 1.18 | 2.6 | 1.0 | 1.07 | .93 | .73 | .757 |
| | ¼ | 6.2 | 1.81 | 2.9 | 1.0 | 1.27 | 1.16 | 2.1 | .81 | 1.07 | .91 | .73 | .759 |

# E-11

## ANGLES
## UNEQUAL LEGS
### PROPERTIES FOR DESIGNING
(continued)

| Size | Thick-ness | Weight per Foot | Area | AXIS X-X | | | | AXIS Y-Y | | | | AXIS Z-Z | |
|------|-----------|-----------------|------|------|------|------|------|------|------|------|------|------|------|
| | | | | I | Z | k | y | I | Z | k | x | k | Tan α |
| In. | In. | Lb. | In.² | In.⁴ | In.³ | In. | In. | In.⁴ | In.³ | In. | In. | In. | |
| 4 x 3 | ⅝ | 13.6 | 3.98 | 6.0 | 2.3 | 1.23 | 1.37 | 2.9 | 1.4 | .85 | .87 | .64 | .534 |
| | ½ | 11.1 | 3.25 | 5.1 | 1.9 | 1.25 | 1.33 | 2.4 | 1.1 | .86 | .83 | .64 | .543 |
| | ⁷⁄₁₆ | 9.8 | 2.87 | 4.5 | 1.7 | 1.25 | 1.30 | 2.2 | 1.0 | .87 | .80 | .64 | .547 |
| | ⅜ | 8.5 | 2.48 | 4.0 | 1.5 | 1.26 | 1.28 | 1.9 | .87 | .88 | .78 | .64 | .551 |
| | ⁵⁄₁₆ | 7.2 | 2.09 | 3.4 | 1.2 | 1.27 | 1.26 | 1.7 | .73 | .89 | .76 | .65 | .554 |
| | ¼ | 5.8 | 1.69 | 2.8 | 1.0 | 1.28 | 1.24 | 1.4 | .60 | .90 | .74 | .65 | .558 |
| 3½ x 3 | ½ | 10.2 | 3.00 | 3.5 | 1.5 | 1.07 | 1.13 | 2.3 | 1.1 | .88 | .88 | .62 | .714 |
| | ⁷⁄₁₆ | 9.1 | 2.65 | 3.1 | 1.3 | 1.08 | 1.10 | 2.1 | .98 | .89 | .85 | .62 | .718 |
| | ⅜ | 7.9 | 2.30 | 2.7 | 1.1 | 1.09 | 1.08 | 1.9 | .85 | .90 | .83 | .62 | .721 |
| | ⁵⁄₁₆ | 6.6 | 1.93 | 2.3 | .95 | 1.10 | 1.06 | 1.6 | .72 | .90 | .81 | .63 | .724 |
| | ¼ | 5.4 | 1.56 | 1.9 | .78 | 1.11 | 1.04 | 1.3 | .59 | .91 | .79 | .63 | .727 |
| 3½x2½ | ½ | 9.4 | 2.75 | 3.2 | 1.4 | 1.09 | 1.20 | 1.4 | .76 | .70 | .70 | .53 | .486 |
| | ⁷⁄₁₆ | 8.3 | 2.43 | 2.9 | 1.3 | 1.09 | 1.18 | 1.2 | .68 | .71 | .68 | .54 | .491 |
| | ⅜ | 7.2 | 2.11 | 2.6 | 1.1 | 1.10 | 1.16 | 1.1 | .59 | .72 | .66 | .54 | .496 |
| | ⁵⁄₁₆ | 6.1 | 1.78 | 2.2 | .93 | 1.11 | 1.14 | .94 | .50 | .73 | .64 | .54 | .501 |
| | ¼ | 4.9 | 1.44 | 1.8 | .75 | 1.12 | 1.11 | .78 | .41 | .74 | .61 | .54 | .506 |
| 3 x 2½ | ½ | 8.5 | 2.50 | 2.1 | 1.0 | .91 | 1.00 | 1.3 | .74 | .72 | .75 | .52 | .667 |
| | ⁷⁄₁₆ | 7.6 | 2.21 | 1.9 | .93 | .92 | .98 | 1.2 | .66 | .73 | .73 | .52 | .672 |
| | ⅜ | 6.6 | 1.92 | 1.7 | .81 | .93 | .96 | 1.0 | .58 | .74 | .71 | .52 | .676 |
| | ⁵⁄₁₆ | 5.6 | 1.62 | 1.4 | .69 | .94 | .93 | .90 | .49 | .74 | .68 | .53 | .680 |
| | ¼ | 4.5 | 1.31 | 1.2 | .56 | .95 | .91 | .74 | .40 | .75 | .66 | .53 | .684 |
| 3 x 2 | ½ | 7.7 | 2.25 | 1.9 | 1.0 | .92 | 1.08 | .67 | .47 | .55 | .58 | .43 | .414 |
| | ⁷⁄₁₆ | 6.8 | 2.00 | 1.7 | .89 | .93 | 1.06 | .61 | .42 | .55 | .56 | .43 | .421 |
| | ⅜ | 5.9 | 1.73 | 1.5 | .78 | .94 | 1.04 | .54 | .37 | .56 | .54 | .43 | .428 |
| | ⁵⁄₁₆ | 5.0 | 1.47 | 1.3 | .66 | .95 | 1.02 | .47 | .32 | .57 | .52 | .43 | .435 |
| | ¼ | 4.1 | 1.19 | 1.1 | .54 | .95 | .99 | .39 | .26 | .57 | .49 | .43 | .440 |
| | ³⁄₁₆ | 3.07 | .90 | .84 | .41 | .97 | .97 | .31 | .20 | .58 | .47 | .44 | .446 |

# E-12

## TWO UNEQUAL ANGLES

### PROPERTIES OF SECTIONS

#### SHORT LEGS BACK TO BACK

| Size | Thickness | Weight per Ft. 2 Angles | Area of 2 Angles | AXIS X-X | | | | RADII OF GYRATION ABOUT AXIS Y-Y Back to Back of Angles, Inches | | | | | |
|---|---|---|---|---|---|---|---|---|---|---|---|---|---|
| | | | | I | Z | k | y | 0 | 1/4 | 3/8 | 1/2 | 5/8 | 3/4 |
| In. | In. | Lb. | In.² | In.⁴ | In.³ | In. | In. | | | | | | |
| 6 x 4 | 5/8 | 40.0 | 11.72 | 15.0 | 5.1 | 1.13 | 1.03 | 2.78 | 2.87 | 2.92 | 2.97 | 3.01 | 3.06 |
| | 9/16 | 36.2 | 10.62 | 13.8 | 4.6 | 1.14 | 1.01 | 2.77 | 2.86 | 2.91 | 2.96 | 3.00 | 3.05 |
| | 1/2 | 32.4 | 9.50 | 12.5 | 4.2 | 1.15 | .99 | 2.76 | 2.85 | 2.90 | 2.95 | 2.99 | 3.04 |
| | 7/16 | 28.6 | 8.36 | 11.2 | 3.7 | 1.16 | .96 | 2.75 | 2.84 | 2.88 | 2.93 | 2.98 | 3.03 |
| | 3/8 | 24.6 | 7.22 | 9.8 | 3.2 | 1.17 | .94 | 2.74 | 2.83 | 2.87 | 2.92 | 2.97 | 3.02 |
| 6 x 3½ | 1/2 | 30.6 | 9.00 | 8.5 | 3.2 | .97 | .83 | 2.83 | 2.92 | 2.97 | 3.02 | 3.07 | 3.12 |
| | 3/8 | 23.4 | 6.84 | 6.7 | 2.5 | .99 | .79 | 2.81 | 2.90 | 2.95 | 3.00 | 3.05 | 3.09 |
| | 5/16 | 19.6 | 5.74 | 5.7 | 2.1 | 1.00 | .76 | 2.80 | 2.89 | 2.94 | 2.99 | 3.03 | 3.08 |
| 5 x 3½ | 3/4 | 39.6 | 11.62 | 11.1 | 4.4 | .98 | 1.00 | 2.34 | 2.43 | 2.48 | 2.53 | 2.58 | 2.63 |
| | 5/8 | 33.6 | 9.84 | 9.7 | 3.8 | .99 | .95 | 2.31 | 2.40 | 2.45 | 2.50 | 2.55 | 2.60 |
| | 1/2 | 27.2 | 8.00 | 8.1 | 3.1 | 1.01 | .91 | 2.29 | 2.38 | 2.43 | 2.48 | 2.53 | 2.58 |
| | 7/16 | 24.0 | 7.06 | 7.3 | 2.8 | 1.01 | .88 | 2.28 | 2.37 | 2.41 | 2.46 | 2.51 | 2.56 |
| | 3/8 | 20.8 | 6.10 | 6.4 | 2.4 | 1.02 | .86 | 2.27 | 2.36 | 2.40 | 2.45 | 2.50 | 2.55 |
| | 5/16 | 17.4 | 5.12 | 5.4 | 2.0 | 1.03 | .84 | 2.26 | 2.35 | 2.38 | 2.43 | 2.48 | 2.53 |
| 5 x 3 | 1/2 | 25.6 | 7.50 | 5.2 | 2.3 | .83 | .75 | 2.36 | 2.46 | 2.50 | 2.55 | 2.60 | 2.65 |
| | 3/8 | 19.6 | 5.72 | 4.1 | 1.8 | .84 | .70 | 2.34 | 2.43 | 2.48 | 2.53 | 2.58 | 2.63 |
| | 5/16 | 16.4 | 4.80 | 3.5 | 1.5 | .85 | .68 | 2.33 | 2.42 | 2.47 | 2.52 | 2.57 | 2.62 |
| 4 x 3½ | 1/2 | 23.8 | 7.00 | 7.6 | 3.0 | 1.04 | 1.00 | 1.76 | 1.85 | 1.89 | 1.94 | 1.99 | 2.04 |
| | 7/16 | 21.2 | 6.18 | 6.8 | 2.7 | 1.05 | .98 | 1.75 | 1.84 | 1.89 | 1.94 | 1.98 | 2.03 |
| | 3/8 | 18.2 | 5.34 | 6.0 | 2.3 | 1.06 | .96 | 1.74 | 1.83 | 1.88 | 1.92 | 1.97 | 2.02 |
| | 5/16 | 15.4 | 4.50 | 5.1 | 2.0 | 1.07 | .93 | 1.73 | 1.81 | 1.86 | 1.91 | 1.96 | 2.00 |
| | 1/4 | 12.4 | 3.62 | 4.2 | 1.6 | 1.07 | .91 | 1.72 | 1.80 | 1.85 | 1.90 | 1.94 | 1.99 |
| 4 x 3 | 5/8 | 27.2 | 7.96 | 5.7 | 2.7 | .85 | .87 | 1.84 | 1.94 | 1.99 | 2.03 | 2.08 | 2.14 |
| | 1/2 | 22.2 | 6.50 | 4.8 | 2.2 | .86 | .83 | 1.82 | 1.92 | 1.96 | 2.01 | 2.06 | 2.11 |
| | 7/16 | 19.6 | 5.74 | 4.4 | 2.0 | .87 | .80 | 1.81 | 1.90 | 1.95 | 1.99 | 2.04 | 2.09 |
| | 3/8 | 17.0 | 4.96 | 3.8 | 1.7 | .88 | .78 | 1.80 | 1.89 | 1.94 | 1.98 | 2.03 | 2.08 |
| | 5/16 | 14.4 | 4.18 | 3.3 | 1.5 | .89 | .76 | 1.79 | 1.88 | 1.93 | 1.97 | 2.02 | 2.07 |
| | 1/4 | 11.6 | 3.38 | 2.7 | 1.2 | .90 | .74 | 1.78 | 1.87 | 1.92 | 1.96 | 2.01 | 2.06 |
| 3½ x 3 | 1/2 | 20.4 | 6.00 | 4.7 | 2.2 | .88 | .88 | 1.56 | 1.65 | 1.70 | 1.75 | 1.80 | 1.85 |
| | 7/16 | 18.2 | 5.30 | 4.2 | 2.0 | .89 | .85 | 1.54 | 1.63 | 1.68 | 1.73 | 1.78 | 1.83 |
| | 3/8 | 15.8 | 4.60 | 3.7 | 1.7 | .90 | .83 | 1.53 | 1.62 | 1.67 | 1.72 | 1.77 | 1.82 |
| | 5/16 | 13.2 | 3.86 | 3.2 | 1.4 | .90 | .81 | 1.52 | 1.61 | 1.66 | 1.71 | 1.76 | 1.81 |
| | 1/4 | 10.8 | 3.12 | 2.6 | 1.2 | .91 | .79 | 1.52 | 1.61 | 1.65 | 1.70 | 1.75 | 1.80 |
| 3½ x 2½ | 1/2 | 18.8 | 5.50 | 2.7 | 1.5 | .70 | .70 | 1.62 | 1.71 | 1.76 | 1.81 | 1.86 | 1.91 |
| | 7/16 | 16.6 | 4.86 | 2.5 | 1.4 | .71 | .68 | 1.61 | 1.70 | 1.75 | 1.80 | 1.85 | 1.90 |
| | 3/8 | 14.4 | 4.22 | 2.2 | 1.2 | .72 | .66 | 1.61 | 1.69 | 1.74 | 1.79 | 1.84 | 1.89 |
| | 5/16 | 12.2 | 3.56 | 1.9 | 1.0 | .73 | .64 | 1.60 | 1.68 | 1.73 | 1.77 | 1.82 | 1.88 |
| | 1/4 | 9.8 | 2.88 | 1.6 | 0.8 | .74 | .61 | 1.58 | 1.67 | 1.71 | 1.76 | 1.81 | 1.86 |
| 3 x 2½ | 1/2 | 17.0 | 5.00 | 2.6 | 1.5 | .72 | .75 | 1.35 | 1.45 | 1.50 | 1.55 | 1.60 | 1.65 |
| | 7/16 | 15.2 | 4.42 | 2.4 | 1.3 | .73 | .73 | 1.34 | 1.44 | 1.49 | 1.54 | 1.59 | 1.64 |
| | 3/8 | 13.2 | 3.84 | 2.1 | 1.2 | .74 | .71 | 1.34 | 1.43 | 1.48 | 1.53 | 1.58 | 1.63 |
| | 5/16 | 11.2 | 3.24 | 1.8 | 1.0 | .74 | .68 | 1.32 | 1.41 | 1.46 | 1.51 | 1.56 | 1.60 |
| | 1/4 | 9.0 | 2.62 | 1.5 | 0.8 | .75 | .66 | 1.31 | 1.40 | 1.45 | 1.50 | 1.55 | 1.60 |

# TWO UNEQUAL ANGLES

## PROPERTIES OF SECTIONS

### LONG LEGS BACK TO BACK

### (continued)

| Size | Thick-ness | Weight per Ft. 2 Angles | Area of 2 Angles | AXIS X-X | | | | RADII OF GYRATION ABOUT AXIS Y-Y | | | | | |
|---|---|---|---|---|---|---|---|---|---|---|---|---|---|
| | | | | | | | | Back to Back of Angles, Inches | | | | | |
| | | | | I | Z | k | y | 0 | $\frac{1}{4}$ | $\frac{3}{8}$ | $\frac{1}{2}$ | $\frac{5}{8}$ | $\frac{3}{4}$ |
| In. | In. | Lb. | In.² | In.⁴ | In.³ | In. | In. | | | | | | |
| 6 x 4 | $\frac{5}{8}$ | 40.0 | 11.72 | 42.1 | 10.6 | 1.90 | 2.03 | 1.53 | 1.62 | 1.66 | 1.71 | 1.76 | 1.80 |
| | $\frac{9}{16}$ | 36.2 | 10.62 | 38.5 | 9.7 | 1.90 | 2.01 | 1.52 | 1.61 | 1.66 | 1.70 | 1.75 | 1.79 |
| | $\frac{1}{2}$ | 32.4 | 9.50 | 34.8 | 8.7 | 1.91 | 1.99 | 1.52 | 1.60 | 1.65 | 1.69 | 1.74 | 1.78 |
| | $\frac{7}{16}$ | 28.6 | 8.36 | 31.0 | 7.7 | 1.92 | 1.96 | 1.50 | 1.59 | 1.63 | 1.68 | 1.72 | 1.77 |
| | $\frac{3}{8}$ | 24.6 | 7.22 | 26.9 | 6.6 | 1.93 | 1.94 | 1.50 | 1.58 | 1.62 | 1.67 | 1.71 | 1.76 |
| 6 x 3½ | $\frac{1}{2}$ | 30.6 | 9.00 | 33.2 | 8.5 | 1.92 | 2.08 | 1.27 | 1.36 | 1.40 | 1.45 | 1.49 | 1.55 |
| | $\frac{3}{8}$ | 23.4 | 6.84 | 25.7 | 6.5 | 1.94 | 2.04 | 1.26 | 1.34 | 1.39 | 1.43 | 1.48 | 1.53 |
| | $\frac{5}{16}$ | 19.6 | 5.74 | 21.8 | 5.5 | 1.95 | 2.01 | 1.26 | 1.33 | 1.38 | 1.42 | 1.46 | 1.51 |
| 5 x 3½ | $\frac{3}{4}$ | 39.6 | 11.62 | 27.8 | 8.6 | 1.55 | 1.75 | 1.40 | 1.49 | 1.54 | 1.59 | 1.64 | 1.69 |
| | $\frac{5}{8}$ | 33.6 | 9.84 | 24.1 | 7.3 | 1.56 | 1.70 | 1.37 | 1.46 | 1.51 | 1.56 | 1.60 | 1.65 |
| | $\frac{1}{2}$ | 27.2 | 8.00 | 20.0 | 6.0 | 1.58 | 1.66 | 1.36 | 1.44 | 1.49 | 1.54 | 1.58 | 1.63 |
| | $\frac{7}{16}$ | 24.0 | 7.06 | 17.8 | 5.3 | 1.59 | 1.63 | 1.35 | 1.43 | 1.47 | 1.52 | 1.57 | 1.62 |
| | $\frac{3}{8}$ | 20.8 | 6.10 | 15.6 | 4.6 | 1.60 | 1.61 | 1.34 | 1.42 | 1.46 | 1.51 | 1.55 | 1.60 |
| | $\frac{5}{16}$ | 17.4 | 5.12 | 13.2 | 3.9 | 1.61 | 1.59 | 1.33 | 1.41 | 1.45 | 1.50 | 1.54 | 1.59 |
| 5 x 3 | $\frac{1}{2}$ | 25.6 | 7.50 | 18.9 | 5.8 | 1.59 | 1.75 | 1.11 | 1.21 | 1.25 | 1.30 | 1.35 | 1.40 |
| | $\frac{3}{8}$ | 19.6 | 5.72 | 14.7 | 4.5 | 1.61 | 1.70 | 1.09 | 1.18 | 1.23 | 1.27 | 1.32 | 1.37 |
| | $\frac{5}{16}$ | 16.4 | 4.80 | 12.5 | 3.8 | 1.61 | 1.68 | 1.09 | 1.17 | 1.22 | 1.26 | 1.31 | 1.36 |
| 4 x 3½ | $\frac{5}{8}$ | 29.4 | 8.60 | 12.7 | 4.7 | 1.22 | 1.29 | 1.46 | 1.55 | 1.60 | 1.65 | 1.70 | 1.75 |
| | $\frac{1}{2}$ | 23.8 | 7.00 | 10.6 | 3.9 | 1.23 | 1.25 | 1.44 | 1.53 | 1.58 | 1.63 | 1.67 | 1.72 |
| | $\frac{7}{16}$ | 21.2 | 6.18 | 9.5 | 3.4 | 1.24 | 1.23 | 1.44 | 1.52 | 1.57 | 1.62 | 1.66 | 1.71 |
| | $\frac{3}{8}$ | 18.2 | 5.34 | 8.4 | 3.0 | 1.25 | 1.21 | 1.43 | 1.52 | 1.56 | 1.61 | 1.66 | 1.70 |
| | $\frac{5}{16}$ | 15.4 | 4.50 | 7.1 | 2.5 | 1.26 | 1.18 | 1.42 | 1.50 | 1.55 | 1.59 | 1.64 | 1.69 |
| | $\frac{1}{4}$ | 12.4 | 3.62 | 5.8 | 2.1 | 1.27 | 1.16 | 1.41 | 1.49 | 1.54 | 1.58 | 1.63 | 1.67 |
| 4 x 3 | $\frac{5}{8}$ | 27.2 | 7.96 | 12.1 | 4.6 | 1.23 | 1.37 | 1.22 | 1.31 | 1.36 | 1.41 | 1.46 | 1.51 |
| | $\frac{1}{2}$ | 22.2 | 6.50 | 10.1 | 3.8 | 1.25 | 1.33 | 1.20 | 1.29 | 1.33 | 1.38 | 1.43 | 1.48 |
| | $\frac{7}{16}$ | 19.6 | 5.74 | 9.0 | 3.4 | 1.25 | 1.30 | 1.18 | 1.27 | 1.32 | 1.36 | 1.41 | 1.46 |
| | $\frac{3}{8}$ | 17.0 | 4.96 | 7.9 | 2.9 | 1.26 | 1.28 | 1.18 | 1.26 | 1.31 | 1.35 | 1.40 | 1.45 |
| | $\frac{5}{16}$ | 14.4 | 4.18 | 6.8 | 2.5 | 1.27 | 1.26 | 1.17 | 1.25 | 1.30 | 1.35 | 1.39 | 1.44 |
| | $\frac{1}{4}$ | 11.6 | 3.38 | 5.5 | 2.0 | 1.28 | 1.24 | 1.16 | 1.25 | 1.29 | 1.34 | 1.38 | 1.43 |
| 3½ x 3 | $\frac{1}{2}$ | 20.4 | 6.00 | 6.9 | 2.9 | 1.07 | 1.13 | 1.25 | 1.34 | 1.38 | 1.43 | 1.48 | 1.53 |
| | $\frac{7}{16}$ | 18.2 | 5.30 | 6.2 | 2.6 | 1.08 | 1.10 | 1.23 | 1.32 | 1.37 | 1.41 | 1.46 | 1.51 |
| | $\frac{3}{8}$ | 15.8 | 4.60 | 5.4 | 2.3 | 1.09 | 1.08 | 1.22 | 1.31 | 1.36 | 1.40 | 1.45 | 1.50 |
| | $\frac{5}{16}$ | 13.2 | 3.86 | 4.7 | 1.9 | 1.10 | 1.06 | 1.22 | 1.30 | 1.35 | 1.39 | 1.44 | 1.49 |
| | $\frac{1}{4}$ | 10.8 | 3.12 | 3.8 | 1.6 | 1.11 | 1.04 | 1.21 | 1.29 | 1.34 | 1.38 | 1.43 | 1.48 |
| 3½x2½ | $\frac{1}{2}$ | 18.8 | 5.50 | 6.5 | 2.8 | 1.09 | 1.20 | .99 | 1.08 | 1.13 | 1.18 | 1.23 | 1.29 |
| | $\frac{7}{16}$ | 16.6 | 4.86 | 5.8 | 2.5 | 1.09 | 1.18 | .98 | 1.07 | 1.12 | 1.17 | 1.22 | 1.27 |
| | $\frac{3}{8}$ | 14.4 | 4.22 | 5.1 | 2.2 | 1.10 | 1.16 | .97 | 1.07 | 1.11 | 1.16 | 1.21 | 1.26 |
| | $\frac{5}{16}$ | 12.2 | 3.56 | 4.4 | 1.9 | 1.11 | 1.14 | .96 | 1.05 | 1.10 | 1.15 | 1.20 | 1.24 |
| | $\frac{1}{4}$ | 9.8 | 2.88 | 3.6 | 1.5 | 1.12 | 1.11 | .95 | 1.04 | 1.09 | 1.13 | 1.18 | 1.23 |

# E-13

## ELASTIC SECTION MODULUS TABLE

$Z_x$

### FOR SHAPES USED AS BEAMS

| Elastic Modulus | Shape | Elastic Modulus | Shape | Elastic Modulus | Shape |
|---|---|---|---|---|---|
| 243 | W 27 × 94 | 112 | W 14 × 74 | 53.8 | C 15 × 50 |
| 236 | S 24 × 105.9 | 110 | W 21 × 55 | 51.9 | W 12 × 40 |
| | | 108 | W 18 × 60 | 50.8 | S 12 × 50 |
| 221 | W 24 × 94 | 107 | W 12 × 79 | 49.1 | W 10 × 45 |
| 220 | W 18 × 114 | 104 | W 16 × 64 | 48.6 | W 14 × 34 |
| | | 103 | S 18 × 70 | | |
| 212 | W 27 × 84 | 103 | W 14 × 68 | 47.2 | W 16 × 31 |
| 202 | W 18 × 105 | | | 46.5 | C 15 × 40 |
| 199 | S 24 × 100 | 98.4 | W 18 × 55 | 46.0 | W 12 × 36 |
| 198 | W 21 × 96 | 97.5 | W 12 × 72 | 45.4 | S 12 × 40.8 |
| | | 94.4 | W 16 × 58 | 42.2 | W 10 × 39 |
| 197 | W 24 × 84 | | | 42.0 | C 15 × 33.9 |
| 189 | W 14 × 119 | 93.3 | W 21 × 49 | | |
| 187 | S 24 × 90 | 92.2 | W14 × 61 | 41.9 | W 14 × 30 |
| 185 | W 18 × 96 | 89.4 | S 18 × 54.7 | 39.5 | W 12 × 31 |
| | | | | | |
| 176 | W 24 × 76 | 89.1 | W 18 × 50 | 38.3 | W 16 × 26 |
| 176 | W 14 × 111 | 88.0 | W 12 × 65 | 38.2 | S 12 × 35 |
| 175 | S 24 × 79.9 | | | 36.4 | S 12 × 31.8 |
| 169 | W 21 × 82 | 81.6 | W 21 × 44 | | |
| 166 | W 16 × 96 | 80.8 | W 16 × 50 | 35.1 | W 14 × 26 |
| 164 | W 14 × 103 | 79.0 | W 18 × 45 | 35.0 | W 10 × 33 |
| 161 | S 20 × 95 | 78.1 | W 12 × 58 | 34.2 | W 12 × 27 |
| 157 | W 18 × 85 | 77.8 | W 14 × 53 | 31.1 | W 8 × 35 |
| | | 75.1 | MC 18 × 58 | 30.8 | W 10 × 29 |
| 153 | W 24 × 68 | 73.7 | W 10 × 66 | 29.4 | S 10 × 35 |
| 152 | S 20 × 85 | 72.5 | W 16 × 45 | | |
| 151 | W 21 × 73 | 70.7 | W 12 × 53 | 28.9 | W 14 × 22 |
| 151 | W 16 × 88 | 70.2 | W 14 × 48 | 27.4 | W 8 × 31 |
| 151 | W 14 × 95 | 69.7 | MC 18 × 51.9 | 27.0 | C 12 × 30 |
| 142 | W 18 × 77 | | | 26.6 | M 10 × 29.1 |
| | | 68.4 | W 18 × 40 | 26.5 | W 10 × 25 |
| 140 | W 21 × 68 | 67.1 | W 10 × 60 | | |
| 138 | W 14 × 87 | 64.8 | S 15 × 50 | 25.3 | W 12 × 22 |
| 131 | W 14 × 84 | 64.7 | W 12 × 50 | 24.7 | S 10 × 25.4 |
| | | | | 24.3 | W 8 × 28 |
| 130 | W 24 × 61 | 64.6 | W 16 × 40 | 24.1 | C 12 × 25 |
| 129 | W 18 × 70 | 64.3 | MC 18 × 45.8 | 23.6 | M 10 × 22.9 |
| 128 | S 20 × 75 | 62.7 | W 14 × 43 | | |
| 128 | W 16 × 78 | 61.6 | MC 18 × 42.7 | 21.5 | C 12 × 20.7 |
| 127 | W 21 × 62 | 60.4 | W10 × 54 | 21.5 | W 10 × 21 |
| 121 | W 14 × 78 | 59.6 | S 15 × 42.9 | | |
| 118 | S 20 × 65.4 | 58.2 | W 12 × 45 | 21.3 | W 12 × 19 |
| 118 | W 18 × 64 | | | | |
| 116 | W 16 × 71 | 57.9 | W 18 × 35 | 21.1 | M 14 × 17.2 |
| 116 | W 12 × 85 | 56.5 | W 16 × 36 | 20.8 | W 8 × 24 |
| | | 54.7 | W 14 × 38 | 20.7 | C 10 × 30 |
| 114 | W 24 × 55 | 54.6 | W 10 × 49 | | |

# ELASTIC SECTION MODULUS TABLE

# E-13

$$Z_x$$

## FOR SHAPES USED AS BEAMS

### (continued)

| Elastic Modulus | Shape | Elastic Modulus | Shape | Elastic Modulus | Shape |
|---|---|---|---|---|---|
| 18.8 | W 10 × 19 | 10.6 | C 9 × 13.4 | 5.08 | W 6 × 8.5 |
| 18.2 | C 10 × 25 | 10.5 | W 10 × 11.5 | 5.06 | C 6 × 10.5 |
| | | 10.5 | S 7 × 15.3 | 4.92 | S 5 × 10 |
| 17.6 | W 12 × 16.5 | 10.2 | W 6 × 16 | | |
| 17.0 | W 8 × 20 | 10.0 | W 6 × 15.5 | | |
| 16.7 | W 6 × 25 | 9.90 | W 8 × 13 | 4.62 | M 8 × 6.5 |
| 16.2 | W 10 × 17 | | | 4.42 | MC 10 × 6.5 |
| 16.2 | S 8 × 23 | 9.23 | MC 12 × 10.6 | 4.38 | C 6 × 8.2 |
| 15.8 | C 10 × 20 | 9.09 | C 8 × 13.75 | 3.56 | C 5 × 9 |
| | | 8.77 | S 6 × 17.25 | | |
| 14.8 | W 12 × 14 | 8.14 | C 8 × 11.5 | 3.44 | M 7 × 5.5 |
| 14.4 | S 8 × 18.4 | | | 3.39 | S 4 × 9.5 |
| 14.1 | W 8 × 17 | 7.76 | M 10 × 9 | 3.00 | C 5 × 6.7 |
| 13.8 | W 10 × 15 | 7.80 | W 8 × 10 | 3.04 | S 4 × 7.7 |
| 13.5 | C 10 × 15.3 | 7.78 | C 7 × 14.75 | | |
| 13.5 | C 9 × 20 | 7.37 | S 6 × 12.5 | 2.40 | M 6 × 4.4 |
| 13.4 | W 6 × 20 | 7.25 | W 6 × 12 | 2.29 | C 4 × 7.25 |
| 12.1 | S 7 × 20 | 6.93 | C 7 × 12.25 | 1.95 | S 3 × 7.5 |
| | | | | 1.93 | C 4 × 5.4 |
| | | | | 1.68 | S 3 × 5.7 |
| 12.0 | M 12 × 11.8 | 6.40 | MC 10 × 8.4 | 1.38 | C 3 × 6 |
| 11.8 | W 8 × 15 | 6.08 | C 7 × 9.8 | 1.12 | C 3 × 5 |
| 11.3 | C 9 × 15 | 6.09 | S 5 × 14.75 | | |
| 11.0 | C 8 × 18.75 | 5.80 | C 6 × 13 | 1.17 | C 3 × 4.1 |

# E-14

## ALLOWABLE STRESS, KSI, FOR COMPRESSION MEMBERS OF 36 KSI YIELD POINT STEEL

| Main and secondary members $KL/k$ not over 120 | | | | | | Main members $KL/k$ 121 to 200 | | | | Secondary members $L/k$ 121 to 200 | | | |
|---|---|---|---|---|---|---|---|---|---|---|---|---|---|
| $\frac{KL}{k}$ | $\frac{P}{A}$ | $\frac{KL}{k}$ | $\frac{P}{A}$ | $\frac{KL}{k}$ | $\frac{P}{A}$ | $\frac{KL}{k}$ | $\frac{P}{A}$ | $\frac{KL}{k}$ | $\frac{P}{A}$ | $\frac{L}{k}$ | $\frac{P}{A}$ | $\frac{L}{k}$ | $\frac{P}{A}$ |
| 1 | 21.56 | 41 | 19.11 | 81 | 15.24 | 121 | 10.14 | 161 | 5.76 | 121 | 10.19 | 161 | 7.25 |
| 2 | 21.52 | 42 | 19.03 | 82 | 15.13 | 122 | 9.99 | 162 | 5.69 | 122 | 10.09 | 162 | 7.20 |
| 3 | 21.48 | 43 | 18.95 | 83 | 15.02 | 123 | 9.85 | 163 | 5.62 | 123 | 10.00 | 163 | 7.16 |
| 4 | 21.44 | 44 | 18.86 | 84 | 14.90 | 124 | 9.70 | 164 | 5.55 | 124 | 9.90 | 164 | 7.12 |
| 5 | 21.39 | 45 | 18.78 | 85 | 14.79 | 125 | 9.55 | 165 | 5.49 | 125 | 9.80 | 165 | 7.08 |
| 6 | 21.35 | 46 | 18.70 | 86 | 14.67 | 126 | 9.41 | 166 | 5.42 | 126 | 9.70 | 166 | 7.04 |
| 7 | 21.30 | 47 | 18.61 | 87 | 14.56 | 127 | 9.26 | 167 | 5.35 | 127 | 9.59 | 167 | 7.00 |
| 8 | 21.25 | 48 | 18.53 | 88 | 14.44 | 128 | 9.11 | 168 | 5.29 | 128 | 9.49 | 168 | 6.96 |
| 9 | 21.21 | 49 | 18.44 | 89 | 14.32 | 129 | 8.97 | 169 | 5.23 | 129 | 9.40 | 169 | 6.93 |
| 10 | 21.16 | 50 | 18.35 | 90 | 14.20 | 130 | 8.84 | 170 | 5.17 | 130 | 9.30 | 170 | 6.89 |
| 11 | 21.10 | 51 | 18.26 | 91 | 14.09 | 131 | 8.70 | 171 | 5.11 | 131 | 9.21 | 171 | 6.85 |
| 12 | 21.05 | 52 | 18.17 | 92 | 13.97 | 132 | 8.57 | 172 | 5.05 | 132 | 9.12 | 172 | 6.82 |
| 13 | 21.00 | 53 | 18.08 | 93 | 13.84 | 133 | 8.44 | 173 | 4.99 | 133 | 9.03 | 173 | 6.79 |
| 14 | 20.95 | 54 | 17.99 | 94 | 13.72 | 134 | 8.32 | 174 | 4.93 | 134 | 8.94 | 174 | 6.76 |
| 15 | 20.89 | 55 | 17.90 | 95 | 13.60 | 135 | 8.19 | 175 | 4.88 | 135 | 8.86 | 175 | 6.73 |
| 16 | 20.83 | 56 | 17.81 | 96 | 13.48 | 136 | 8.07 | 176 | 4.82 | 136 | 8.78 | 176 | 6.70 |
| 17 | 20.78 | 57 | 17.71 | 97 | 13.35 | 137 | 7.96 | 177 | 4.77 | 137 | 8.70 | 177 | 6.67 |
| 18 | 20.72 | 58 | 17.62 | 98 | 13.23 | 138 | 7.84 | 178 | 4.71 | 138 | 8.62 | 178 | 6.64 |
| 19 | 20.66 | 59 | 17.53 | 99 | 13.10 | 139 | 7.73 | 179 | 4.66 | 139 | 8.54 | 179 | 6.61 |
| 20 | 20.60 | 60 | 17.43 | 100 | 12.98 | 140 | 7.62 | 180 | 4.61 | 140 | 8.47 | 180 | 6.58 |
| 21 | 20.54 | 61 | 17.33 | 101 | 12.85 | 141 | 7.51 | 181 | 4.56 | 141 | 8.39 | 181 | 6.56 |
| 22 | 20.48 | 62 | 17.24 | 102 | 12.72 | 142 | 7.41 | 182 | 4.51 | 142 | 8.32 | 182 | 6.53 |
| 23 | 20.41 | 63 | 17.14 | 103 | 12.59 | 143 | 7.30 | 183 | 4.46 | 143 | 8.25 | 183 | 6.51 |
| 24 | 20.35 | 64 | 17.04 | 104 | 12.47 | 144 | 7.20 | 184 | 4.41 | 144 | 8.18 | 184 | 6.49 |
| 25 | 20.28 | 65 | 16.94 | 105 | 12.33 | 145 | 7.10 | 185 | 4.36 | 145 | 8.12 | 185 | 6.46 |
| 26 | 20.22 | 66 | 16.84 | 106 | 12.20 | 146 | 7.01 | 186 | 4.32 | 146 | 8.05 | 186 | 6.44 |
| 27 | 20.15 | 67 | 16.74 | 107 | 12.07 | 147 | 6.91 | 187 | 4.27 | 147 | 7.99 | 187 | 6.42 |
| 28 | 20.08 | 68 | 16.64 | 108 | 11.94 | 148 | 6.82 | 188 | 4.23 | 148 | 7.93 | 188 | 6.40 |
| 29 | 20.01 | 69 | 16.53 | 109 | 11.81 | 149 | 6.73 | 189 | 4.18 | 149 | 7.87 | 189 | 6.38 |
| 30 | 19.94 | 70 | 16.43 | 110 | 11.67 | 150 | 6.64 | 190 | 4.14 | 150 | 7.81 | 190 | 6.36 |
| 31 | 19.87 | 71 | 16.33 | 111 | 11.54 | 151 | 6.55 | 191 | 4.09 | 151 | 7.75 | 191 | 6.35 |
| 32 | 19.80 | 72 | 16.22 | 112 | 11.40 | 152 | 6.46 | 192 | 4.05 | 152 | 7.69 | 192 | 6.33 |
| 33 | 19.73 | 73 | 16.12 | 113 | 11.26 | 153 | 6.38 | 193 | 4.01 | 153 | 7.64 | 193 | 6.31 |
| 34 | 19.65 | 74 | 16.01 | 114 | 11.13 | 154 | 6.30 | 194 | 3.97 | 154 | 7.59 | 194 | 6.30 |
| 35 | 19.58 | 75 | 15.90 | 115 | 10.99 | 155 | 6.22 | 195 | 3.93 | 155 | 7.53 | 195 | 6.28 |
| 36 | 19.50 | 76 | 15.79 | 116 | 10.85 | 156 | 6.14 | 196 | 3.89 | 156 | 7.48 | 196 | 6.27 |
| 37 | 19.42 | 77 | 15.69 | 117 | 10.71 | 157 | 6.06 | 197 | 3.85 | 157 | 7.43 | 197 | 6.26 |
| 38 | 19.35 | 78 | 15.58 | 118 | 10.57 | 158 | 5.98 | 198 | 3.81 | 158 | 7.39 | 198 | 6.24 |
| 39 | 19.27 | 79 | 15.47 | 119 | 10.43 | 159 | 5.91 | 199 | 3.77 | 159 | 7.34 | 199 | 6.23 |
| 40 | 19.19 | 80 | 15.36 | 120 | 10.28 | 160 | 5.83 | 200 | 3.73 | 160 | 7.29 | 200 | 6.22 |

# E-15

## PIPE
### Dimensions and properties

| Nominal Diameter In. | Outside Diameter In. | Inside Diameter In. | Thick- ness In. | Weight per Foot Lb. Plain Ends | ASA Schedule No. | $I$ In.⁴ | $A$ In.² | $k$ In. |
|---|---|---|---|---|---|---|---|---|
| **Dimensions** | | | | | | **Properties** | | |
| Standard Weight | | | | | | | | |
| ⅛ | .405 | .269 | .068 | .24 | 40 | .001 | .072 | .12 |
| ¼ | .540 | .364 | .088 | .42 | 40 | .003 | .125 | .16 |
| ⅜ | .675 | .493 | .091 | .57 | 40 | .007 | .167 | .21 |
| ½ | .840 | .622 | .109 | .85 | 40 | .017 | .250 | .26 |
| ¾ | 1.050 | .824 | .113 | 1.13 | 40 | .037 | .333 | .33 |
| 1 | 1.315 | 1.049 | .133 | 1.68 | 40 | .087 | .494 | .42 |
| 1¼ | 1.660 | 1.380 | .140 | 2.27 | 40 | .195 | .669 | .54 |
| 1½ | 1.900 | 1.610 | .145 | 2.72 | 40 | .310 | .799 | .62 |
| 2 | 2.375 | 2.067 | .154 | 3.65 | 40 | .666 | 1.075 | .79 |
| 2½ | 2.875 | 2.469 | .203 | 5.79 | 40 | 1.530 | 1.704 | .95 |
| 3 | 3.500 | 3.068 | .216 | 7.58 | 40 | 3.017 | 2.228 | 1.16 |
| 3½ | 4.000 | 3.548 | .226 | 9.11 | 40 | 4.788 | 2.680 | 1.34 |
| 4 | 4.500 | 4.026 | .237 | 10.79 | 40 | 7.233 | 3.174 | 1.51 |
| 5 | 5.563 | 5.047 | .258 | 14.62 | 40 | 15.16 | 4.300 | 1.88 |
| 6 | 6.625 | 6.065 | .280 | 18.97 | 40 | 28.14 | 5.581 | 2.25 |
| *8 | 8.625 | 7.981 | .322 | 28.55 | 40 | 72.49 | 8.399 | 2.94 |
| *10 | 10.750 | 10.020 | .365 | 40.48 | 40 | 160.7 | 11.91 | 3.67 |
| *12 | 12.750 | 12.000 | .375 | 49.56 | ... | 279.3 | 14.58 | 4.38 |
| Extra Strong | | | | | | | | |
| ⅛ | .405 | .215 | .095 | .31 | 80 | .001 | .093 | .11 |
| ¼ | .540 | .302 | .119 | .54 | 80 | .004 | .157 | .15 |
| ⅜ | .675 | .423 | .126 | .74 | 80 | .009 | .217 | .20 |
| ½ | .840 | .546 | .147 | 1.09 | 80 | .020 | .320 | .25 |
| ¾ | 1.050 | .742 | .154 | 1.47 | 80 | .045 | .433 | .32 |
| 1 | 1.315 | .957 | .179 | 2.17 | 80 | .106 | .639 | .41 |
| 1¼ | 1.660 | 1.278 | .191 | 3.00 | 80 | .242 | .881 | .52 |
| 1½ | 1.900 | 1.500 | .200 | 3.63 | 80 | .391 | 1.068 | .61 |
| 2 | 2.375 | 1.939 | .218 | 5.02 | 80 | .868 | 1.477 | .77 |
| 2½ | 2.875 | 2.323 | .276 | 7.66 | 80 | 1.924 | 2.254 | .92 |
| 3 | 3.500 | 2.900 | .300 | 10.25 | 80 | 3.894 | 3.016 | 1.14 |
| 3½ | 4.000 | 3.364 | .318 | 12.51 | 80 | 6.280 | 3.678 | 1.31 |
| 4 | 4.500 | 3.826 | .337 | 14.98 | 80 | 9.610 | 4.407 | 1.48 |
| 5 | 5.563 | 4.813 | .375 | 20.78 | 80 | 20.67 | 6.112 | 1.84 |
| 6 | 6.625 | 5.761 | .432 | 28.57 | 80 | 40.49 | 8.405 | 2.20 |
| 8 | 8.625 | 7.625 | .500 | 43.39 | 80 | 105.7 | 12.76 | 2.88 |
| 10 | 10.750 | 9.750 | .500 | 54.74 | 60 | 211.9 | 16.10 | 3.63 |
| 12 | 12.750 | 11.750 | .500 | 65.42 | ... | 361.5 | 19.24 | 4.34 |
| Double-Extra Strong | | | | | | | | |
| ½ | .840 | .252 | .294 | 1.71 | ... | .024 | .504 | .22 |
| ¾ | 1.050 | .434 | .308 | 2.44 | ... | .058 | .718 | .28 |
| 1 | 1.315 | .599 | .358 | 3.66 | ... | .140 | 1.076 | .36 |
| 1¼ | 1.660 | .896 | .382 | 5.21 | ... | .341 | 1.534 | .47 |
| 1½ | 1.900 | 1.100 | .400 | 6.41 | ... | .568 | 1.885 | .55 |
| 2 | 2.375 | 1.503 | .436 | 9.03 | ... | 1.311 | 2.656 | .70 |
| 2½ | 2.875 | 1.771 | .552 | 13.70 | ... | 2.871 | 4.028 | .84 |
| 3 | 3.500 | 2.300 | .600 | 18.58 | ... | 5.992 | 5.466 | 1.05 |
| 4 | 4.500 | 3.152 | .674 | 27.54 | ... | 15.28 | 8.101 | 1.37 |
| 5 | 5.563 | 4.063 | .750 | 38.55 | ... | 33.64 | 11.34 | 1.72 |
| 6 | 6.625 | 4.897 | .864 | 53.16 | ... | 66.33 | 15.64 | 2.06 |
| 8 | 8.625 | 6.875 | .875 | 72.42 | ... | 162.0 | 21.30 | 2.76 |
| *10 | 10.750 | 8.750 | 1.000 | 104.13 | 140 | 367.8 | 30.63 | 3.47 |
| *12 | 12.750 | 10.750 | 1.000 | 125.49 | 120 | 641.7 | 36.91 | 4.17 |

*Sections are not readily available but may be secured on special order.

When ordering pipe for structural purposes refer to notes on page 3-44.

The schedule number is not a structural term but refers to the pressure-stress ratio of pipe. For further discussion on schedule numbers, see American Standard ASA B36.10-1959.

# Index